CARBON-13 NUCLEAR MAGNETIC RESONANCE SPECTROSCOPY

Carbon-13 Nuclear Magnetic Resonance Spectroscopy

Second Edition

GEORGE C. LEVY
The Florida State University

ROBERT L. LICHTER
Hunter College, City University of New York

GORDON L. NELSON
General Electric Company

A Wiley-Interscience Publication
JOHN WILEY & SONS
New York / Chichester / Brisbane / Toronto / Singapore

Library of Congress Cataloging in Publication Data:

Levy, George C
  Carbon-13 nuclear magnetic resonance spectroscopy.

  "A Wiley-Interscience publication."
  Edition of 1972 published under title: Carbon-13
nuclear magnetic resonance for organic chemists.
  Includes index.
  1. Nuclear magnetic resonance spectroscopy.
2. Carbon—Isotopes. I. Lichter, Robert L., joint
author. II. Nelson, Gordon L., 1943-   joint author.
III. Title.
QD272.S6L474 1980   547.3'0877   80-17289
ISBN 0-471-53157-X

To Linda

To Derek and Allison

To Marge

and to the memory of
Charles A. Levy

Preface

In the eight years since publication of the first edition of *Carbon-13 Nuclear Magnetic Resonance For Organic Chemists* the methodology and application of carbon nuclear magnetic resonance (nmr) to chemistry problems have grown explosively. In 1980 ^{13}C nmr at natural isotopic abundance can be as routine as proton nmr was in 1972. At the same time ^{13}C nmr is constantly showing us applications that are not conceivable with the use of proton magnetic resonance. In the 1980s ^{13}C nmr techniques will prove valuable for analysis of intractable organic solids, elucidation of *in vivo* intracellular metabolism, sensitive detection of short-lived organic reaction intermediates, and other problems not yet amenable to chemical analysis.

The first edition introduced ^{13}C nmr spectroscopy to organic chemists. The objective of the second edition remains similar, with the added responsibility of presenting the large body of currently evolving techniques that are extending the applicability of ^{13}C nmr. This edition is still directed toward the organic chemist; however, we include some detailed treatments of ^{13}C nmr concepts that are appropriate for physical chemists and others.

The structure of this book roughly parallels that of the first edition with greatly expanded treatment of spin relaxation, natural products, and synthetic and biopolymers. All the material is updated, with extensive referencing. The direct photocopy process has allowed inclusion of 1979 and, in some cases, early 1980 references.

Chapters 1 and 2 introduce necessary theoretical and experimental concepts and provide a brief overall view of ^{13}C spectral characteristics. Chapters 3 to 6 discuss ^{13}C methods, results, and applications for aliphatic and aromatic compounds, functional groups, organic intermediates, and organometallic compounds. Chapter 7 describes ^{13}C nmr of synthetic high-molecular-weight polymers, both in solution and as bulk solid samples. Chapter 8 details concepts and results of ^{13}C spin-relaxation processes and is intended to provide the chemist with a qualitative understanding of this topic while at the same time presenting theoretical considerations to some extent. Application of ^{13}C nmr to the various classes of natural products and biopolymers is treated in Chapter 9. Special methods and applications of ^{13}C nmr are discussed in Chapter 10, including studies of reaction mechanisms and nmr in liquid crystals and in solids. The techniques of two-dimensional FT ^{13}C nmr are briefly outlined, as are a number of specialized ^{13}C nmr pulse schemes.

Several scientists have provided unpublished materials, especially Drs. Felix Wehrli, Dr. W. Hull, and Dr. B. Knüttel. Dr. John Grutzner commented critically on aspects of the manuscript. Sandy DeChello, Leslie B. Heinz, and Barbara Maybin prepared the photoready text. Drafting and graphics were carried out by Stephen Leukanech and Richard Roche. We appreciate the support and the encouragement of our editor, Dr. Theodore Hoffman. One of us (R. L. L.) acknowledges the hospitality afforded by the Chemistry Department of Florida State University during part of the preparation of this book.

Finally, the authors acknowledge the General Electric Corporate Research and Development Laboratory for its support and the many scientists who made helpful suggestions during the preparation of the first edition of this book.

GEORGE C. LEVY
ROBERT L. LICHTER
GORDON L. NELSON

Tallahassee, Florida
New York, New York
Pittsfield, Massachusetts
May 1980

Acknowledgments

John Wiley and Sons (Figures 1.7, 1.11, 7.11, 7.12, 10.4)

Pergamon Press (Figures 9.2, 10.12)

National Academy of Sciences (Figures 9.4, 9.8, 9.9)

Heyden and Sons (Figure 9.1)

Verlag Chemie (Figures 6.1, 10.2)

National Research Council of Canada (Figure 9.6)

North-Holland Publishing Co. (Figure 10.8)

University Park Press (Figure 9.16)

Information Retrieval Ltd. (Figure 9.17)

American Chemical Society (Figures 9.3, 9.5, 9.10-9.13, 9.15, 9.18-9.20)

Academic Press, Inc. (Figures 1.10, 7.4-7.8, 8.5, 8.8, 10.6)

IPC Science and Technology Press, Ltd. (Figures 10.9-10.11)

G.C.L.
R.L.L.
G.L.N.

Contents

Compounds. Diazo Compounds and Ketenes. Isocyanates.
Thio- and Selenocarbonyls.

Chapter 1

Introduction, Theory, and Methods

HISTORY OF ^{13}C NMR

Nuclear magnetic resonance (nmr) in the condensed phase was first realized experimentally in late 1945 in the laboratories of Bloch[1] and Purcell[2] and their colleagues. A great deal of foresight and imagination would have been required to predict where these first measurements on the protons of water and paraffin wax were going to lead. It was not until 1951 that Arnold, Dharmatti, and Packard[3] reported the "resolution" of the spectrum of ethanol into three peaks (CH$_3$, CH$_2$, and OH). Since 1953, however, when the first commercial "high-resolution" nmr spectrometer was sold by Varian Associates, proton nmr spectroscopy has become a mature scientific discipline and has achieved the status of a routine analytical tool, supporting research in diverse fields but especially in organic chemistry.

The first nmr observations of carbon-13 (^{13}C) nuclei were reported in 1957.[4] These experiments indicated that the direct observation of carbon nuclei had many advantages over the equivalent proton studies. However, the extreme difficulty of the experiments, coupled with poor spectral resolution and the requirement of working with highly soluble, very low molecular weight materials, severely restricted the early application of ^{13}C nmr [or carbon magnetic resonance (cmr*) spectroscopy]. Nevertheless, by the mid-1960s, research groups headed by D. M. Grant, P. C. Lauterbur, J. D. Roberts, J. B. Stothers, and others succeeded in studying many classes of organic compounds with ^{13}C nmr techniques.

The first great breakthrough in experimental ^{13}C nmr, wideband proton decoupling, was available as early as 1965, but the advancement of ^{13}C nmr spectroscopy to the status of a practical analytical research tool for organic chemists required several instrumental and technical developments that became available only after 1970. Figures 1.1-1.3 demonstrate the evolution of ^{13}C nmr spectroscopy.

* The abbreviation "cmr" is used occasionally in this text; however, its general use is currently discouraged.

1

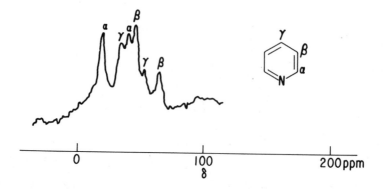

FIGURE 1.1.[5]   Carbon-13 nmr spectrum of pyridine. Chemical
shift scale in parts per million upfield from
*CS_2. Spectrum obtained without ^{1}H decoupling;
before 1958.

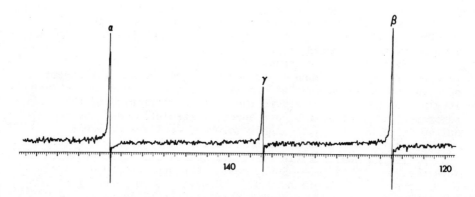

FIGURE 1.2.   Carbon-13 nmr spectrum of pyridine (^{1}H decoupled).
Six 10-Hz/sec, 1000-Hz-wide scans of 2 g of
pyridine in approximately 3 ml of total volume.
Spectrum accumulated on a Varian C-1024 time averaging computer. Total acquisition time = 600 sec.
Chemical shifts in parts per million downfield
from TMS (tetramethylsilane).

2

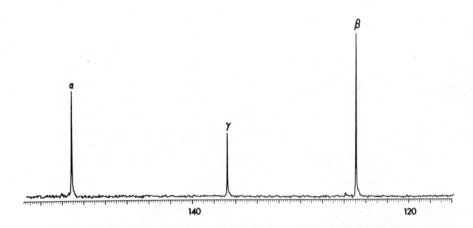

FIGURE 1.3.   *Fourier transform ^{13}C nmr spectrum of pyridine. Same solution and chemical-shift convention as in Figure 1.2 but only 6 sec of total acquisition time.*

^{13}C NMR: THEORY AND EXPERIMENT

Practical problems aside, carbon nmr had greater potential than did ^{1}H nmr for the study of organic systems. Indeed, the enhanced effective resolution possible in ^{13}C nmr was recognized early. Carbon resonances of organic compounds are found over a chemical-shift range of 600 ppm, compared with the <20-ppm range for proton nuclei. With modern instrumental methods it is also possible to have narrower resonance lines in cmr than in ^{1}H spectra.* Thus it is not unusual in cmr to be able to identify individual resonances for each carbon in a compound whose molecular weight is 300 to 500. In such complex molecules proton nmr is largely useful only for "fingerprint" identification.

Additional advantages of carbon nmr relate to the direct observation of molecular backbones, of carbon-containing functional groups with no attached protons (e.g., carbonyls and nitriles), and of carbon reaction sites of interest. However, adoption of carbon nmr for routine use by organic chemists was slow because of the experimental difficulties associated with the methods, and not because of any fundamental limitations on applicability. Only in the past decade, after intensive research in instrumental methods, has cmr begun to reach full maturity.

* In most cases ^{1}H spectra are extensively spin coupled, thus leading to broadened resonance bands for most protons. As is shown later, in cmr the situation is more easily controlled and spin-spin splitting is not usually present.

Sensitivity.   Nuclear magnetic resonance of carbon nuclei and of protons share some characteristics but differ in one important aspect. The most abundant isotope of carbon, atomic weight 12, has no nuclear spin and is not observable in nmr experiments. The ^{13}C isotope has a nuclear spin of 1/2 (as does ^{1}H); however, the natural abundance of ^{13}C is 1.1%. This natural abundance is low enough to make observation of $^{13}C-^{13}C$ spin-spin coupling interactions unlikely in unenriched compounds and yet large enough to be considered practical for nmr. A high natural abundance of ^{13}C would have caused immeasurable grief for early ^{1}H nmr spectroscopists since the proton spectra would have been complicated by $^{13}C-^{1}H$ coupling.

Another factor further lowers the effective sensitivity of ^{13}C nmr experiments. The magnetogyric ratio, γ, of ^{13}C nuclei is about 1/4 that of ^{1}H nuclei. Here γ is a function of the magnetic moment and spin quantum number of the nucleus. Since the sensitivity of a nucleus in a magnetic resonance experiment at constant field is proportional to the cube of γ, a ^{13}C nucleus gives rise to $(1/4)^{3}$, or 1/64 the signal that a proton nucleus would yield on excitation. The result of the 1.1% natural isotopic abundance and lower γ is a lowering of the sensitivity by a factor of ~6000 in a ^{13}C nmr experiment relative to a proton nmr experiment.

The NMR Experiment.   Let us briefly review what constitutes an nmr experiment in the context of ^{13}C nmr. In a static magnetic field, B_0, an isolated nucleus of spin 1/2 has two available energy states, corresponding to alignment with and against the field (Figure 1.4). Absorption of energy (resonance) occurs when the sample is irradiated with electromagnetic energy at a radiofrequency ν_0. The magnetic component of the irradiation, called B_1, induces transitions between the two energy levels. For carbon nuclei, when the static magnetic field B_0 is 2.35 T, ν_0 = 25.2 MHz (for protons, ν_0 = 100.1 MHz at 2.35 T).

The energy difference between the upper and lower states (corresponding to ν_0) is very small, only a few millicalories per mole. This means that at ordinary temperatures only a very small excess of nuclei is in the lower energy state at thermal equilibrium, <1 in 10^{5}, according to the Boltzmann distribution law. This small excess gives rise to an observable nmr signal. The sensitivity of the nmr method is low relative to spectroscopies dealing with larger energy transitions [e.g., ultraviolet (uv) spectroscopy]. The utility of the nmr method results from the sensitive variation of resonance frequency with chemical structure and molecular environment. Electron clouds surrounding individual nuclei are polarized by chemical bonding. These electrons "shield" the nuclei from the static magnetic field to different extents. This gives rise to chemical shifts for the same nuclear species in different molecular environments.

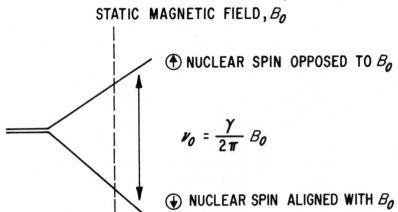

FIGURE 1.4.   *Energy levels for a nucleus with spin = 1/2.*

In a typical nmr experiment the excitation frequency ν is swept through the range of frequencies covering all possible resonance lines of interest. As ν reaches a resonance frequency that will lead to energy absorption or emission by a nucleus, transitions between the two energy levels are stimulated. The probabilities of transitions are proportional to the populations of the energy levels; thus the probability of an upward transition is slightly greater than that of a downward transition. If the B_1 irradiation passes through a resonance line very rapidly, perturbation of the populations in the two energy levels for that nucleus does not become a significant factor. However, in most swept nmr experiments a relatively slow sweep rate is used; therefore, the energy-level populations may equalize while the excitation occurs. If the populations of the energy levels become equivalent, no further absorption is observed and the sample is said to be saturated.

Nuclear Spin Relaxation.   There is a process, spin-lattice relaxation, that allows the nuclei to reestablish a Boltzmann population distribution. The average lifetime of an ensemble of identical nuclei (not being irradiated) in either the upper or lower energy state is described by T_1, the spin-lattice relaxation time. When the nuclei are irradiated (and thus the Boltzmann distribution is disturbed) T_1 defines the time required for the nuclear spins to exponentially return toward Boltzmann populations. After $1 \cdot T_1$, $1/e$ (=36.8%) of the initial magnetization remains; it takes $5 \cdot T_1$ for the relaxation to result in return to >99% of the initial Boltzmann distribution. (The preceding describes an ensemble of nuclear spins without considering effects of double irradiation, e.g., the nuclear Overhauser effect, discussed later.)

The nuclei interact with their surroundings (denoted by the term "lattice" for either liquids or solids). Several interaction mechanisms operate to various extents in different molecules and systems. All the mechanisms have one thing in common: the relaxation of a given nucleus is caused by the fluctuations of localized magnetic or electric fields in the sample that result from molecular motion(s). The localized fields can be considered as fluctuating at various frequencies, corresponding to different molecular motional components. Those motional (Fourier) components, fluctuating at frequencies near the excitation (or Larmor) frequency of a nucleus, are most effective in relaxing that nucleus; significantly faster or slower fluctuations have little effect.

A long spin-lattice relaxation time indicates that a given nucleus has no efficient relaxation pathway and is thus easy to saturate. Care must be exercised in the nmr experiment to avoid saturation; rapid sweep rates and/or low power settings of the irradiating field B_1 should be used. Unfortunately, both techniques degrade experimental results. Fast sweep rates limit achievable resolution (in the normal swept experiment), and low-power B_1 irradiation results in lower signal intensities, whereas the baseline noise remains constant.

Spin-Lattice Relaxation Mechanisms.   There are several common mechanisms that lead to spin-lattice relaxation for ^{13}C nuclei:

1. Dipole-Dipole Relaxation. Spin-lattice relaxation for ^{13}C nuclei can arise from fluctuating fields as a result of dipole-dipole interactions with neighboring magnetic nuclei (or with unpaired electrons).

2. Spin-Rotation Relaxation. Small molecules and freely rotating CH_3 groups can be effectively relaxed by a mechanism involving quantum-rotational states of the molecule or the group. In these cases spin-rotation relaxation often competes with dipole-dipole relaxation for protonated carbons, whereas the spin-rotation interaction dominates the relaxation of non-protonated carbons.

3. Chemical-Shift Anisotropy. Significant anisotropy (directionality) in the shielding tensor of a nucleus can give rise to fluctuating magnetic fields when the molecule tumbles in solution (relative to the fixed laboratory magnetic field).

4. Scalar Relaxation. A ^{13}C nucleus that is spin-spin (or scalar) coupled to a nucleus X that is undergoing rapid spin-lattice relaxation may itself be relaxed because of the fluctuating scalar interaction between the two nuclei. This mechanism is normally encountered when the X nucleus has a spin

I, which is >1/2 (and X is relaxed by a mechanism known as *quadrupolar relaxation*), but scalar relaxation of ^{13}C nuclei may also occur when X = ^{1}H. Scalar relaxation of ^{13}C nuclei is generally confined to spin-spin relaxation. (The spin-spin relaxation time, T_2, defines the resonance linewidths but does not involve saturation of energy-level populations; see Chapter 8.) In a few cases scalar spin-lattice relaxation of ^{13}C nuclei does occur. When X is undergoing (quadrupolar) spin-lattice relaxation very rapidly and when the Larmor (resonance) frequency of X is close to the Larmor frequency for ^{13}C, scalar spin-lattice relaxation of ^{13}C nuclei may be competitive with the other mechanisms.

In cases where scalar relaxation of ^{13}C nuclei is limited to spin-spin relaxation, line broadening is often evident in the cmr spectra. For example, ^{13}C nuclei attached to ^{14}N nuclei frequently appear as broad resonances.

^{13}C-^{1}H Dipole-Dipole Relaxation.   In most organic molecules the ^{13}C-^{1}H dipole-dipole relaxation mechanism is dominant for ^{13}C spin-lattice relaxation. This is particularly true for carbons that have directly attached hydrogens (protonated carbons).

Two main factors determine the efficiency of dipole-dipole relaxation in individual cases: the magnetogyric ratio of the nucleus causing relaxation and the proximity of that nucleus to the nucleus being relaxed. In both ^{1}H and ^{13}C nuclear magnetic resonance, protons usually dominate dipole-dipole relaxation because of their high magnetogyric ratio (dipole-dipole relaxation is proportional to γ^2). The distance dependence of ^{13}C-^{1}H dipole-dipole relaxation is proportional to $1/r^6$ where *r* is the internuclear distance; thus protonated carbons are most easily relaxed, and the dipole-dipole mechanism usually dominates the relaxation of these carbons. For ^{13}C nuclei that do not have directly attached hydrogens (nonprotonated carbons), other mechanisms may compete with or replace ^{13}C-^{1}H dipole-dipole relaxation. However, in medium to large organic molecules, [molecular weight (mw)$\gtrsim$ 300] dipolar relaxation may become exclusive even for nonprotonated carbons.

The presence of unpaired spins can lead to very efficient dipole-dipole relaxation (replacing ^{13}C-^{1}H dipole-dipole relaxation) since the magnetic moment of the electron is much greater than that of the proton. Both T_1 and T_2 can be affected by unpaired spins, and spectral lines are often significantly broadened in their presence.

(The preceding sections serve only as an introduction to spin relaxation and relaxation mechanisms. Further discussion and applications of relaxation measurements are found in Chapter 8.)

The Nuclear Overhauser Effect.   This is a by-product of proton irradiation in ^{1}H-decoupled ^{13}C nmr experiments (notation: ^{13}C $\{^1$H$\}$). Irradiation of the protons in a sample disturbs the Boltzmann distribution of the upper and lower ^{1}H energy levels (full ^{1}H saturation equalizes the levels). The ^{13}C nuclei depend chiefly on the ^{1}H nuclei for spin-lattice relaxation. The carbon nuclei react to the equalization of ^{1}H energy-level populations by changing their own energy-level populations. This results in an equilibrium excess of nuclei in the lower ^{13}C energy level relative to that specified by a Boltzmann distribution. Experimentally, this means that more radiofrequency (rf) energy will be absorbed by the ^{13}C nuclei as a result of the larger excess population in the lower energy level.

In a ^{13}C $\{^1$H$\}$ nmr experiment, the theoretical nuclear Overhauser enhancement factor (NOEF) is 1.988. This means that if the NOE is fully operative, each carbon resonance will have a peak area 2.988 times the total resonance signal area observed in the absence of ^{1}H (the additional NOEF, 1.988, is added to the original signal, 1.00, giving a resonance of total intensity 2.988).

The nuclear Overhauser effect is derived from the ^{1}H-induced relaxation of ^{13}C nuclei. It is further dependent on the dominance of the ^{13}C-^{1}H dipole-dipole relaxation mechanism. Appreciable contribution to ^{13}C spin-lattice relaxation from other than the dipole-dipole mechanism results in lower nuclear Overhauser enhancements. The integrated intensities of ^{1}H-decoupled ^{13}C resonances can vary in a single molecule as a result of differing Overhauser enhancements, particularly in small, symmetrical molecules where the dipole-dipole relaxation mechanism may not be dominant even for some protonated carbons. In 1971 Allerhand et al.[6] showed that in larger, relatively rigid molecules, virtually all carbons are relaxed by the dipole-dipole mechanism; thus most carbons yield a full Overhauser enhancement. Even in these molecules, some nonprotonated carbons still may have appreciable contributions from other relaxation mechanisms. Hence their integrated intensities would be lower in ^{13}C $\{^1$H$\}$ experiments. Furthermore, in *very* large molecules, even when ^{13}C-^{1}H dipolar relaxation is operative, the NOEF is reduced. The interpretation problem associated with variable Overhauser enhancements is treated in Chapter 2.

^{1}H Decoupling Methods.   The original ^{13}C $\{^1$H$\}$ experiments suffered because only one ^{1}H frequency could be irradiated at a given instant. Thus a ^{13}C spectrum might have one "singlet" carbon resonance amidst many partially decoupled ^{13}C multiplets. A practical way to achieve complete proton decoupling was first developed by Ernst.[7] Ernst utilized a single ^{1}H decoupling frequency as the center of a finite-excitation band. The single frequency was modulated by a pseudorandom noise generator

yielding effective excitation throughout a preset bandwidth.
The bandwidth can be set broad enough to cover all protons in a
sample. This can be visualized as irradiation of all proton
resonances sequentially within the time required to observe each
individual carbon resonance. Modern ^{13}C nmr instruments use
other modulation schemes for higher-efficiency broad-band decoupling needed at today's higher spectrometer frequencies.

Over an order of magnitude increase in sensitivity is
achieved in wide-band ^1H decoupled ^{13}C experiments, relative to
experiments with no ^1H decoupling. This corresponds to the increase from collapse of C-H spin multiplets and also from the
increased peak area/height resulting from NOEs.

Unfortunately, a price is paid for this increase in sensitivity. Except when ^{31}P or ^{19}F is present, all carbon resonances
appear as singlets; all coupling information is lost. Various
decoupling techniques allow recovery of C-H coupling information.
In a single-frequency off-resonance decoupling experiment the
^1H irradiation is kept at high power levels, but the center
frequency is moved 500 to 1000 Hz away from the protons to be
irradiated, and the excitation bandwidth generator is switched
off. In these experiments one-bond C-H coupling patterns return,
allowing spectral assignments of nonprotonated carbons, CH, CH_2,
and CH_3 observed as singlets, doublets, triplets, and quartets,
respectively. The observed couplings in off-resonance decoupled
experiments are not equal to the actual one-bond coupling constants but are reduced or residual couplings. In off-resonance
^1H-decoupled ^{13}C spectra, peak assignments are more easily made
since overlap between adjacent carbon resonance bands is less
likely. The observed residual coupling is a function of the
actual ^{13}C-^1H coupling, the decoupling power, and the decoupler
offset. Suitable adjustment of the offset allows resolution of
overlapping sets of lines. These experiments show relatively
high sensitivity since NOE is still present and because long-
range C-H couplings are absent.

Figure 1.5 compares the spectrum of phenol obtained in
proton-coupled, wide-band, and off-resonance ^1H-decoupled experiments. Experimental observation time was equivalent in each
case; therefore, signal-to-noise (S:N) improvements in the decoupled spectra are real indications of sensitivity improvement.
In the coupled spectrum (Figure 1.5a) of this symmetrical, low-
molecular-weight compound, two resonance bands overlap. These
bands do not overlap in the off-resonance ^1H-decoupled spectrum.
In large organic molecules it is generally not possible to
separate all carbons in an off-resonance ^1H-decoupled spectrum.
Nevertheless, relatively isolated multiplets can usually be
assigned even in the largest of molecules. Additionally, some
multiplets observed in off-resonance decoupled spectra can be
ambiguous since second-order spin-spin coupling can occur, especially for CH_2 carbons.

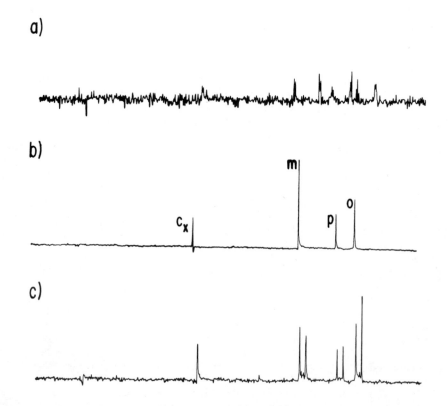

FIGURE 1.5.   *Carbon-13 25.2-MHz spectrum of phenol (2500 Hz spectral width): (a) no ^1H decoupling; (b) wide-band ^1H decoupled; (c) off-resonance ^1H decoupled. Assignments noted in (b).*

A third method of ^1H decoupling is found to be useful for ^{13}C spectral assignments--single-frequency proton decoupling. This method is more nearly analogous to the usual "observe proton A, decouple proton B" experiment (proton homonuclear decoupling). If a given proton resonance can be identified in an ^1H nmr spectrum, it is possible to completely irradiate only those protons at low rf power. The result observed in a ^{13}C experiment is that the carbon attached to that proton collapses to a singlet while other protonated carbons retain some C-H coupling. Thus it is possible to relate ^1H and ^{13}C spectral lines. Figure 1.6 shows the ^{13}C spectrum of 2-methyl-1-butanol with wide-band ^1H decoupling and specific (single-frequency) decoupling of the CH_2OH methylene protons (332 Hz downfield from TMS in the ^1H spectrum).

It is possible to obtain a set of single-frequency decoupled spectra by sequencing through the proton resonance region. A graphical representation then relates ^{1}H and ^{13}C spectral lines (see Figure 1.7). This technique can be extremely useful for structural analysis.

Several types of ^{1}H-decoupling experiment use interrupted irradiation or pulsed irradiation of the protons. Heteronuclear multiple resonance is also possible, with two or more kinds of nuclei being simultaneously decoupled from carbon.

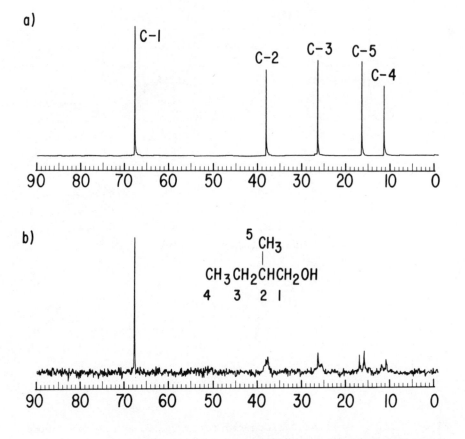

FIGURE 1.6.   Carbon-13 25.1-MHz spectrum of 2-methyl-1-butanol: (a) wide-band ^{1}H decoupled; (b) specific frequency ^{1}H decoupled with irradiation of the C**H**$_2$OH protons.

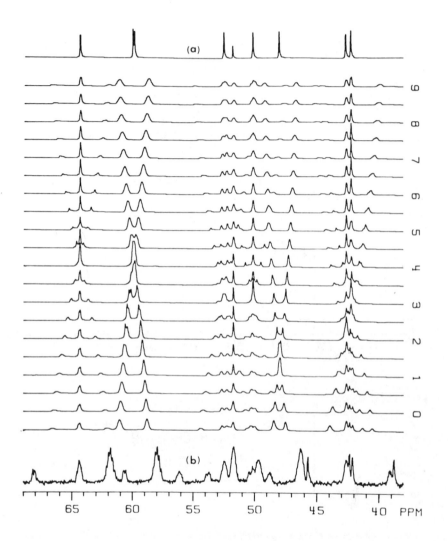

FIGURE 1.7.   Partial ^{13}C spectra for strychnine (39 to 69 ppm):
(a) fully proton decoupled; (b) proton coupled;
others, sequentially single frequency decoupled,
stepping through the proton spectrum at 0.5-ppm
intervals (L. F. Johnson, in _Topics in Carbon-13_
NMR Spectroscopy, Vol. 3, G. C. Levy, ed., Wiley-
Interscience, New York; 1979).

12

NMR INSTRUMENTATION--SECOND-, THIRD-, AND FOURTH-GENERATION NMR SPECTROMETERS

In the late 1960s several breakthroughs in nmr instrumentation occurred, resulting largely from the need for increased flexibility in multinuclei nmr studies. These third-generation spectrometers differed in several important ways from previous research-grade nmr instrumentation such as the second-generation Varian HA-100 and JEOL 4H-100, first sold in the mid-1960s. Third- and fourth-generation spectrometers were designed for ^{13}C and other nuclei nmr studies, whereas the use of unmodified second-generation spectrometers in ^{13}C nmr required some measure of perseverance.

Second-generation nmr spectrometer systems shared several common features:

1. Field/frequency control (lock) on a signal in the analytical sample. The signal is of the same nuclear species as the nucleus whose spectrum is being taken (*homonuclear* lock).
2. Audiofrequency and/or magnetic field sweeps for recording spectra.
3. Heteronuclear decoupling, if present, is not locked to the observe channel frequency.

Third-generation nmr systems (e.g., Varian XL-100, Bruker HFX-90, and JEOL PS-100) differed from older instruments in several important ways:

1. At least three independent rf irradiation channels are supplied, allowing simultaneous observation, lock, and decoupling of three or more nuclear species (e.g., ^{13}C observe, ^{2}H lock, and ^{1}H decouple). The three channels are derived from or are locked to one master source, preventing frequency drifts and so on.
2. Two receiving channels are required to detect and amplify the observe and lock resonance signals.
3. Digital or analog rf sweep and single sideband detection are utilized to optimize performance in the wide sweeps required for multinuclei nmr. In addition, these systems are easily converted to Fourier transform (FT) operation.
4. Wide-gap (>1 in.) 1.8- to 2.3-T high-resolution electromagnet systems generally utilize universal probes (one probe with several receiver inserts, tuned circuits, and preamplifiers for all observe nuclei) and large-diameter (12 to 15 mm) sample tubes.
5. Interfacing of the spectrometer to a laboratory computer allows external control of experimental parameters as well as digital treatment of data. Third-generation spectrometer systems are designed to facilitate interfacing and control.

The fourth-generation spectrometer systems (e.g., Bruker WP-80, WP-200, WM 250, WM 500, WH 400, Varian XL-200, Nicolet NTC-150, NTC 400, JEOL FX-90Q, and FX-200) first developed in the mid-1970s are actually derived from third-generation designs with major modifications.[8]   In the newest configurations the observe (and possibly the decoupling) channel rf is produced by a true rf synthesizer (or an equivalent infinitely variable and *stable* pure frequency generation scheme).   Furthermore, the probehead of the spectrometer may be constructed so as to allow tuning over a wide range of frequencies without change of any components.   With these improvements it becomes possible to change the observation nucleus without removing the sample, or even while maintaining the heteronuclear lock condition!

Fourth-generation instruments generally do not incorporate the additional electronic circuitry required for sweeping radio-frequencies; they are used in the pulsed FT mode only.   One other major change for fourth-generation spectrometers is the *integration* of the digital computer *into* the spectrometer design, rather than as an interfaced accessory (Figure 1.12).

Recent spectrometer designs utilize electromagnets for 15 to 25-MHz ^{13}C nmr or superconducting solenoids for ^{13}C spectroscopy at higher frequencies (50 to 125 MHz).

FOURIER TRANSFORM NMR

The experimental advances in the late 1960s allowed ^{13}C nmr spectroscopy to be utilized in practical studies of organic systems, but those studies still required a fair amount of determination.   The development of FT nmr[9] has made versatile ^{13}C nmr studies not only practical but also nearly comparable with ^{1}H nmr in terms of experimental ease and quality of results.

The great inefficiency of conventional frequency or field sweep nmr [also called *continuous wave* (CW) nmr] is the fact that at any given instant only one frequency is being observed. Thus for ^{13}C nmr at 2.3 T, where the chemical-shift range for most molecules covers $\gtrsim$5000 Hz (~200 ppm), each 1-Hz-wide resonance line would be observed $\gtrsim$1/5000 of the time.   The rest of the time would be spent observing other peaks or looking at the baseline!   Rapid sweeps help sensitivity somewhat by allowing more rf power to be used without saturation occurring. However, even with very rapid sweeps, only one frequency is observed at a given instant.   Rapid sweeps also limit resolution, artificially broadening signals.   The inefficiency of swept nmr is even more acute with high-field ^{13}C spectrometers since the ^{13}C shift range covers 12 to 30 kHz.

The only solution to the inefficiency of single-frequency observation is to excite all ^{13}C nuclei in the sample simultaneously and observe the total response of the sample. In principle, this is possible by utilization of several thousand transmitters and an equal number of receivers spaced across the ^{13}C chemical-shift range. Each receiver would have to be "tuned" to receive signals only from its paired transmitter, ignoring signals resulting from excitation by adjacent trans- mitters. Multiple transmitters/receivers of this type are, of course, impractical. Fortunately, another method of excitation and detection is possible.

A short rf pulse does not produce excitation at one fre- quency but instead excites a finite bandwidth of frequencies. If the pulse is short enough (~5 to 50 μsec), the bandwidth of excitation can be sufficient to simultaneously excite all ^{13}C nuclei in a sample. To effectively excite the sample in this short period of time, very high rf power is required.

The response of a sample to this excitation is absorption of individual frequency components by each nucleus. These fre- quencies, called *precession frequencies*, are detected by the receiver. The pattern detected by the receiver is called a *free induction decay* (FID). The FID corresponds to simultaneous "reradiation" of all of the frequencies absorbed by different kinds of ^{13}C nuclei in the sample. The FID corresponding to absorption of one frequency (e.g., a singlet absorption in a ^{13}C {^{1}H} spectrum) is an exponentially decaying sine wave. The frequency of the sine wave is the difference between the center (carrier) frequency of the rf excitation pulse and the rf fre- quency for that particular absorption (the Larmor frequency for that nucleus). This is shown in Figure 1.8a, where the Larmor

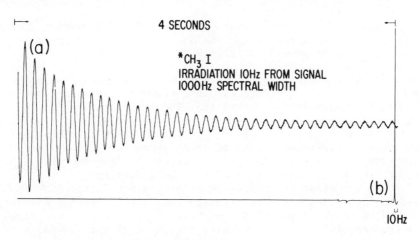

FIGURE 1.8.   (a) Time domain--FID; (b) frequency domain-- spectrum (*CH₃I, ¹H decoupled).

frequency is 10 Hz from the center frequency of the rf carrier. The FID shown in Figure 1.8*a* was recorded for 4 sec; the sine wave contains 40 cycles, and thus the precession frequency is 40/4 = 10 Hz.   The FT spectrum of the FID is shown in Figure 1.8*b*. The rf carrier was located at the far right during irradiation. The single peak is found 10 Hz from the right limit (the full spectral width is 1 kHz in Figure 1.8*b*).

The FID for a single-resonance line is a simple, easily recognized, exponentially decaying sine wave.   When two or more energy absorptions take place, however, the FID becomes more complex (Figure 1.9*a*).   In the general case, to abstract the

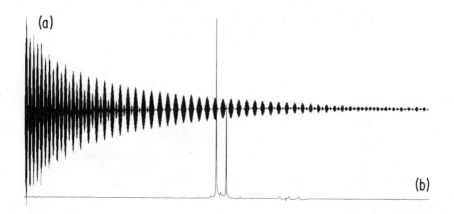

FIGURE 1.9.   *(a)   Free induction decay and (b) frequency domain spectrum of $(CH_2=CH)_4Si$.   (The small triplet is benzene-d_6 solvent.)*

spectral information (Figure 1.9*b*) from the FID, an FT must be performed.   The FT abstracts all the frequency components out of the complex waveform present in the FID.   The duration (in seconds) of the individual frequency component defines the resolution (linewidth in hertz) of each resonance line.   Frequencies that decay rapidly correspond to broad lines, and frequencies that decay slowly yield sharp lines after FT.   The overall shape of the FID is a juxtaposition of all these factors.

It is important to realize that nothing artificial has been done in this procedure.   The simultaneous excitation and detection allowed in an ideal pulsed nmr experiment will produce the same spectrum as will a true slow-passage frequency or field-sweep experiment.   The mathematical operation, FT, is required only to abstract the information from the free induction decay and present it in the same format as in CW experiments.

The sensitivity advantage gained over CW nmr in a single-pulse experiment is proportional to $(F/\Delta)^{1/2}$ where F is the total chemical shift range and Δ is the linewidth of the narrowest signal. For ^{13}C nmr the theoretical increase in sensitivity is of the order $(5 \text{ to } 25 \text{ kHz}/1\text{Hz})^{\frac{1}{2}}$ or 70 to 150.

Both pulsed FT and conventional sweep mode nmr employ time-averaging to improve S:N ratios of spectra. In CW cmr it is not uncommon for several hundred or several thousand 50-to-250-sec scans to be accumulated. In FT cmr as many as one million pulses spaced 0.1 sec to several seconds apart might be used. The S:N ratio of a spectrum improves with the square root of the number of scans or pulses. The advantage of the FT method can be expressed in terms of a much shorter time to achieve a given S:N ratio than that required in a frequency or field-sweep experiment. The sensitivity improvement of FT nmr results in a time improvement of order 10^3 to 10^4 (see Figures 1.2 and 1.3).

Ernst,[10] Kaiser,[11] and others[12] have suggested alternative forms of excitation for use in FT nmr. A number of newer pulse sequences are described in Chapter 10.

Instrumental Requirements for Fourier Transform ^{13}C NMR.   Discussions of instrumental requirements for pulsed FT nmr appear elsewhere.[13,14] Therefore, only the main features of FT instrumentation are covered here. Complete ^{13}C nmr FT instrumentation requires two subsystems: a spectrometer and a data-acquisition/processing package. The spectrometer usually conforms to third- or fourth-generation nmr specifications. In cases where organic applications of ^{13}C are of primary concern, the general requirements of the spectrometer subsystem for pulsed FT operation are not greatly different from those needed in CW nmr. A high-powered amplifier is required to increase the rf input to the probe and rf gates are needed to form the pulse and to protect the receiver while the pulse is switched on.

The greatest difference between CW and FT instrumentation is the requirement for sophisticated data-acquisition/processing equipment. The FID signal that is acquired in an FT experiment contains frequency components as high as the spectrum width. Sampling theory requires measurement of at least two data points per cycle to measure the frequency of a sine wave. For a 5000-Hz spectral width, this means that the data rate must be $\geq 10{,}000$ measurements per second; at high magnetic fields data acquisition is even faster. Measurement of each data point is done by an analog-to-digital converter (ADC), which converts the analog data into digital format for storage and processing by the computer.

Fourier transform nmr systems have minicomputers directly incorporated in the spectrometer. Computer memory requirements for FT nmr systems range from 16K (16,384 computer words) to

40K (40,960 words) or more. The configuration of the computer
hardware (especially the number of words and the word length in
bits) is a limiting factor in the resolution and dynamic range
of FT nmr systems. In FT systems, a part of computer memory is
utilized for programming, generally leaving 8 to 32K words for
placement of data. The mathematical operation of Fourier transformation of the FID yields only half as many data points in the
final absorption nmr spectrum; half are lost in the calculation.
Thus most FT spectra contain 4 to 16K real data points. Inclusion
of the full ^{13}C chemical-shift range in the spectrum results in
a digital limitation on resolution at 20 MHz of 0.25 to 1 Hz,
but at higher frequencies the digital resolution would be more
limited.

The computer word length (in bits) sets the maximum dynamic
range* of the experiment and also determines the mode of scan
averaging. Thus older nmr data systems using 16-bit words can
directly accumulate only 16 (2^4) scans assuming 12 bit ADC and
an FID with S:N ratio exceeding 1.† (those data systems used
software scaling routines to maximize flexibility and dynamic
range).   Modern FT nmr data systems perform 20 to 32-bit data
acquisition, followed by fixed-point or, in some cases, floating-
point fast Fourier transform (FFT) calculations. Dynamic ranges
exceeding 10^5 have been demonstrated.

An increasing number of commercial ^{13}C spectrometers have
the capability to do Fourier transforms of 64K or larger data
tables. The utility of these extra large transforms for ^{13}C nmr
at normal magnetic fields is marginal since other factors limit
the resolution to ≳1 Hz for typical wideband ^{1}H-decoupled experiments. Nevertheless, it can be advantageous in some cases to be
able to calculate these very large FIDs; this is particularly
true for ^{13}C measurements at very high frequencies. In cases
where spectral resolution is actually limited by available computer memory, spectra may usually be obtained that cover only
partial sweep widths. In such experiments the spectral resolution
will be limited only by magnetic field inhomogeneities, natural
linewidths, or decoupling efficiency.

In FT nmr all precession frequencies within the effective
bandwidth of the pulse are excited. The only way to obtain

* "Dynamic range" is defined as the observed ratio between the
largest and smallest real signals after Fourier transformation
of an FID.

† In actuality it is not generally good practice to utilize
the full 12 bits of an ADC unless single scan S:N is very large--
6 to 9 bit digitization is equally good; for low S:N FIDs even
2 to 3 bit digitization is sufficient.

partial sweep widths is to lower the sampling rate of the ADC,
thus limiting the highest frequency to be digitized and stored
in the computer data table. Unfortunately, if frequencies higher
than the maximum specified are present in the FID, they are
converted down to lower frequencies during the digitization
process. Down conversion (*folding* or *aliasing*) of frequencies
is minimized by placing low-frequency pass filters between the
receiver and the ADC. Frequencies much higher than the desired
spectral width are effectively eliminated; frequencies only a
little higher than the filter setting are partially attenuated.

   Spectral folds in phase-corrected FT spectra* usually can
be identified by the fact that the folds appear out of phase
in relation to the rest of the spectrum. A signal whose true
frequency is X Hz higher than the spectrum window (i.e., SW + X)
will be folded to frequency (SW - X) (see Figure 1.10).

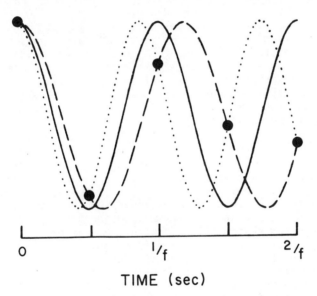

$$0 \qquad {}^{1}/_{f} \qquad {}^{2}/_{f}$$

TIME (sec)

FIGURE 1.10.[13b]   *Illustration of aliasing. The three curves
represent sinewaves of frequencies f Hz (——),
(f-X) Hz (---), and (f+X) Hz (···); the dots
(●) represent data sampling at 2f points/sec.
Note that (f+X) Hz and (f-X) Hz are indistinguishable.*

* All FT systems currently in use yield phase-corrected absorption spectra. Some early FT systems calculated power spectra
containing no phase information and yielding spectral lines
broadened by the contributions of dispersion mode signals.[13b]

There is another kind of spectral folding in FT nmr. The
ADC measures an audio difference frequency representing the
separation (in hertz) between the input carrier rf and the
frequency of the absorption signal. Single-phase spectrometer
receivers do not differentiate between frequencies higher or
lower than the reference frequency, but instead pass $+Y$ Hz or
$-Y$ Hz as Y Hz. If the carrier (reference) frequency is mistaken-
ly set within the range of absorption lines, the spectrum will
show both correct and "folded" resonances on one arbitrary side
of the carrier. This type of spectral fold is more difficult
to suppress than the first type described because low-frequency
filters do not discriminate between positive and negative
frequencies. However, folded resonances will usually appear
misphased in the spectrum, relative to resonance correctly
represented in digitization (Figure 1.11).

The second type of spectral folding is avoided in modern
FT nmr spectrometers that use quadrature-detection (QD) schemes.[15]
The QD method has other advantages also, with respect to efficiency
of pulse excitation and the use of lower ADC rates. (In QD nmr,
the input carrier is placed near the center of the spectrum;
positive and negative frequencies are differentiated.) One
complication can arise in QD nmr. If the irradiation frequency
is misset, so that some lines fall outside of the chosen window, then those lines will be folded into the opposite side of
the spectrum (Figure 1.11c).

Data Systems: A block diagram representative of modern FT
nmr spectrometer designs is given in Figure 1.12. The central
role of the computer in fourth-generation nmr spectrometers is
clearly indicated. The various manufacturers use somewhat
different approaches for the digital part of the spectrometer.
Current commercial designs allow extended word or double-pre-
cision data collection and FFT format, to improve spectral
dynamic range over that obtainable with the shorter 16-bit word
length common until recently. User input/output (I/O) is
commonly via quiet moderate-speed teletype substitutes, although
most spectrometers offer soft copy displays for more rapid user
I/O. Analog plotters are now yielding to digital x,y plotters
with rapid electrostatic plotters available as options.

Nuclear magnetic resonance data systems used on modern
commercial spectrometers can perform foreground/background processing. Thus data acquisition may proceed in the background,
while the user manipulates previously acquired data in the
foreground partition of the computer. The newest nmr computer
designs have more complex configurations or use dual processors
to perform several tasks simultaneously. Unfortunately, eco-
nomic considerations have prevented manufacturers from imple-
menting the fastest new minicomputer hardware and floating-
point arithmetic.

In larger, multispectrometer laboratories it is practical

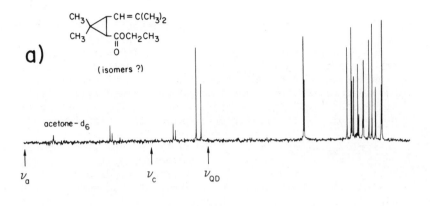

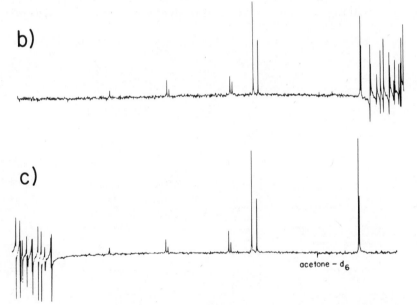

FIGURE 1.11.   *Single phase and quadrature detection FT nmr spectra of ethyl chrysanthemumate. Each spectrum obtained at 67.9 MHz, 8 scans, spectral width 15 kHz, ν_a, ν_b, ν_c, indicate placement of rf excitation in the three cases. Spectrum: (a) single phase detection, no lines aliased; (b) single phase spectrum, high field resonances folded; (c) quadrature detection spectrum with high field resonances folded into the low field region (irradiation near ν_{QD} would avoid aliasing). Note that spectrum (c) demonstrates S:N improvement of $\sqrt{2}$ as predicted for QD FT nmr. (Spectra, courtesy of T. Gedris and R. Rosanske.)*

21

to utilize one of the new super minicomputers in a distributed processing configuration. A data system based on such a computer extends data-processing capabilities significantly, although the costs of software development are not insignificant. One such system has been developed at Florida State University, based on a Data General Eclipse computer and programmed primarily in fast extended FORTRAN. Using Data General FORTRAN 5, an 8K floating-point FFT calculation requires less than 8 sec.[16]

High-Field Spectrometers.   Carbon-13 nmr spectroscopy at high magnetic fields is *generally* characterized by (1) increased sensitivity, (2) increased resolution (spectral dispersion), and (3) simplified ^{1}H-coupled ^{13}C spectra. The advantages of high-field ^{13}C spectroscopy have been particularly useful for synthetic and biopolymer studies (see Chapters 7 and 9) and for elucidating details of molecular dynamics with ^{13}C spin-relaxation measurements (Chapter 8).

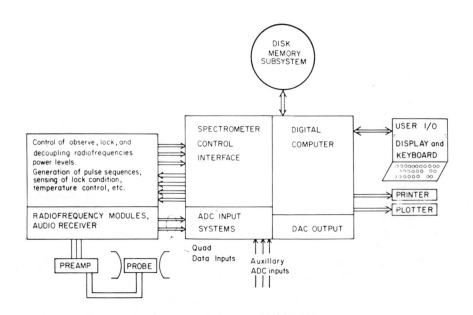

FIGURE 1.12.   Representative block diagram of a modern FT nmr spectrometer emphasizing the central role of the digital computer. Some systems utilize additional microprocessors to control input/output or spectrometer functions.

STATE-OF-THE-ART ^{13}C NMR

The first ^{13}C experiments required samples of several grams
that contained only a few kinds of carbon nuclei. Most compounds were of very low molecular weight and/or a high degree
of symmetry. At that stage of experimental development, spectra clearly indicated the promise of ^{13}C nmr but only teased
the organic chemist in terms of application to most practical
problems. This chapter has briefly covered many of the major
theoretical and experimental techniques developed over the past
decade. Table 1.1 compares current ^{13}C capabilities with
typical earlier experiments. It must be emphasized that the
numbers shown in Table 1.1 are not precise. The differences
in the five columns, however, are in general agreement with
recent experimental results. All values assume large sample
tubes and solubility limitations on sampling. Commercially
available microcells can increase relative sensitivity (two
to fourfold) when the available quantity of sample (and not
sample solubility) is limiting. Many nmr spectrometers are
not capable of running 20 to 25-mm sample tubes. For these
instruments the lower concentration limit should be increased.

One example of an ultrahigh sensitivity ^{13}C spectrum is
shown in Figure 1.13. This spectrum illustrates the "state-
of-the-art" for samples available in relatively large quantities,
where either solubility is limited or where the application
otherwise calls for low sample concentrations.

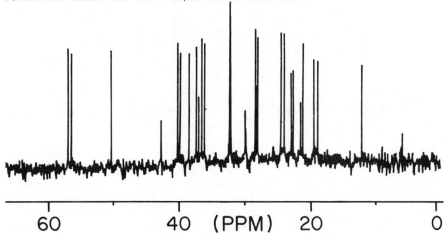

FIGURE 1.13.   *90.6 MHz ^{13}C nmr spectrum of 1 mM cholesteryl*
acetate (aliphatic region shown): 20,000 scans,
20 mm sample, Bruker WH-360 widebore spectrometer.
(Spectrum courtesy of F. Wehrli and M. Mattingly,
Bruker Instruments, Inc.)

An example of the other limiting case, where sample availability is limiting, is shown in Figure 1.14. Here 200 µg of cholesterol was run in a special-design 5 mm microcell probe (0.1 ml
total volume). This probe utilizes a largely transverse spinning
angle with a solenoid coil to achieve better sensitivity per unit
of sample. While ^{13}C nmr with $\gtrsim$ 100 µgm samples cannot be considered routine, it is clear that combined use of this type of
probe technology with higher field spectrometers will make practical such studies.

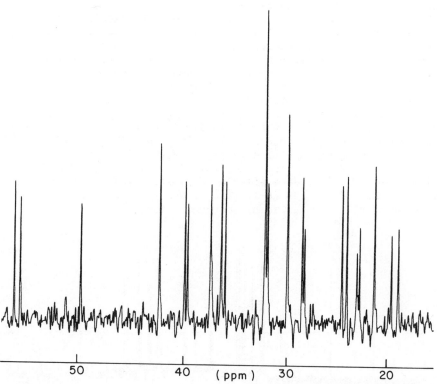

FIGURE 1.14.   50 MHz ^{13}C nmr spectrum of 200 µg of cholesterol
(natural isotopic abundance!) obtained in a tilt-
micro probe on a JEOL FX-200 spectrometer. The
spectrum was obtained from 50,000 scans (14 hrs);
saturated aliphatic region shown. (Spectrum,
courtesy of M. J. Albright, JEOL, Inc.)

TABLE 1.1.   NATURAL-ABUNDANCE ^{13}C NMR CAPABILITIES

CRITERION	1959-1963[a]	1965-1969[b]	1970-1972[c]	1975-1977	1980[d]
Typical molar concentration (assumes no molecular symmetry)	>5	1-2	0.5-1.0	0.1-1.0	0.05-0.1
Best resolution (organic molecules) (in parts per million)	>5	<0.05	<0.05	<0.02	<0.01
Practical lower limit on molar concentration[e]	>1	0.1	0.01	0.002	~0.0005
Resolution in experiments at low concentration[f] (in parts per million)	--	~0.4	<0.1	<0.1	~0.01
Experimental time required for lower concentration limit.[f,g] (in hours)	--	20	12	18	18
Experimental time required to run 0.5 to 1-M solution[h]	--	>1 hr	~1 min	≤5 sec	<1 sec

[a]Dispersion mode, rapid passage sweeps.
[b]Absorption mode time-averaged intermediate and slow sweeps with wide-band proton decoupling, homonuclear field/frequency control.
[c]Pulsed FT data acquisition, wide-band proton decoupling, heteronuclear field/frequency control.
[d]Pulsed FT QD, wide-band proton decoupling, heteronuclear field/frequency control, 20-mm sample tube at 50 to 90 MHz.
[e]To attain S:N ratios ≈12 (signal height over peak-to-peak noise divided by 2.5).
[f]Assumes entire ^{13}C chemical-shift range covered.
[g]After initial setup, run unattended.
[h]To achieve S:N ratios exceeding 25, for practical organic analyses.

REFERENCES

1. F. Bloch, W. W. Hansen, and M. E. Packard, *Phys. Rev.*, **69**, 127 (1946).
2. E. M. Purcell, H. C. Torrey, and R. V. Pound, *ibid.*, **69**, 37 (1946).
3. J. T. Arnold, S. S. Dharmatti, and M. E. Packard, *J. Chem. Phys.*, **19**, 507 (1951).
4. P. C. Lauterbur, *ibid.*, **21**, 217 (1957); C. H. Holm, *ibid.*, 707.
5. P. C. Lauterbur, *Ann. N. Y. Acad. Sci.*, **70**, 841 (1958).
6. A. Allerhand, D. Doddrell, and R. Komoroski, *J. Chem. Phys.*, **55**, 189 (1971).
7. R. R. Ernst, *ibid.*, **45**, 3845 (1966).
8. Early modifications of third-generation spectrometers: (a) D. D. Traficante, J. A. Sims, and M. Mulcay, *J. Magn. Resonance*, **15**, 484 (1974); (b) H. C. Dorn, L. Simeral, J. J. Natterstad, and G. E. Maciel, *ibid.*, **18**, 1 (1975); (c) C. S. Peters, R. Codrington, H. C. Walsh, and P. D. Ellis, *ibid.*, **11**, 437 (1973).
9. R. R. Ernst and W. A. Anderson, *Rev. Sci. Inst.*, **37**, 93 (1966).
10. R. R. Ernst, *J. Magn. Resonance*, **3**, 10 (1970).
11. R. Kaiser, *ibid.*, **3**, 28 (1970); *ibid.*, **15**, 44 (1974).
12. D. Ziessow and B. Blümich, *Ber. der Bunsenges.*, **78**, 1168 (1974); J. Dadok and R. F. Sprecher, *J. Magn. Resonance*, **13**, 243 (1974); R. K. Gupta, J. A. Ferretti, and E. D. Becker, *ibid.*, 275; B. L. Tomlinson and H. D. W. Hill, *J. Chem. Phys.*, **59**, 1775 (1973).
13. (a) D. Shaw, *Fourier Transform NMR Spectroscopy*, Elsevier, London, 1976; (b) T. C. Farrar and E. D. Becker, *Pulse and Fourier Transform NMR*, Academic, New York, 1971.
14. D. I. Hoult, in *Topics in Carbon-13 NMR Spectroscopy*, Vol. 3, G. C. Levy, *Ed.*, Wiley-Interscience, New York, 1979, Chapter 1.
15. (a) J. D. Ellet, *et al.*, *Adv. Magn. Resonance*, **5**, 117 (1971); (b) A. G. Redfield and R. K. Gupta, *ibid.*, **82**; (c) E. O. Stejskal and J. Schaefer, *J. Magn. Resonance*, **14**, 160 (1974); (d) D. M. Wilson, R. W. Olson, and A. L. Burlingame, *Rev. Sci. Inst.*, **45**, 1095 (1974).
16. This software-hardware project is described in *Computer Networks in the Chemical Laboratory*, G. C. Levy and D. Terpstra, Eds., Wiley-Interscience, New York, in press.
17. J. N. Shoolery, in *Topics in Carbon-13 NMR Spectroscopy*, Vol. 3, G. C. Levy, *Ed.*, Wiley-Interscience, New York, 1979, Chapter 1 (see also ref. 15d).

BIBLIOGRAPHY

Selected General NMR Reference Books

E. D. Becker, *High Resolution NMR, Theory and Applications,*
Academic Press, New York, 1980. Complete coverage at inter-
mediate level. Clear explanations.
F. A. Bovey, *NMR Spectroscopy*, Academic, New York, 1968. A
lower-level approach than that of Becker.
L. M. Jackman and S. Sternhell, *Applications of Nuclear Magnetic
Resonance Spectroscopy in Organic Chemistry*, 2nd ed., Pergamon,
New York, 1969.
J. W. Emsley, J. Feeney, and L. H. Sutcliffe, *High Resolution
Nuclear Magnetic Resonance Spectroscopy*, Pergamon, New York,
1965.

Monographs on ^{13}C NMR

J. B. Stothers, *Carbon-13 NMR Spectroscopy*, Academic, New York,
1972. Comprehensive treatise. Literature coverage through
mid-1970.
E. Breitmaier and W. Voelter, ^{13}*C NMR Spectroscopy, Methods and
Applications*, Verlag Chemie, Weinheim, West Germany, 1974.
Expensive. Second edition published in Europe in 1978.
J. T. Clerc, E. Pretsch, and S. Sternhell, ^{13}*C-Nuclear Resonance
Spectroscopy* [in German], Akad. Verlag, Frankfurt, West Ger-
many, 1973.
F. Wehrli and T. Wirthlin, *Interpretation of Carbon-13 NMR Spectra*,
Heyden, London, 1976. Excellent text.
R. J. Abraham and P. Loftus, *Proton and Carbon-13 NMR Spectra*,
Heyden, London, 1978. Useful, low-level text. Extensive ma-
terials from literature are used without references.
E. Buncel and C. C. Lee, Eds., *Isotopes in Organic Chemistry*,
Vol. 3, Elsevier, London, 1977. Mechanistic and spectroscopic
studies, biosynthesis, and stereochemistry.
G. C. Levy, Ed., *Topics in Carbon-13 NMR Spectroscopy*, Wiley-
Interscience, New York. Advanced treatments of new developments covering applications of ^{13}C nmr in chemistry and biology;
three volumes published to date.
Vol. 1 (1974): theory of chemical shifts; substituent effects;
^{13}C relaxation; synthetic high-molecular-weight
polymers; high-magnetic-field cmr; studies of
reaction mechanisms.
Vol. 2 (1976): natural products; peptides; biopolymers; ^{13}C biosynthetic studies; organometallics; structure
assignments with ^{13}C spin-relaxation data; the
computer in FT nmr; theory of spin-spin coupling.

Vol. 3 (1979): experimental techniques; ^{13}C spin relaxation; ^{13}C
nmr of heterocyclics; synthetic and biopolymers in
the "solid" state, ^{13}C CIDNP.

Special Topics

D. Shaw, *Fourier Transform NMR Spectroscopy*, Elsevier, London,
1976. Expensive but modern treatment of FT nmr theory and
techniques.

K. Mullen and P. S. Pregosin, *Fourier Transform NMR Techniques:
A Practical Approach*, Academic, London, 1976. Low-level
approach.

Data Banks, Spectra Catalogs

L. F. Johnson and W. C. Jankowski, *Carbon-13 NMR Spectra*, Wiley-
Interscience, New York, 1972. Five hundred spectra.

V. Formacek, L. Desnoyer, H. P. Kellerhals, T. Keller, and J. T.
Clerc, ^{13}C *Data Bank*, Vol. 1, Bruker Physik (West Germany),
1976.

NIH-EPA Chemical Information System (CIS) CNMR Search System,
described in D. L. Dalrymple, C. L. Wilkins, G. W. A. Milne, and
S. R. Heller, *Org. Magn. Resonance*, **11**, 535 (1978).

E. Breitmaier, G. Hass, and W. Voelter, *Atlas of Carbon-13 NMR
Data*, **Vols.** 1-3, through 1979, Heyden, London.

P. L. Fuchs and C. A. Bunnell, *Carbon-13 NMR Based Organic Spectral Problems*, Wiley, New York, 1979.

Chapter 2

^{13}C NMR Spectral Characteristics

In proton nmr the three types of spectral information--chem-
ical shifts, coupling constants, and peak area measurements--
are of nearly comparable value in organic structure determinations.
Although the same types of information are available in cmr, chemical shifts generally are the most useful parameters. Quantitative coupling information is not usually obtained in routine ^{13}C
nmr studies. Peak intensities can be profitably utilized. However, the direct relationship between area measurement and the
number of nuclei absorbing energy is often lost unless the experiment is specifically designed to retain this information.
Carbon-13 spin-relaxation parameters may be an additional aid in
structure or dynamic ^{13}C studies. This chapter presents some
generalizations about ^{13}C nuclear magnetic resonance characteristics and how they can be used in structure determinations and
other studies.

CHEMICAL SHIFTS

Known carbon chemical shifts in diamagnetic compounds (compounds containing no unpaired electrons) cover a range of
approximately 600 ppm. For almost all organic molecules, complete
^{13}C spectra appear between deshielded carbonyl groups and shielded
methyl groups in a range of just over 200 ppm. Early cmr studies
referenced ^{13}C chemical shifts to an external capillary of ^{13}C-
enriched CS_2 (*CS_2). According to that convention, positive
chemical shifts were *upfield* from CS_2. During the 1970s tetramethylsilane (TMS), the same standard used in ^{1}H nmr, became the
nearly universal internal ^{13}C shift standard. Shifts downfield
from the TMS methyl carbon resonance signal are stated as positive,
in analogy with the proton δ scale. Chemical shifts in this book
are referenced to TMS. In practice, solvent resonances are often
used as secondary standards with chemical shifts adjusted to the
TMS scale. Table 2.1 lists a number of solvent ^{13}C chemical
shifts for use in reporting ^{13}C chemical shifts. It is important
to realize that solvent effects and use of different standards
reduce precision in comparisons of ^{13}C chemical shifts obtained
in different studies.

29

TABLE 2.1.   TMS-BASED ^{13}C CHEMICAL SHIFTS FOR COMMON STANDARDS
AND SOLVENTS

COMPOUND, SOLVENT	CHEMICAL SHIFT[a]	
	PROTIO COMPOUND	DEUTERIO COMPOUND
Acetonitrile (methyl carbon)	1.7	1.3[b]
Toluene (methyl carbon)	21.3	20.4[b]
Cyclohexane	27.5	26.1
Acetone (methyl carbon)	30.4	29.2
Hexamethylphosphortriamide	36.9	35.8[b]
Dimethylsulfoxide	40.5	39.6
Methanol	49.9	49.0[b]
Methylene chloride	54.0	53.6
Nitromethane	62.8	--
Dioxane	67.4	66.5[b]
Chloroform	77.2	76.9
Carbon tetrachloride	96.0	
Benzene	128.5	128.0
Acetic acid (*CO)	178.3	
CS$_2$	192.8	
CS$_2$ capillary	193.7	

[a]   In parts per million from internal TMS; ±0.05 ppm at 38°C.

[b]   F. Wehrli and T. Wirthlin, *Interpretation of Carbon-13 NMR Spectra*, Heyden, London, 1976.

    In the past few years extensive efforts have been made to theoretically calculate ^{13}C chemical shifts.[2,3] Unfortunately, available theoretical treatments still do not fully explain trends in ^{13}C shielding; the most accurate calculations are at present limited to small molecules containing only first- and second-row elements. The methods of current theoretical treatments are not detailed in this text; however, it is appropriate to list the various contributions to carbon chemical shifts: (1) hybridization, (2) inductive effects, (3) electric field effects, (4) "steric" effects, (5) hyperconjugative effects, (6) mesomeric (resonance) effects, (7) neighbor anisotropic effects,

(8) heavy halogen effects, and (9) isotopic effects. These
effects give rise to the major empirical trends observed in all
nmr spectra. The particular sensitivity of ^{13}C shieldings to
hybridization and substituent electronegativity (with subsequent
inductive effects) has given rise to the popular misconception
that carbon chemical shifts depend entirely on electron density
considerations. Large steric and electric field effects couple
with not-well-understood universal β-substituent effects to produce
complex patterns for ^{13}C shieldings. In some cases mesomeric
effects or heavy halogen (bromine and iodine) substituents may
result in shielding changes larger than 50 ppm. By contrast,
anisotropic neighboring groups, so important in ^1H nmr, are relatively unimportant in ^{13}C nmr, amounting generally to a few parts
per million or less.

In one approximate description three terms contribute to ^{13}C
shifts: a diamagnetic shielding term, a paramagnetic shielding
term, and a term containing the contributions from anisotropy
in the magnetic susceptibility of neighboring atoms or groups.
The diamagnetic term is due to currents involving electrons on
the nucleus being considered (and nearest neighbors). The paramagnetic term arises from contributions of higher electronic
states (especially low-level "excited states") to the description
of the ground state. The third term accounts for electron cir-
culations on neighboring atoms and specific anisotropic groups
such as phenyl rings, carbonyl groups, and acetylenes.

The anisotropic term depends only on the spatial relationship between the nucleus whose shift is being considered and the
anisotropic group. It remains essentially constant (in parts per
million) regardless of the nucleus being considered. Since ^{13}C
chemical shifts occur over a far greater range than do proton
shifts, the anisotropic term is relatively insignificant in cmr.
Available calculations agree that although the contribution of the
diamagnetic term is not insignificant, the paramagnetic term
normally dominates ^{13}C chemical shifts.

For the cmr user, semiempirical relationships developed over
the years are perhaps most helpful. It has been found experimentally that the effects of substituents on ^{13}C chemical shifts
are largely additive; thus predictions of cmr shifts by semiempirical methods are usually successful. Some general classes of
compounds and their chemical-shift ranges are shown in Figure 2.1.

Some of the shielding trends shown in Figure 2.1 have direct
analogies with ^1H nmr chemical shifts, although magnified here
by a factor of 15 to 20. For example, sp^3 carbons of saturated
hydrocarbons are highly shielded, whereas sp^2 carbons are among
the least shielded. Within the class of sp^3 carbons, methyl
resonances are the most shielded, again in analogy with proton
nmr. Acetylenic sp carbons are found at positions intermediate
between those of sp^3 and sp^2 carbons, as in proton spectra. Not
surprisingly, the effect of electronegative substituents is to

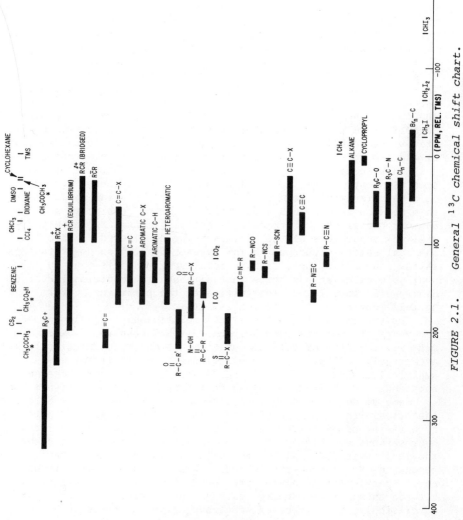

FIGURE 2.1.   General ^{13}C chemical shift chart.

32

deshield a given carbon; for a single fluorine substituent, the change at the substituted carbon is on the order of 60-70 ppm!

Of course, some of the shielding trends given in Figure 2.1 have no direct analogy in proton spectra. For example, carbon-containing functional groups such as carbonyls, nitriles, and derivatives may be examined directly in cmr. Since the shielding of similar carbons may in fact be proportional to electron charge (this is true for certain types of molecular systems if comparisons are carefully constructed) studies of charged organic intermediates such as carbocations can be quite fruitful.

The shielding trends of cmr are outlined in the following chapters.

"Steric" Shifts.   Remote substituent effects in ^{13}C nmr have been observed to be sensitive to relative stereochemistry. For example, γ-gauche alkyl substituents give rise to upfield shifts of 2 to 6 ppm, depending on molecular conformational possibilities. Even δ-gauche alkyl groups can conformationally affect the resonance position of a carbon, in this case deshielding it by 1 to 2 ppm. Specific shifts combined with the use of related model compounds are powerful tools in the analysis of ^{13}C nmr spectra of complex organic molecules. These considerations are dealt with in Chapters 3 and 9.

Chemical-Shift Reagents and ^{13}C NMR.   Lanthanide chemical-shift reagents[4] that have been used to "simplify" ^{1}H nmr spectra of compounds containing coordinating functional groups (hydroxyl, carbonyl, ether, etc.) can also be used in cmr studies. Since ^{13}C resonances enjoy a far greater freedom from overlap than do proton resonances, chemical-shift reagents do not yield the same level of benefits as they do in proton nmr. Nevertheless, it is possible to aid ^{13}C spectral assignments with pseudocontact shift measurements made with these reagents.[5] It has been noted that the assumption of insignificant contact shifts will often not apply[6] for ^{13}C, although this complication can be obviated with the use of Yb^{3+} shift reagents.[6a,7] Furthermore, the difficulties of dealing with nonrigid molecules have been treated.[8] Another significant use of shift reagents is in $^{13}C\{^{1}H\}$ decoupling.

SPIN-SPIN COUPLING INVOLVING CARBON

There are three types of carbon spin-spin interaction: coupling between two ^{13}C nuclei, coupling between a ^{13}C nucleus and a proton, and coupling between a ^{13}C nucleus and a magnetic nucleus other than ^{1}H or ^{13}C (^{13}C-X coupling). Because of the low probability of having two ^{13}C nuclei in the same molecule, ^{13}C-^{13}C coupling is not usually observed in ^{13}C spectra of natural-abundance compounds. ^{13}C-^{1}H Coupling constants are not obtained from

^{13}C experiments using wide-band ^{1}H decoupling and thus are not usually determined in the course of structure determinations. One-bond and long-range ^{13}C-^{1}H coupling constants can be obtained in ^{13}C nmr experiments without ^{1}H decoupling. However, the technique is of limited utility for complex molecules, except with use of two-dimensional FT nmr techniques (see Chapter 10 and ref. 9).

For simple molecules, ^{1}H-coupled ^{13}C spectra are often well resolved. Additionally, gated decoupling in these cases ({^{1}H}ON *between* data acquisitions) can regain some sensitivity through production of NOE. Some ^{13}C-H scalar couplings may be obtainable from ^{13}C satellites in ^{1}H spectra.

Molecules that contain magnetic nuclei other than ^{1}H and ^{13}C can give rise to spin-spin coupling that is observable in ^{13}C nmr, assuming a reasonable isotopic abundance for X and the lack of significant quadrupole broadening if X has a nuclear spin >1/2. Many studies of ^{13}C-X coupling involve the isotopically abundant, spin-1/2 nuclei ^{19}F and ^{31}P; some have shown C-X coupling to other nuclei. A few representative C-X couplings are shown in Table 2.2.[10]

The theoretical framework of spin-spin coupling is complex.[11] However, it is possible to estimate one-bond ^{13}C-^{1}H and ^{13}C-^{13}C couplings in organic molecules from semiempirical arguments. In hydrocarbons, for example, one-bond ^{13}C-^{1}H couplings (notation: $^{1}J_{CH}$) have been related to the amount of s character on the carbon nucleus. Thus sp^3-hybridized carbons have the lowest one-bond C-H coupling ($^{1}J_{CH}$ ~125 Hz), whereas $^{1}J_{CH}$ for sp carbons is highest (~250 Hz). Substitution on carbon by electronegative atoms increases $^{1}J_{CH}$ significantly. The twin dependence of $^{1}J_{CH}$ on carbon hybridization and substitution by electronegative atoms is evident in Table 2.3.

Various theoretical calculations have been employed for predictions of ^{13}C-^{1}H, ^{13}C-^{13}C, and ^{13}C-X spin-spin coupling constants. A recently proposed self-consistent field (SCF) finite-perturbation method that utilizes the INDO molecular orbital approximation has been used successfully in treating a large number of one-bond C-H and C-C couplings.[12] One-bond C-C coupling constants increase crudely with increasing s character of the two ^{13}C nuclei. This is true for C-Si and C-Sn one-bond coupling. Table 2.4 lists representative ^{13}C-^{13}C coupling constants. Note that cyclopropyl ring coupling constants show the unusual nature of cyclopropane bonding with "sp^2" ^{13}C-^{1}H couplings, $^{1}J_{CH}$ = 161 Hz and C-C couplings of 10 to 13 Hz corresponding to "sp^5" C-C bonds.

TABLE 2.2. REPRESENTATIVE C-X COUPLINGS[a], [10]

		$(Ph)_n X$			
X	n	$^1J_{CX}$	$^2J_{CX}$ (ortho)	$^3J_{CX}$ (meta)	$^4J_{CX}$ (para)
F	1	-245.3	+ 21.0	+ 7.7	+ 3.3
P	3	12.4	19.55	6.7	~ 0
P^+	4	88.4	10.9	12.8	2.9
Hg	2	1,186	88	101.6	17.8
B^-	4	49.5		2.6	
H	1	157.7	+ 1.2	+ 7.6	- 1.2

			$(C_n H_{2n+1})_m X$		
X	m	n	$^1J_{CX}$	$^2J_{CX}$	$^3J_{CX}$
F	1	6	-166.6	+ 19.9	5.25
P	3	4	- 10.9	+ 11.7	12.5
P^+	4	4	+ 47.6	- 4.3	15.4
Hg	2	4	+656	- 26.3	
Sn	4	2	+307.4 (^{117}Sn) +321.5 (^{119}Sn)	- 23.5	
H	1	2	+125	- 4.5	

[a] In Hz; specified signs refer to reduced coupling constants. $^iJ_{CX}$ refers to the coupling constant between X and the carbon i bonds away.

TABLE 2.3.   ONE-BOND CARBON-HYDROGEN COUPLING CONSTANTS[a]

COMPOUND	$^1J_{CH}$
CH_3CH_3	125.0
$CH_2=CH_2$	156.2
C_6H_6	158.5
$HC\equiv CH$	249.0
CH_3NH_2	133.0
CH_3OCH_3	140.0
CH_3F	149.0
CH_2F_2	184.5
CHF_3	239.1
CH_2O	172.0
HCO_2^-	194.8
HCO_2H	222.0
$HC\equiv N$	269.0
$HC\equiv \overset{+}{N}H$	320.0

[a] In Hertz. Data from ref. 12.

^{13}C PEAK AREA; INTEGRATION

In ^{13}C nmr several factors lead to difficulties when meaning-ful peak area measurements are desired. Peak area measurements may be obtained; however, correlation between integrated peak areas and the number of carbon nuclei in each peak requires care-ful design of the experiment. Area measurements are not usually obtained in routine ^{13}C spectra. There are two main reasons for the loss of correlation between the number of carbon nuclei

TABLE 2.4.   ONE-BOND CARBON-CARBON COUPLING CONSTANTS[a]

COMPOUND	$^1J_{CC}$	COMPOUND	$^1J_{CC}$
CH_3CH_3	34.6	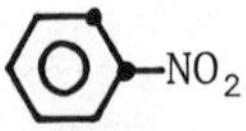 —NO_2	61.3
$CH_3CH_2CH_3$	37.7	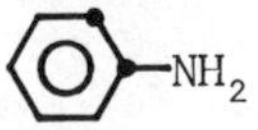 —NH_2	55.4
CH_3CCl_3	56.1	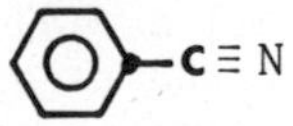 —$C \equiv N$	80.3
CH_3CO_2H	56.7	$(CH_3)_2C=C=CH_2$	99.5
CH_3CO_2-	51.6	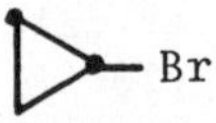 — Br	13.3
$CH_3C \equiv CH$	56.5	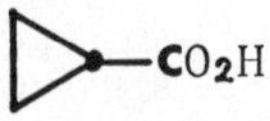 —CO_2H	54
$CH_2=CH_2$	67.6	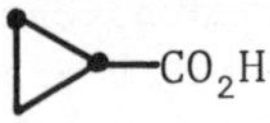 —CO_2H	10.0
$CH_3C \equiv N$	67.4		
$HC \equiv CH$	171.4		

[a]In Hertz.   Data from refs. 12 to 14.

comprising a peak and the integrated peak area: possible differential saturation effects from variable spin-lattice relaxation times and variable nuclear Overhauser enhancements.

Swept (CW) experiments normally utilize relatively slow (e.g., 50 to 250-sec) sweeps under repetitive, time-averaging conditions. Carbon-13 nuclei that have particularly long spin-lattice relaxation times (T_1) will remain partially saturated from previous sweeps because it takes five times T_1 to restore the Boltzmann distribution to >99% of its thermal equilibrium value. The signal from a partially saturated nucleus is of lower intensity than the signal from a nucleus whose T_1 is much shorter than the time required for each sweep. In FT nmr the situation is generally much worse because the time for each "sweep" (pulse) is much shorter. In a typical FT experiment the pulse repetition rate may be 0.1 to 1 sec. Carbon-13 spin-lattice relaxation times in most organic compounds are longer than 0.1 sec. If T_1 is close to the experimental repetition rate, even relatively small differences in T_1 values among the different nuclei in a molecule will lead to significant differences in peak areas. When accurate peak areas are required, the sweep time or pulse repetition rate must be ≳5 times the longest T_1 value for a nucleus whose peak area is to be measured.

Carbon-13 peak areas may not properly indicate relative numbers of ^{13}C nuclei even when precautions are taken to ensure that all ^{13}C nuclei in a sample relax completely between scans or pulses. In the normal mode of operation utilizing ^{1}H decoupling, ^{13}C resonances are augmented by NOEs. The Overhauser enhancement is not necessarily equal for each carbon in a molecule. In fact, it is unusual for all carbons in a low-molecular-weight organic molecule to achieve the theoretical maximum Overhauser enhancement.

In ^{13}C{^{1}H} experiments the maximum *integrated intensity ratio* for decoupled and undecoupled (i.e., coupled) resonances (I_D/I_O) is 2.99, corresponding to a maximum NOE factor, NOEF $\equiv$ [(I_D/I_O)-1] of 1.99[15] (this theoretical limit depends on the observed and decoupled nuclear species). It is often possible to achieve good quantitative results in ^{13}C{^{1}H} experiments, especially for protonated carbons. In molecules containing more than a few carbons (molecular weights ≳200) the NOE is usually (but not always) complete for all protonated carbons, indicating the dominance of dipole-dipole spin-lattice relaxation.[16] Nonprotonated carbons achieve NOEF values of 0.8 to ~2.0 in small organic molecules (integrated intensity ratios of 1.8 to ~3.0). In larger organic molecules all carbons typically yield full NOEs at least when spectra are obtained at low or moderate magnetic fields.[16,17] The problems of peak area measurement are summarized in Figures 2.2 to 2.4. Integrated peak intensities for the spectra shown in Figures 2.2 to 2.4 are given in Table 2.5.

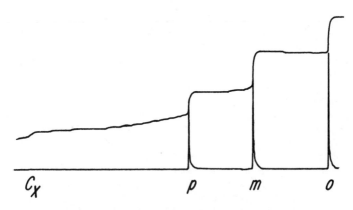

C_X p m o

FIGURE 2.2.   *Normal FT ^{13}C spectrum, pulse interval 2 sec (Note: the nonprotonated carbon (C_X) resonance is broadened because of scalar T2 relaxation with ^{14}N of the nitro group; integration data, summarized in* **Table 2.4**).

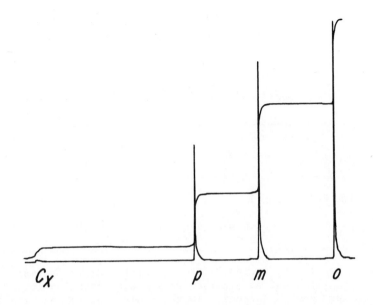

FIGURE 2.3.   Fourier transform ^{13}C spectrum obtained with 12-sec
pulse interval (10-sec delay) (see notes, Figure
2.2).

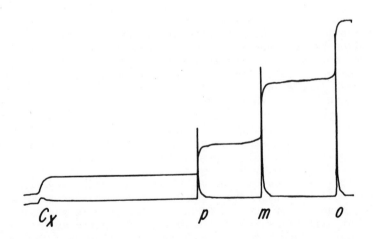

FIGURE 2.4.   Fourier transform ^{13}C spectrum obtained under long
(300-sec) pulse-interval conditions (see notes,
Figure 2.2).

TABLE 2.5.   NORMALIZED INTEGRATED PEAK INTENSITIES FOR NITROBENZENE

PULSE INTERVAL, (sec)	NORMALIZED INTEGRATIONS[a]			c_x
2	5.63	5.76	3.62	0.87
12	5.78	5.84	3.37	1.57
300	5.94	5.94	3.06	2.03
Spin-Lattice Relaxation Time, T_1[b]				
	6.9	6.9	4.8	56

[a] From the spectra in Figures 2.2 to 2.4; integrations normalized from *ortho, meta,* and *para* carbon peak areas, assuming full NOE for these protonated carbons. In the table, one carbon with full NOE = 2.99; one carbon with no NOE = 1.00

[b] T_1 in seconds.

    Figure 2.2 represents a normal ^{1}H-decoupled FT ^{13}C spectrum of nitrobenzene. The spectrum was accumulated in 15 pulses, spaced 2 sec apart. As shown in Table 2.5, the relative peak areas do not closely correlate with the number of carbons.
    If a short delay (10 sec) is inserted after the data acquisition following each pulse, the protonated carbon peak areas roughly correspond to the respective number of ^{13}C nuclei (Figure 2.3 and Table 2.5). The T_1 for the nonprotonated carbon is 56 sec, and a long delay is required to eliminate the effect of partial saturation for that carbon. Figure 2.4 gives the FT spectrum of nitrobenzene under long (300-sec) pulse delay conditions. The integrated peak intensities of all four resonances in Figure 2.4 thus have physical significance. The lower peak area for the (^{14}N quadrupole-broadened) nonprotonated carbon signal additionally results from a lower NOE.
    There are instrumental sources of error in peak area measurements that must be considered. The first contribution involves the rf pulse amplifier used in FT spectrometers. To achieve accurate intensity measurements in FT nmr, the pulse amplifier must be powerful enough to give uniform excitation over the entire spectral width. For ^{13}C nmr at 25.2 MHz, a pulse amplifier yielding an excitation field B_1 greater than 10 kHz would be adequate to effectively cover the usual 200-ppm ^{13}C chemical-shift range; at higher frequencies B_1 would have to be correspondingly higher. Some ^{13}C FT systems currently in use do not have sufficiently

powerful rf amplifiers, especially for larger-sized samples. In these cases, smaller spectral windows or shorter pulses yielding <90 ° nuclear flip angles (see Chapter 8) must be used if accurate intensities are to be obtained. Quadrature-detection schemes make more efficient use of available pulse power since they allow irradiation in the center of the nmr spectrum.

Another potential problem arises from the limited size of FT computer data tables. Thus a digitized nmr spectrum may have individual peaks defined by as few as two or three data points, and this can result in imprecise representation of these resonances. Fortunately, FT nmr peak integrations are less sensitive to digitization errors than are peak height measurements. Except in cases where a high degree of accuracy is required (±5%), resonances defined by two or three data points may be satisfactory.

REPRESENTATIVE ^{13}C SPIN-LATTICE RELAXATION TIMES

Detailed discussion of ^{13}C spin-relaxation is deferred until Chapter 8. However, to obtain meaningful ^{13}C peak integrations, it is necessary to predict approximate ^{13}C relaxation behavior. Table 2.6 lists ^{13}C T_1s and NOEs for representative compounds. Two trends are easily observed from these data:

1. Carbons that have one or more directly attached hydrogens (protonated carbons) have shorter T_1s than nonprotonated carbons;
2. Values of T_1 in large molecules or in associated molecules (e.g., 2-aminoethanol and cyclopentanol) are shorter than in small-sized, nonassociating molecules (e.g., benzene and cyclopentanone).

Another trend noted from the data in Table 2.6 is that NOEFs, where reported, are at or near the theoretical maximum, 1.99. For most carbons in *very large* molecules, (e.g., oligonucleotides, oligopeptides, synthetic polymers, and highly associated steroids), lower than maximum NOEs are usually observed. Also, T_1 and NOE data can depend on the magnetic field of the spectrometer. These considerations are detailed in Chapters 8 and 9.

QUANTITATIVE ANALYSIS WITH ^{13}C NMR

Despite potential difficulties in obtaining meaningful peak integrations, the superior spectral dispersion of ^{13}C nmr makes it suitable for quantitative analysis. In fact, quantitative analysis by ^{13}C nmr can be uniquely attractive when care is exercised in experimental design. Spin-relaxation reagents[23,24] have been used to shorten and equalize ^{13}C T_1s and to suppress NOEs by circumventing normal ^{13}C-^{1}H dipolar relaxation. Unfortunately, use of these reagents can actually degrade results for larger organic molecules, where their general use is not recommended.[25]

TABLE 2.6. REPRESENTATIVE ^{13}C SPIN-RELAXATION DATA[a]

COMPOUND	CARBON	T_1(sec)	NOEF	REFERENCE
$CHCl_3$		32		
(benzene)		29	1.6	16a
$CH_3CH_2CH_2CH_2OH$	1	3.5		
	2	4.0		
	3	6.0		16b
	4	6.0		
$H_2NCH_2CH_2OH$ (2, 1)	1	0.61	2.0	18
	2	0.71	2.0	
(phenyl benzoate; positions 3, 2, 4, 2', 3', 4')	2,3	3.0	2.0	
	2',3'	3.5	2.0	16a
	4,4'	1.4	2.0	
(cyclopentanone)	2,5	12.2	2.0	19
	3,4	12.0	2.0	
(cyclopentanol)	1	2.5	2.0	
	2,5	4.1	2.0	19
	3,4	3.3	2.0	
(phenylacetylene, $-C{\equiv}CH$, α β)	1	107	0.8	
	2,3	14.0	2.0	
	4	8.2	1.9	16b
	α	132	0.9	
	β	9.3	2.0	
(piperidine, NH)	2	8.0		
	3	7.6		20
	4	6.9		
(dimethylcyclohexane, CH_3, CH_3)	1	70		
	2	11.9		
	3	11.3		20
	4	10.4		
	CH_3	8.4		

42

TABLE 2.6.   Con't.

COMPOUND	CARBON	T_1(sec)	NOEF	REFERENCE

Adenosine 5'-mono
phosphate

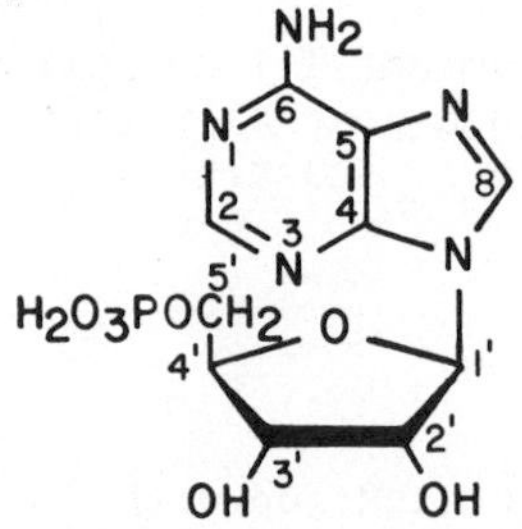 1'	0.19		
2'	0.23		
3'	0.22		
4'	0.19		
5'	0.11		17
2	0.15		
4	5.3		
5	4.7		
6	2.4		
8	0.16		

Methyl methoxypodocarpate

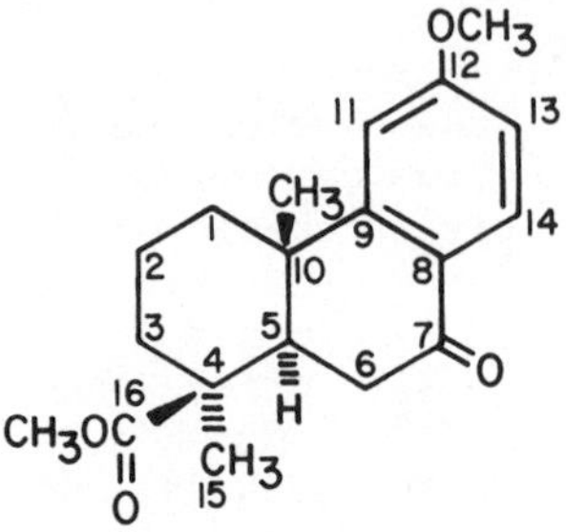 1,2,6	1.1 ± 0.2	1.9	
11,14,5	1.0 ± 0.2	2.0	
13	1.5	2.0	
4	16.0	1.9	
10	15.0	1.9	21
7[b]	34.5	1.8	
8[b]	48.0	1.7	
9[b]	29.0	1.9	
12[b]	33.0	1.8	
16[b]	48.0	1.8	

Codeine

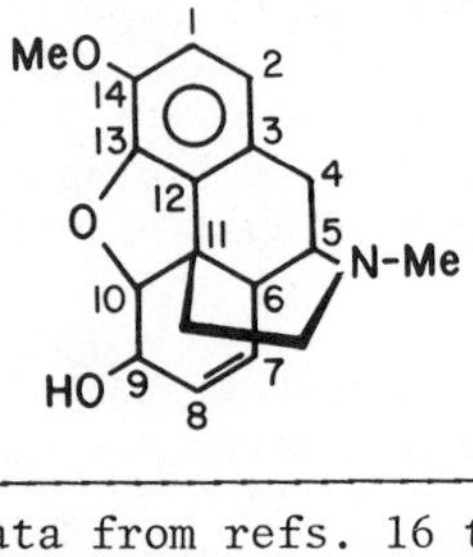 1,2,5,6,9,10	0.15 ± 0.1		
3	1.82		
4	0.11		
7,8	0.15,0.18		22
11	1.58		
12	5.1		
13	8.9		
14	5.2		

[a]Data from refs. 16 to 22. Most data obtained at low magnetic
fields (1.4 to 2.3T) and at or near room temperature.
[b]At high fields these carbons were measured to have shorter T_1s
and lower NOEs because of competitive chemical-shift anisotropy
relaxation (see Chapter 8).

43

Nuclear Overhauser effects may be effectively suppressed by using gated decoupling[26] with moderate delays between scans. For molecules with unfavorable relaxation characteristics, joint use of relaxation agents and gated ^{1}H decoupling with short scan delays gives excellent results.[27]

Various "inert" relaxation agents have been used for organic samples, including trisacetylacetonatochromium(III),Cr(acac)₃. A better agent is trisdipivaloylmethanatochromium(III),Cr(dpm)₃,[28] which is more soluble in nonpolar solutions and generally inert to hydrogen bonding interactions with hydroxylic substrates [in contrast to the behavior of Cr(acac)₃[23c,29]].

ORGANIC STRUCTURAL ASSIGNMENTS WITH ^{13}C NMR

Carbon-13 nmr spectra used alone or in combination with ^{1}H spectra give the chemist unparalleled insight into the details of organic molecular structure. Assignment of ^{13}C resonances for a moderately complex molecule may use some or all of the following techniques:

1. Chemical-Shift Correlations for Similar Structural Features or for Model Compounds. This may include chemical modifications, such as sample derivatization, selective deuterium incorporation, or effects on ^{13}C shifts from solvent change or ionization.

2. Spin-Coupling Information. Multiplicities in off-resonance decoupling; single-frequency decoupling, ^{13}C-^{1}H, ^{13}C-^{13}C, and ^{13}C-X couplings if observable, or special decoupling methods, and so on.

3. Spin-Relaxation Parameters. Values of T_1 and NOE can be used to identify and distinguish nonprotonated carbons and sometimes CH, CH₂, and CH₃ groups; T_2s usually evaluated from line-widths can identify unique features (e.g., bonding to nitrogen) or detect chemical exchanges.

4. Miscellaneous Techniques. These include use of lanthanide shift reagents or selective spin-relaxation agents, labeling with spin ½ nuclei (^{13}C, ^{15}N).

Progress has been reported in computer-aided interpretation of nmr spectra with the use of pattern recognition and large available ^{13}C data bases. For example, fifteen *exo* and *endo* substituted norbornanes were characterized by their chemical shifts in a multivariate data analysis. Regularities found in the data were used to correctly classify a test set of 28 compounds and also to indicate some erroneous spectral assignments.[30]

PROBLEMS

The following three problems (as well as those presented in later chapters) are provided to give readers experience in cmr spectral interpretation and to help them evaluate their understanding of material covered in the text. Answers to some of the problems are provided in the appendix

The following conventions are used for presentation of problem spectra. Peaks are labeled with successive numbers from low to high field. Tetramethylsilane, if present, is not numbered. Off-resonance ^1H decoupled spectra are seldom included, but the information from those experiments is often noted by single-letter codes next to the peak numbers. The letters s, d, t, and q represent singlet, doublet, triplet, and quartet, respectively, corresponding to nonprotonated carbons, CH, CH$_2$, and CH$_3$ carbons. Finally, most of the problem spectra are FT spectra recorded at high pulse-repetition rates; thus peak intensities may be mis-leading. (These peak intensities can be helpful--nonprotonated carbons often give signals of reduced intensity.)

PROBLEM 2.1.   *This aromatic compound ($C_8H_5NO_2$) has only six kinds of carbon. In this case peak intensities are sig-nificant; the smaller peaks (1, 2, 5, and 6) are due to nonprotonated carbons. What is this com-pound? (Hint: this molecule contains a cyano group but no nitro group!)*

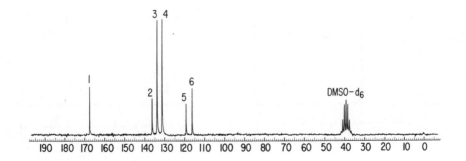

PROBLEM 2.2. *A compound with the molecular formula $C_4H_6O_2$ has the ^{13}C spectrum shown in the accompanying figure. (The top trace shows an off-resonance ^1H-decoupled spectrum.) Using the information in Figure 2.1, deduce the structure. (Note that peak intensities in this spectrum are particularly misleading.)*

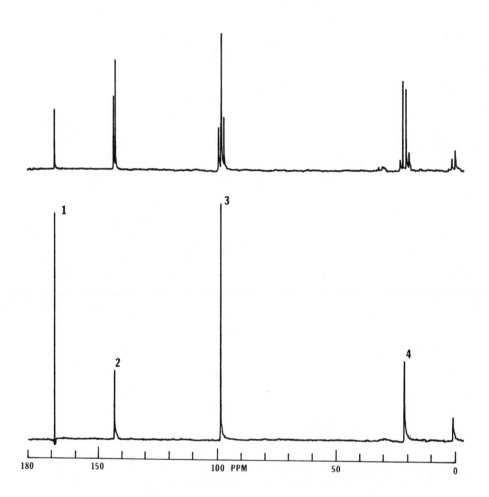

PROBLEM 2.3.   *An epoxy compound ($C_7H_{10}O_3$) has the ^{13}C spectrum shown in the accompanying figure. Make structure and resonance assignments from the spectrum and off-resonance $\{^1H\}$ information.*

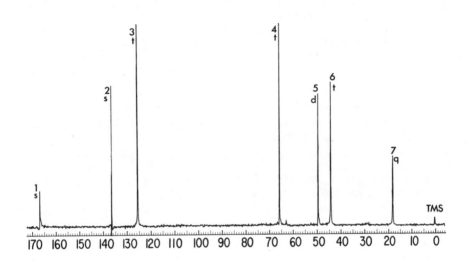

REFERENCES

1. G. C. Levy and J. D. Cargioli, *J. Magn. Resonance,* **6**, 143 (1972).
2. P. D. Ellis and R. Ditchfield, in *Topics in Carbon-13 NMR Spectroscopy,* Vol. 1, G. C. Levy, Ed., Wiley-Interscience, New York, 1974, Chapter 1 and references cited therein.
3. M. J.-Heravi and G. A. Webb, *Org. Magn. Resonance,* **11**, 524 (1978); M. Kondo, I. Ando, R. Chujo, and A. Nishioka, *J. Magn. Resonance,* **24**, 315 (1976); J. Mason, *Org. Magn. Resonance,* **10**, 188 (1977); M. Kondo and I. Ando, *Bull. Chem. Soc. Jap.,* **21**, 2072 (1978); A. R. Garber, P. D. Ellis, K. Seidman, and K. Schade, *J. Magn. Resonance,* **34**, 1 (1979).
4. R. E. Sievers, Ed., *Nuclear Magnetic Resonance Shift Reagents,* Academic, New York, 1973.
5. (a) J. Briggs, F. A. Hart, G. P. Moss, and E. W. Randall, Chem. Commun., **1971**, 364; (b) E. Wenkert, D. W. Cochran, E. W. Hagaman, R. B. Lewis, and F. M. Schell, *J. Am. Chem. Soc.,* **93**, 6271 (1971); (c) O. A. Gansow, M. R. Willcott, and

R. E. Lenkinski, *ibid.*, **93**, 4295 (1971); (d) J. C. Duggan, W. H. Urry, and J. Schaefer, *Tetrahedron Lett.*, **1971**, 4197, W. B. Smith and D. L. Deavenport, *J. Magn. Resonance*, **6**, 256 (1972).

6. (a) O. A. Gansow, et al., *J. Am. Chem. Soc.*, **95**, 3390 (1973); (b) G. E. Hawkes, C. Marzin, S. R. Johns, and J. D. Roberts, *ibid.*, **95**, 1661, (1973); (c) K. Tori, Y. Yoshimura, M. Kainosho, and K. Ajisaka, *Tetrahedron Lett.*, **1973**, 3127.

7. J. Reuben, *J. Magn. Resonance*, **11**, 105 (1973).

8. (a) G. R. Sullivan, *J. Am. Chem. Soc.*, **98**, 7162 (1976); (b) K. L. Williamson, et al., *ibid.*, **96**, 1471 (1974).

9. References summarized in: D. Terpstra, in *Topics in Carbon-13 NMR Spectroscopy*, Vol. 3, G. C. Levy, Ed., Wiley-Interscience, New York, 1979; R. Freeman and G. A. Morris, *Bull. Mang. Resonance*, **1**, 27 (1979).

10. (a) F. J. Weigert and J. D. Roberts, *J. Am. Chem. Soc.*, **91**, 4940 (1970); (b) H. Günther, H. Seel and M.-E. Günther, *Org. Magn. Resonance*, **9**, 281 (1977).

11. J. Kowalewski, *Progr. Nucl. Magn. Resonance Spectrosc.*, **11**, 1 (1977).

12. G. E. Maciel, J. W. McIver, Jr., N. S. Ostlund, and J. A. Pople, *J. Am. Chem. Soc.*, **92**, 1, 11 (1970).

13. F. J. Weigert and J. D. Roberts, *J. Am. Chem. Soc.*, **94**, 6021 (1972).

14. G. A. Gray, P. D. Ellis, D. D. Traficante, and G. E. Maciel, *J. Magn. Resonance*, **1**, 41 (1968).

15. K. F. Kuhlmann and D. M. Grant, *J. Am. Chem. Soc.*, **90**, 7355 (1968).

16. (a) G. C. Levy, *Acc. Chem. Res.*, **6**, 161 (1973); (b) J. R. Lyerla, Jr. and G. C. Levy, *Topics in Carbon-13 NMR Spectroscopy*, G. C. Levy, Ed., Vol. 1, Wiley-Interscience, New York, 1974, Chapter 3.

17. A. Allerhand, D. D. Doddrell, and R. A. Komoroski, *J. Chem. Phys.*, **55**, 189 (1971).

18. U. Edlund, C. E. Holloway, and G. C. Levy, *J. Am. Chem. Soc.*, **98**, 5069 (1976).

19. R. A. Komoroski and G. C. Levy, *J. Phys. Chem.*, **80**, 2410 (1976).

20. J. B. Lambert and D. A. Netzel, *J. Am. Chem. Soc.*, **98**, 3783 (1976).

21. G. C. Levy and U. Edlund, *ibid.*, **97**, 5031 (1975).

22. F. W. Wehrli, *Adv. Molec. Relaxation Processes*, **6**, 139 (1974).

23. (a) O. A. Gansow, A. R. Burke, and G. N. LaMar, *J. Chem. Soc.*, **1972**, 456; (b) O. A. Gansow, A. R. Burke, and W. D. Vernon, *J. Am. Chem. Soc.*, **94**, 2850 (1972); (c) G. C. Levy and J. D. Cargioli, *J. Magn. Resonance*, **10**, 231 (1973).

24. R. Freeman, K. G. R. Pachler, and G. N. LaMar, *J. Chem. Phys.*, **55**, 4586 (1971).
25. G. C. Levy and U. Edlund, *J. Am. Chem. Soc.*, **97**, 4482 (1975).
26. K. F. Kuhlmann and D. M. Grant, *J. Chem. Phys.*, **55**, 2998 (1971); R. Freeman, H. D. W. Hill, and R. Kaptein, *J. Magn. Resonance*, **7**, 327 (1972); S. J. Opella, D. J. Nelson, and O. Jardetzky, J. Chem. Phys., **164**, 2533 (1976); R. K. Harris and R. H. Newman, *J. Magn. Resonance*, **24**, 449 (1976).
27. J. N. Shoolery, *Prog. NMR Spectrosc.*, **11**, 79 (1977).
28. G. C. Levy, U. Edlund, and J. G. Hexem, *J. Magn. Resonance*, **19**, 259 (1975).
29. G. C. Levy, U. Edlund, and C. E. Holloway, *ibid.*, **24**, 375 (1976).
30. M. Sjostrom and U. Edlund, *ibid.*, **25**, 285 (1977).

Chapter 3

Aliphatic Compounds

ALKANES AND DERIVATIVES

Most introductions to organic chemistry begin discussion with
the simplest of organic molecules, the paraffins. It is thus
appropriate that the discussion of specific cmr applications to
organic chemistry begin with saturated hydrocarbons. Readers
with proton nmr experience will recognize that proton nmr spectra
of saturated hydrocarbons can be used only with difficulty to
distinguish between isomeric paraffins. In cmr this is definitely
not the case. Figure 3.1 shows the ^{1}H and ^{13}C nmr spectra of 3-
methylheptane. Whereas the ^{1}H spectrum shows only gross structural features, the ^{13}C spectrum shows eight distinct lines
No other C_8 hydrocarbon has precisely this cmr spectrum.[1]

The presence of attached and nearby carbon atoms has a profound effect on ^{13}C chemical shifts. Table 3.1[2] summarizes ^{13}C
chemical shifts for several simple liquid alkanes. By focusing
on C-1, it can be seen that both α- and β-carbons deshield by
perhaps 9 ppm; γ-carbons, on the other hand, shield by 2 ppm.
Grant and Paul, and others[2,3] first noted for some 17 simple
paraffins that the ^{13}C chemical shifts are not only distinct but
quite additive and may be derived in a regular fashion based on
the number of α-, β-, γ-, δ-, and ε-carbon atoms. Theoretical
explanations can be advanced,[4,5] but empirical correlations provide the investigator with a most powerful analytical tool. Subsequently, Lindeman and Adams[1] extended the work of Grant and
Paul. Although more recent elaborations of this approach have
been reported,[6] the Lindeman-Adams method remains the most con-
venient. Employing chemical-shift data for some 59 paraffins (C_5
to C_9) the ^{13}C chemical shift of the Kth carbon can be represented
by Equation 3.1, where B_S, A_{SM}, γ_S, and Δ_S are constants; N_{KP}

$$\Delta_C (K) = B_S + \Sigma_{M=2}^{4} D_M A_{SM} + \gamma_S N_{K3} + \Delta_S N_{K4} \tag{3.1}$$

is the number of carbon atoms P bonds away from the Kth carbon;
D_M is the number of carbon atoms bonded to the Kth carbon atom
that has M attached carbons; and S is the number of carbon atoms

bonded to the *K*th carbon atom. Table 3.2 lists the values of the constants B_S, A_{SM}, γ_S, Δ_S as a function of S and M. By using Equation 3.1, chemical shifts can be predicted with a standard

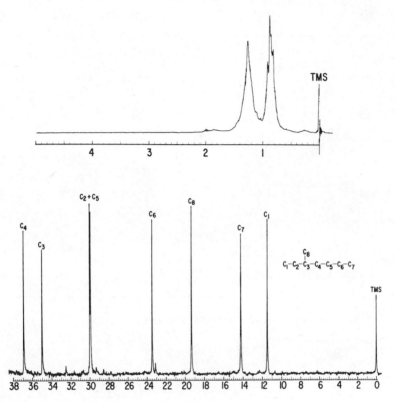

FIGURE 3.1.   *Proton nmr spectrum (above) and* ^{13}C *nmr spectrum of 3-methylheptane.*

error of 0.8 ppm.

For example, the ^{13}C chemical shift of carbon 5 of 3-methyl-heptane (Figure 3.1) can be predicted as follows:

1. The number of carbon atoms bonded to carbon 5 is 2; therefore, $S = 2$.

2. Carbon 5 is bonded to two carbon atoms that are bonded to two carbon atoms; therefore, $D_2 = 2$. There are no attached carbons bearing higher substitution; therefore, $D_3 = 0$, and $D_4 = 0$.

3. The number of carbon atoms three bonds away from carbon 5 is 2; therefore, $N_{K3} = 2$, and likewise $N_{K4} = 1$.

TABLE 3.1. ^{13}C CHEMICAL SHIFTS (IN PARTS PER MILLION) FOR SELECTED ALKANES[a,2]

COMPOUND	C-1	C-2	C-3	C-4	C-5
Methane	-2.3				
Ethane	5.7				
Propane	15.4	15.9			
n-Butane	13.0	24.8			
n-Hexane	13.7	22.7	31.8		
n-Decane	13.9	22.8	32.2	29.7	30.1
2-Methylpropane	24.1	25.0			
2-Methylbutane	21.8	29.7	31.6	11.3	
2-Methylpentane	22.3	27.6	41.6	20.5	13.9

[a]Gas-phase values are slightly different.[2b]

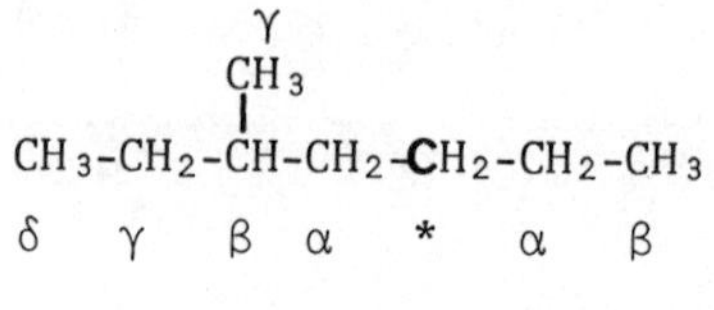

$$
\begin{array}{c}
\gamma \\
CH_3 \\
|
\end{array}
$$
$$
CH_3-CH_2-CH-CH_2-\mathbf{C}H_2-CH_2-CH_3
$$
$$
\delta \quad \gamma \quad \beta \quad \alpha \quad * \quad \alpha \quad \beta
$$

   4. By referring to Table 3.2, the appropriate values may now be substituted into Equation 3.1 yielding

$$\delta_C(K) = 15.34 + 2(9.75) + (-2.69)2 + 0.25(1)$$
$$= 29.7 \text{ ppm (observed 29.7)}$$

Chemical shifts for each carbon can be calculated in a like manner. Table 3.3 gives a comparison of calculated and observed chemical shifts for the carbons of 3-methylheptane. Figure 3.2 graphically summarizes the ^{13}C chemical shifts for paraffins classified on the basis of the number of α-, β-, and γ-carbons. For carbon-5 of 3-methylheptane, there are two α-carbons, two β-carbons, and two γ-carbons. From Figure 3.2 the chemical shift is about 29 ppm.

TABLE 3.2.   [13]C CHEMICAL SHIFTS (IN PARTS PER MILLION) FOR
PARAFFINS[1]

PARAMETER	VALUE	PARAMETER	VALUE
B_1	6.80	Δ_2	0.25
A_{12}	9.56	B_3	23.46
A_{13}	17.83	A_{32}	6.60
A_{14}	25.48	A_{33}	11.14
γ_1	-2.99	A_{34}	14.70
Δ_1	0.49	γ_3	-2.07
B_2	15.34	B_4	27.77
A_{22}	9.75	A_{42}	2.26
A_{23}	16.70	A_{43}	3.96
A_{24}	21.43	A_{44}	7.35
γ_2	-2.69	γ_4	0.68

TABLE 3.3   COMPARISON OF CALCULATED AND OBSERVED CHEMICAL SHIFTS
FOR 3-METHYLHEPTANE

	C-1	C-2	C-3	C-4	C-5	C-6	C-7	C-8
Observed	11.3	29.7	34.7	36.5	29.7	23.3	14.1	19.3
Calculated	10.9	29.6	34.6	36.2	29.7	22.9	13.9	19.1

Although the preceding correlations were derived for paraffins,
they are useful to some degree for alicyclics and substituted
alkanes as well. As is seen in Chapter 7, such correlations even
find use in analysis of polymer chemical shifts.

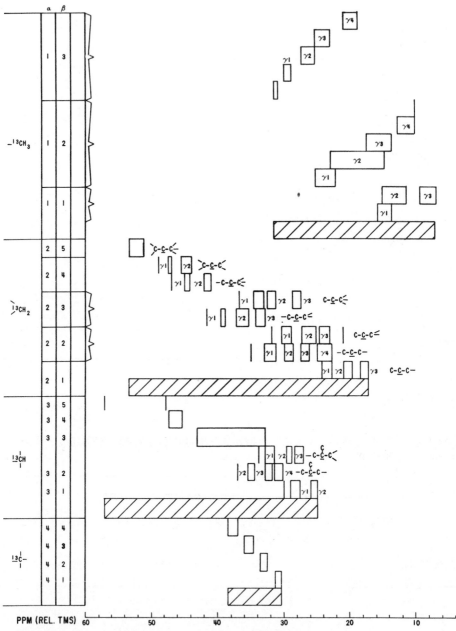

FIGURE 3.2.[1]   Graphical display of ^{13}C chemical shifts for
classes of paraffin carbons.

PROBLEM 3.1.   *Calculate the chemical shifts for 2-methylbutane and 2-methylpentane using Equation 3.1 and Table 3.2. Compare your results with the observed values given in Table 3.1.*

PROBLEM 3.2.   *Use Figure 3.2 to determine the structure and assign the spectrum of the C_8H_{18} hydrocarbon in the accompanying figure.*

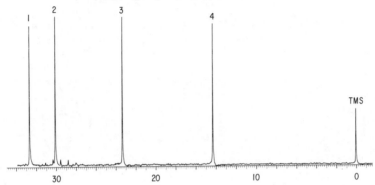

Stereochemistry and Remote Substituent Effects: "Steric" Shifts. As mentioned previously, the relatively additive effect of alkyl groups on ^{13}C chemical shifts was noted rather early in the development of ^{13}C nmr as a structural tool. In particular, the shielding induced by a γ-alkyl group was found to be particularly characteristic. Early attempts[4] at identifying the origin of this effect invoked a "steric compression" between proximate C-H bonds; however, it is now recognized that this cannot be the sole source of the effect because such shielding is often observed even when no such interaction is possible.[5-7] Furthermore, γ-heteroatoms are known to shield ^{13}C nuclei, and this effect is generally larger for a more remote *trans* substituent than for a proximate *gauche* substituent.[8] A more compelling argument against a steric origin for these effects is the deshielding arising from geometrically proximate δ substituents.[9-12] Some examples are shown on p.56;[11] it is interesting that in the norborneols the type of originally proposed C-H perturbations are much less likely with the hydroxyl group, although the effects are comparable to those of the methyl group. Nonetheless, although the mechanistic origins of both the γ and δ effects remain unclear[13,14] --indeed, deshielding γ effects have been reported[15]--their existence and especially their additivity remain empirically useful.

Examples of δ-Substituent Effects:

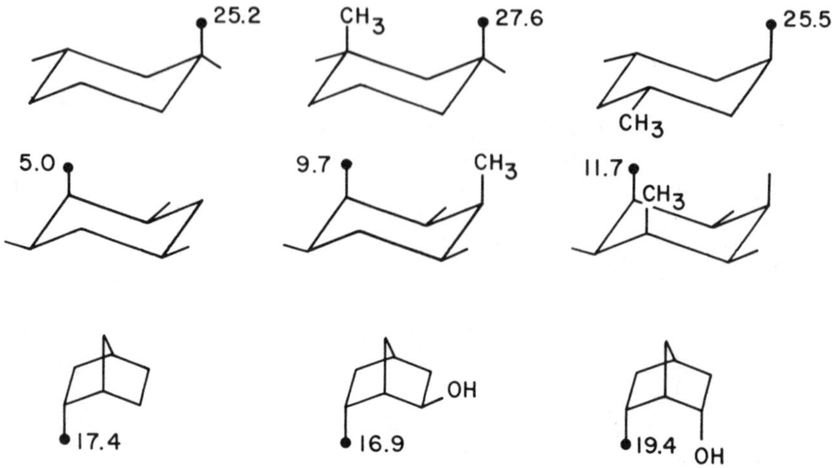

(Chemical shifts in ppm)

Alicyclics offer the problem of not only substitution iso-merism but conformational isomerism as well. Through the work of Dalling and Grant, and others[16] chemical-shift data for the cyclohexane system have been factored on the basis of the geometrical and conformational features found for this class of compounds. Table 3.4[16] gives the parameters obtained (cyclohexane = 27.3 ppm). Chemical shifts for cyclohexyl systems can be predicted with reasonable success using these parameters. Shifts of ring carbons γ-*gauche* to methyl (e.g., the 3- and 5-carbons of 1-*axial* methylcyclohexane) are shielded compared with similar carbon nuclei not so disposed, the "steric" shift referred to earlier in this chapter and in Chapter 2. For the 1-, 3-, and 5-carbons of *axial* methylcyclohexanes, that shielding amounts to 5 ppm. Such interactions have proved valuable in conformational analysis. Sterically crowded alkyl groups can give rise to such shifts, but other substituents may not always behave in this manner.

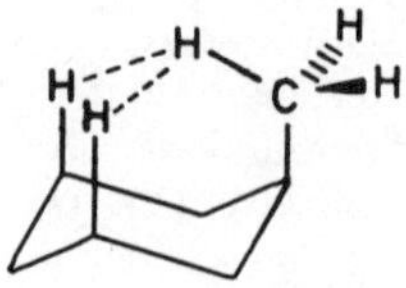

TABLE 3.4.   METHYL-SUBSTITUENT PARAMETERS FOR CYCLOHEXANES[a,16]

	+ 6.0		+ 1.4
	+ 9.0		+ 5.4
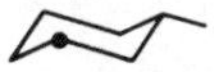	0.0		− 6.4
	− 0.2		0.0
	− 3.8		− 1.3
	− 2.9		− 2.9

[a] Nucleus whose shift is considered is represented by (•); shifts in parts per million relative to cyclohexane.

The complete spectral assignment of cholesterol has been made by employing methyl groups and other substituents and observing particularly the carbons that undergo an upfield steric shift. Using low-temperature ^{13}C nmr at 63 MHz, Anet and co-workers[17] found that indeed for methylcyclohexane itself, the C-3 and C-5 carbons of the axial form are 6 ppm upfield from those in the equatorial form. Christl, Reich, and Roberts[18] gave a similar analysis for the cyclopentyl system. Here, however, the 1,3 interaction is found to be much less pronounced, about 2.6 ppm.

Saturated ring compounds are common in organic chemistry. The chemical shifts for cycloalkanes through C-10 are given in Table 3.5.[19] A comparison of the effects of heteroatoms as a function of ring size is given in Table 3.6.[8,19c] Strain is indicated by significant chemical-shift change only in the three- and four-membered rings. For cyclopropane, the CH_2 carbons are found at very high field, in analogy with the protons in cyclopropyl compounds. The chemical shifts for hydrocarbons related to the norbornane skeleton are given in Table 3.7[20], and those for 2-substituted norbornanes, in Table 3.8.

PROBLEM 3.3.   Using Equation 3.1, calculate the chemical shifts for 2-methylnorbornane. (Hint: In counting the number of α-, β-, γ-, and δ-carbons, include the number of different routes to the same atom.)

PROBLEM 3.4.   Examine the ^{13}C chemical shifts of 2-methyl and 7-methylnorbornane given in Table 3.7. Which carbons are experiencing steric shifts?

PROBLEM 3.5.   *Determine the structure and assign each resonance to the appropriate carbon for the compound whose ^{13}C spectrum appears in the accompanying figure (molecular formula, C_8H_{18}; peak 2 remains a singlet in off-resonance {[1]H} experiments).*

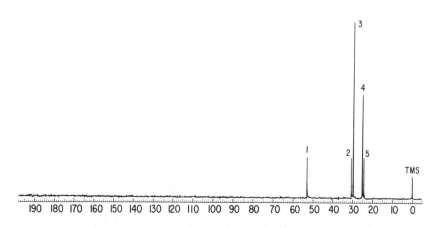

TABLE 3.5.   [13]C CHEMICAL SHIFTS (IN PARTS PER MILLION) OF CYCLOALKANES[18,19]

C_3H_6	-2.9	C_7H_{14}	28.5
C_4H_8	23.3	C_8H_{16}	26.9
C_5H_{10}	26.5	C_9H_{18}	26.1
C_6H_{12}	27.3	$C_{10}H_{20}$	25.3

Substituted Alkanes.   If all the factors that can influence ^{13}C chemical shifts are taken into account, the possibility of using additivity parameters for the calculation of even very approximate chemical shifts would seem remote for paraffins and even more so for substituted alkanes. Substituent parameters can be derived; their success arises from the fact that the influence of substituents is primarily over a short range and is largely electronic in nature.

Replacement of a methyl group by a polar substituent gives rise to changes, particularly at the carbon bearing the substituent (C-1). Table 3.9 summarizes the changes expected at C-1,

C-2, and C-3 for a number of substituents. Positive values indicate downfield shifts. The values are approximate but will give
essentially correct shifts for many situations. As can be seen
from Table 3.9, the changes are not related to electronegativity
in a monotonic manner.

Example: Let us predict the ^{13}C chemical shifts for 2-pentanol.
From Table 3.1 the chemical shifts for the appropriate methyl-substituted compound, 2-methylpentane, are obtained. If shifts for
the methyl-substituted material are unavailable, an estimate may
be obtained by using Equation 3.1 or Figure 3.2. The appropriate
parameters are added to the chemical shift of each carbon as
given in Table 3.9. Table 3.10 gives the results and a comparison
with the observed values.

As in proton nmr, one learns to expect particular shift
changes to be associated with substitution by particular groups.
For most purposes of assignment, such considerations are sufficient.
Carbon-13 chemical shifts can be remarkably regular in this regard.

TABLE 3.6. ^{13}C CHEMICAL SHIFTS (IN PARTS PER MILLION). EFFECT
OF RING SIZE ON ^{13}C SHIFTS OF SATURATED MONOHETEROCYCLES[8,19c]

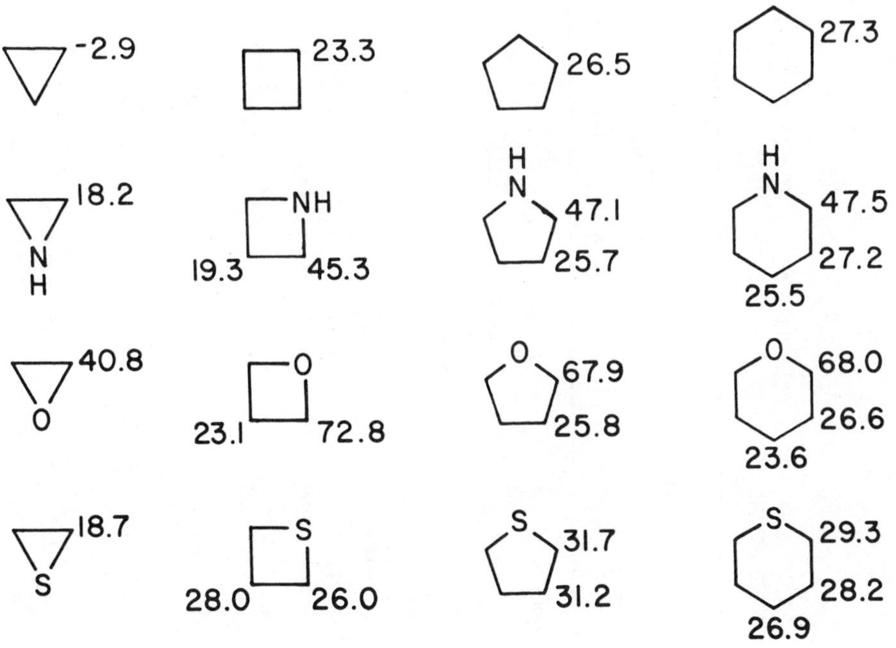

TABLE 3.7.   ¹³C CHEMICAL SHIFT (IN PARTS PER MILLION) OF HYDROCARBONS RELATED TO
NORBORNANE[20]

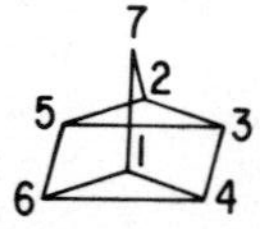

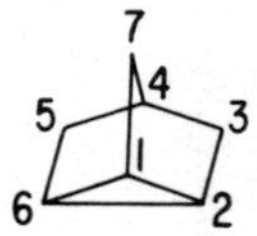

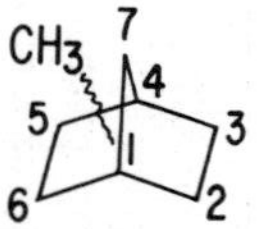

COMPOUND	C-1	C-2	C-3	C-4	C-5	C-6	C-7	Me
Quadricyclene	23.2	15.0	15.0	23.2	15.0	15.0	32.2	
Tricyclene	9.9	9.9	33.2	29.7	33.2	9.9	33.2	
Norbornane	36.8	30.1	30.1	36.8	30.1	30.1	38.7	
1-Methyl norbornane	44.1	37.1	31.6	38.2	31.6	37.1	45.6	21.0
exo-2-Methyl norbornane	43.5	36.8	40.2	37.3	30.3	29.0	35.0	22.3
endo-2-Methyl norbornane	42.2	34.6	40.7	38.2	30.6	22.4	38.9	17.4
7-Methyl norbornane	41.0	27.2	27.2	41.0	31.0	31.0	44.3	12.7

TABLE 3.8. ^{13}C CHEMICAL SHIFTS (IN PARTS PER MILLION) OF exo- AND endo-2-SUBSTITUTED NORBORNANES[20]

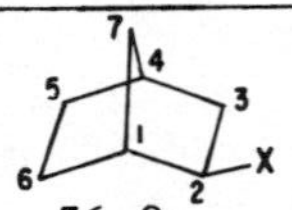

X	C-1	C-2	C-3	C-4	C-5	C-6	C-7	X
H	36.8	30.1	30.1	36.8	30.1	30.1	38.7	
CH₃	43.5	36.8	40.2	37.3	30.3	29.0	35.0	22.3
NH₂	45.7	55.4	42.5	36.4	28.9	27.0	34.3	
OH	44.5	74.4	42.4	35.8	28.8	24.9	34.6	
F	42.4	95.9	40.1	34.9	28.3	22.6	35.3	
CN	42.3	31.1	36.4	36.5	28.6	28.5	37.4	123.4
COOH	41.4	46.8	34.5	36.6	29.8	29.1	36.9	181.4
COOCH₃	41.9	46.5	34.3	36.4	29.0	28.7	36.6	175.7
CH₃	42.2	34.6	40.7	38.2	30.6	22.4	38.9	17.4
NH₂	43.6	53.4	40.6	38.0	30.7	20.6	39.0	
OH	43.1	72.5	39.6	37.7	30.3	20.4	37.8	
CN	40.2	30.2	35.6	37.0	29.4	25.2	38.7	122.6
COOH₃	41.0	46.3	32.2	37.7	29.5	25.3	40.6	180.5
COOCH₃	40.8	46.0	32.3	37.5	29.4	25.1	40.4	174.6

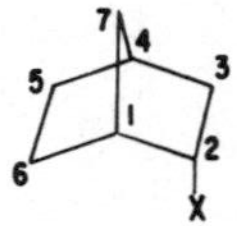

TABLE 3.9.　APPROXIMATE CHANGES IN ^{13}C CHEMICAL SHIFTS (IN PPM)
ON REPLACEMENT OF A METHYL GROUP BY A POLAR SUBSTITUENT

SUBSTITUENT[a]	C-1	C-2	C-3
-OR	+48	-2	-2
-OH	+40	+1	-3
-OCOR	+43	-3	-1
-NO$_2$	+54	-6	
-NH$_2$	+20	+2	-1
-NR$_2$	+33	-3	
-NH$_3^+$	+17	-2	-3
-I	-15	+1	+1
-Br	+12	+2	-1
-Cl	+23	+2	-1
-F	+60	-1	-2
-COX	+15	-5	0
-COOR	+10	-1	-1
-COOH	+12	-3	-1
-CN	-2	-1	-1
-SH	+2	+2	-2

[a]R = alkyl; X = Cl or NR$_2$.　These values correspond generally to
n-alkyl groups; sec-alkyl α and β substituent effects are typically less deshielding by several ppm.

TABLE 3.10.　OBSERVED AND PREDICTED ^{13}C CHEMICAL SHIFTS (IN PPM)
FOR 2-PENTANOL

	C-1	C-2	C-3	C-4	C-5
Observed	23.6	67.3	41.9	19.4	14.3
Predicted	23.3	67.6	42.6	19.5	13.7

TABLE 3.11. ^{13}C CHEMICAL SHIFTS (IN PARTS PER MILLION) FOR SUBSTITUTED CYCLOPROPANES[19b]

SUBSTITUENT	C-1	C-2
H	-2.9	-2.9
I	-20.2	10.3
Br	13.9	8.9
Cl	27.3	8.9
COCl	23.7	12.3
NH₂	23.9	7.3

TABLE 3.12. ^{13}C CHEMICAL SHIFTS (IN PARTS PER MILLION) OF 1-SUBSTITUTED BICYCLO[2.2.2] OCTANES[21b]

SUBSTITUENT	C-1	C-2	C-3	C-4
H	23.9	26.0	26.0	23.9
OCH₃	72.3	29.3	26.9	24.2
F	92.4	31.2	27.3	24.2
Cl	66.2	36.2	28.2	23.3
Br	62.8	37.5	29.0	22.7

TABLE 3.13. ^{13}C CHEMICAL SHIFTS OF 1-SUBSTITUTED NORBORNANES[21b]

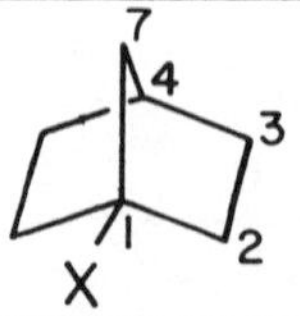

SUBSTITUENT	C-1	C-2	C-3	C-4	C-7
H	36.4	29.8	29.8	36.5	38.4
Cl	69.9	38.4	30.9	34.8	46.8
OH	82.8	35.4	30.3	34.8	43.9
OCH$_3$	87.7	31.0	29.9	33.9	40.0
COOH	52.1	32.9	30.0	37.8	42.3
C$_6$H$_5$	51.3	37.2	30.9	37.3	42.8
NHCOCH$_3$	62.6	33.7	29.9	35.3	41.7

Thus far only replacement of a methyl group by a substituent has been considered. Although reasonable estimates of chemical shifts for methyl substituted alkanes are obtained from Equation 3.1 or Figure 3.2, in many cases the experimental ^{13}C shifts for the appropriate proton-substituted compound are known. What is of interest is the effect of substitution by X on the ^{13}C shifts at C-1, C-2, and C-3. It has already been shown that replacement of a proton by a methyl group deshields C-1 by ~8 ppm, C-2 by ~10 ppm, and shields C-3 by ~2 ppm. Estimates of ^{13}C shifts for replacement of C-H by C-X can thus be made by adding these values to the appropriate parameters of Table 3.9. For example, an estimate of the ^{13}C shift at C-2 of 2-fluoronorbornane (Table 3.8) can be made by adding 61 ppm (from Table 3.9) and 8 ppm (from discussion in text) to the observed shift of 30.1 ppm in norbornane (= ~99 ppm). This compares with 95.9 ppm observed at C-2 for *endo*-2-fluoronorbornane. A comparison can be made for chemical shifts at C-X relative to nornornane for other substituents as listed in Table 3.8. Tables 3.11 to 3.13 present ^{13}C chemical shift data for substituted cyclopropanes,[19b] bicyclooctanes,[21b] and norbornanes.[21b] In each case the effects of substitution can be seen. These data have been discussed and compared with correspondingly substituted adamantanes.[21b,22] Other studies of

polycyclic hydrocarbons have been reported.[23] Some representative
values are shown below. A relationship between proton and carbon
chemical shifts in substituted cyclobutanes has been suggested.[24]

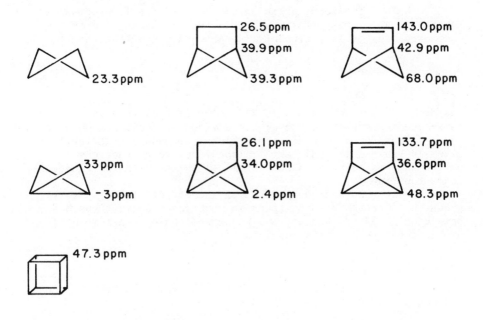

PROBLEM 3.6.   *From Tables 3.8, 3.11, and 3.12 derive additivity
parameters for the substituted carbons and for
carbons 2 and 3 (for the substituents listed,
C-H to C-X).*

     Despite the high degree of regularity for ^{13}C chemical shifts,
subtle differences often may be important. A nearby center of
molecular asymmetry can induce magnetic nonequivalence of otherwise equivalent carbons.[25] This is particularly true for an
isopropyl group with an adjacent asymmetric carbon. For isopropyl-
alkylcarbinols, the ^{13}C chemical-shift difference between the
two isopropyl methyl carbons increases from 0.2 to 6.9 ppm as the
size of the alkyl group increases from methyl to *tert*-butyl.
Although such effects may be strongly attenuated if the number of

intervening carbons is increased, an observable effect has been
observed for a case (an alkane) in which the isopropyl group was
separated from the asymmetric center by three carbons.

Individual Substituents.   Extensive studies have been made for
several functional groups.   Reports include not only chemical
shifts but additivity parameters as well.

Alcohols: The correlation parameters originally proposed by
Roberts et al.[26] for primary alcohols have been extended to
secondary and tertiary alcohols.[27]   An equation of the form

$$\delta_{C-X}^{ROH} = A_i \delta_{C-X}^{RCH_3} + B_i$$

was derived for each carbon (C-X) of the alcohol relative to the
corresponding carbon of the alkane derived by replacing OH with
CH_3. Values for A_i and B_i (i = primary, secondary, tertiary) are
given in Table 3.14.   Ejchart[27] also discussed corrections required when calculating alkane chemical shifts with the Lindeman-
Adams equation.
     The conformational dependence of the effect of the hydroxyl
group on ^{13}C resonance positions in norbornane derivatives has
been elucidated.[28]   Borneol and isoborneol have been completely
assigned and ambiguities resolved.[29]

TABLE 3.14.   SUBSTITUENT PARAMETERS (IN PARTS PER MILLION) FOR
ALCOHOL AND AMINE ^{13}C CHEMICAL SHIFTS

	PRIMARY			SECONDARY			TERTIARY		
	C-1	C-2	C-3	C-1	C-2	C-3	C-1	C-2	C-3
				ALCOHOLS[27]					
A_i	0.709	0.963	0.963	0.786	0.958	0.982	0.755	1.029	1.137
B_i	46.45	1.81	-2.28	45.65	2.07	-0.48	48.37	-0.06	-1.03
				AMINES[32b]					
A_i	0.846	0.955	0.941	0.900	0.942	0.951	0.914	0.999	0.934
B_i	23.09	3.00	-0.07	22.88	2.07	-0.68	22.62	0.45	-0.43

Amines: Comprehensive surveys of amine ^{13}C chemical shifts
have been reported by Sarneski et al.[30a] and by Eggert and
Djerassi.[30b] Substituent parameters were obtained (Table 3.14)
based on the chemical shift of the corresponding carbon in the
alkane derived by replacing the amino group with the appropriate
saturated carbon. Both groups of investigators discussed the
effects of conformation on the chemical shifts and on the substituent parameters.

Protonation of amines changes the resonance positions of the
α and β carbons, but contrary to early suggestions, these changes
may correspond to either shielding or deshielding of the nuclei.[30a]
Protonation shifts may be calculated by using an empirical equation
requiring 15 parameters.[30a]

Table 3.15[30] presents ^{13}C chemical shift data for several mono-
and difunctional amines. *N*-Alkyl substituents deshield the carbon
of attachment and slightly shield carbons α to C-NR$_2$ (C-2). These
β and γ effects, respectively, are useful for ^{13}C resonance
assignments even in complex tertiary amines by converting them to
their quaternary salts.[31]

Alicyclic amine ^{13}C chemical shifts behave similarly to those
of the acyclic compounds, but substituent effects depend on both
ring and substituent geometry. Table 3.16 presents ^{13}C data for
several piperidines, abstracted from a much more comprehensive
tabulation.[8] Effects of substituent geometry may be seen by comparing 8(eq)- and 8(ax)-methyl-*trans*-decahydroquinolines.[8,32a]
Comprehensive studies of conformational equilibria in substituted
cis- and *trans*-decahydroquinolines using ^{13}C nmr have been reported.[32] Vierhapper and Eliel[33] have also demonstrated that the
N-H conformational equilibrium in *trans*-decahydroquinoline is unaffected by 8-methyl substitution but is markedly influenced when
a *tert*-butyl group is in that position.

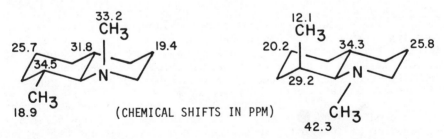

(CHEMICAL SHIFTS IN PPM)

Carboxylic Acids and Carboxylate Anions: Hagen and Roberts[34]
measured the ^{13}C chemical shifts for each carbon of simple aliphatic carboxylic acids and carboxylate anions (formic to valeric acid)
in aqueous solution. The following substituent parameters were
obtained for C*-COOH relative to C*-H:

	C-1	C-2	C-3
Acids	+22	+2.7	-2.3
Anions	+25	+4.5	-1.7

A more comprehensive approach, requiring up to 16 parameters, was developed by Rabenstein and Sayer.[35] The trend observed follows closely those produced by hydroxyl, amino, acetyl, chloro, and bromo groups. The effects at C-1 and C-2 are deshielding, whereas those at C-3 are shielding. Perhaps surprisingly, ionization of an aliphatic acid results in a deshielding of the carbonyl carbon.

Saturated Heterocycles: The effect of heteroatoms incorporated into a ring can affect ^{13}C resonance positions by introducing additional conformational factors. This subject has been extensively reviewed.[8] Some of these influences were given in Table 3.6 for alicyclic compounds with one heteroatom. Representative additional data are given in Table 3.17.[8,36,37]

Halides: Carbon-13 chemical shifts of halomethanes are given in Table 3.18.[38] These values have been recalculated from shifts reported with respect to methane gas at zero density. Resonance positions of alkyl halides display solvent, concentration, and, for gases, pressure dependences. The more heavily fluorinated and chlorinated compounds exhibit solvent effects that may reflect hydrogen-bonding interactions.[39] The shielding induced by increased substitution with bromine and iodine is striking. This "heavy-atom effect" has been treated theoretically by a variety of methods[7a,c] but as of this writing has not been unambiguously accounted for. It is interesting to note that the deshielding effect of multiple chlorine substituents is approximately additive (~25 ppm per chlorine), but with bromine and iodine, multiple substitution gives accelerated shielding.

Spin-Spin Coupling. The discussion of cmr properties of alkanes should not be concluded without some consideration of spin-spin coupling, both directly bonded and long-range C-H and C-C coupling. In early cmr experiments C-H coupling was intentionally suppressed or only residual coupling was retained through off-resonance decoupling techniques. However, a considerable body of J_{CH} data had been accumulated over the years from ^{13}C satellite peaks in proton nmr spectra.[40] With current instrumentation and appropriate multiple-pulse sequences that allow retention of NOE and selective observation of individual resonances, C-H couplings can be routinely obtained in most compounds. The magnitude of the couplings between ^{13}C atoms and the attached protons, $^{1}J_{CH}$, and the internal structure of the coupled peaks are sensitive to the structure of the molecule and the local environment of the carbon atom. Such coupling can be of significant value in structural analyses. The theory of nmr spin-spin coupling has been comprehensively reviewed.[41]

There is considerable variation of $^{1}J_{CH}$ with angular distortion in hydrocarbons.[42,43] Table 3.19[42] presents several examples.

TABLE 3.15. ^{13}C CHEMICAL SHIFTS (IN PARTS PER MILLION) FOR MONO-
AND DIFUNCTIONAL AMINES[30]

	C-1	C-2	C-3	C-4	C-5	C-6
CH_3(2)–N(H)–CH_2(1)–CH_2(3)–NH_2	55.1	36.9	41.9			
CH_3–N(H)–CH_2(1)–CH_2–N(H)–CH_3(2)	51.2	36.6				
H_2N–CH_2(1)–CH_2(2)–CH_2(3)–N(4)(CH_3)(CH_3)	40.8	32.2	57.8	45.4		
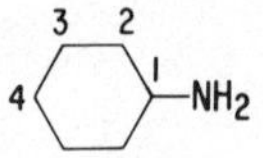 cyclohexyl–NH_2 (1,2,3,4)	50.4	36.7	25.7	25.1		
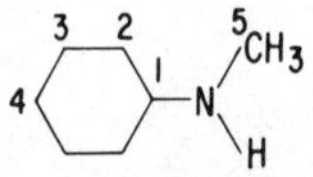 cyclohexyl–N(CH_3(5))(H) (1,2,3,4)	58.6	33.3	25.1	26.3	33.5	
CH_3(2)–CH(1)–CH_2(3)–CH_3(4), NH–CH_3(5)	56.9	19.4	29.5	11.0	33.6	
(CH_3(1)–CH_2(2))$_2$NH	44.2	15.4				
H_2N–CH_2(1)–CH_2(2)–CH_2(3)–CH_2–CH_2–NH_2	42.2	33.4	25.1			
CH_3(3)–N(H)–CH(1)(CH_3(2))(CH_3)	50.9	22.6	33.6			
H_2N–CH_2(1)–CH_2(2)–CH_2(3)–CH_2(4)–CH_2(5)–CH_3(6)	42.6	34.6	27.1	32.3	23.2	14.2
N(CH_2(1)–CH_2(2)–CH_2(3)–CH_3(4))$_3$	54.3	30.3	21.0	14.2		
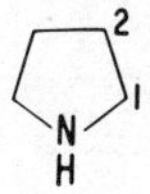 pyrrolidine (1,2), N–H	47.1	25.7				

TABLE 3.16. TABLE 3.16. ^{13}C CHEMICAL SHIFTS (IN PARTS PER MILLION) FOR PIPERIDINES[a,8]

	C-1	C-2	C-3	C-4	C-5	C-6
piperidine	47.7	27.5	26.1			
4-CH₃ piperidine	47.4	37.3	33.5	22.5		
2-CH₃ piperidine	52.5	33.3	26.0	27.4	47.9	23.1
1-CH₃ (4-CH₃) piperidine	56.7	26.3	24.3	47.0		
dimethyl piperazine	52.1	53.3	19.9			
bicyclic	48.7	27.7	21.9			
adamantane N	59.7	33.7	37.2			
adamantane NH	47.2	37.6	27.5	37.0		

[a]Cyclohexane = 27.7 ppm.

TABLE 3.17. ^{13}C CHEMICAL SHIFTS (IN PARTS PER MILLION) OF SATURATED HETEROCYCLES[8,36,37]

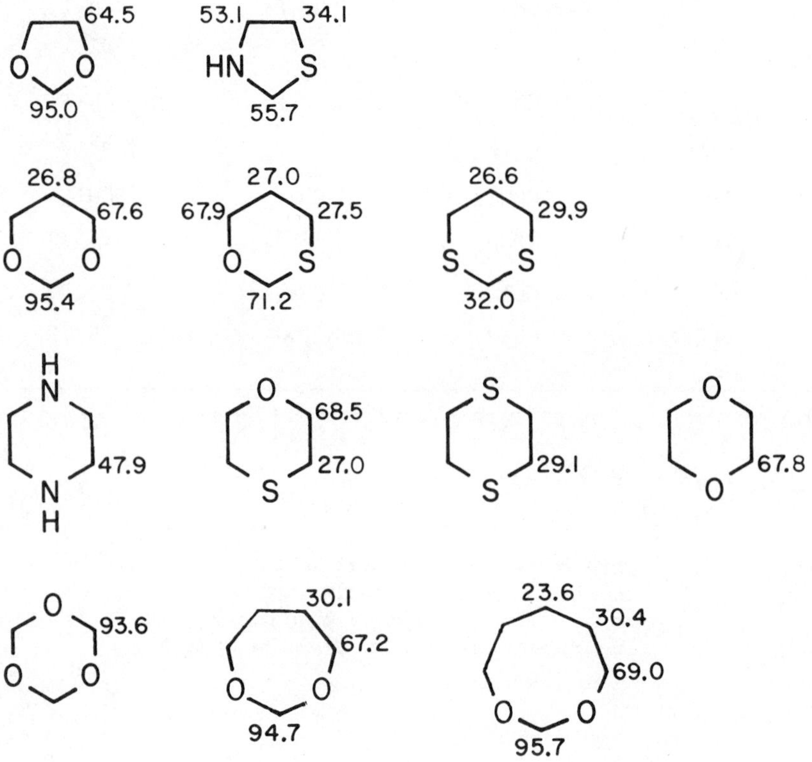

64.5   53.1   34.1

O   O   HN   S

95.0   55.7

26.8   27.0   26.6

67.6   67.9   27.5   29.9

O   O   O   S   S   S

95.4   71.2   32.0

H
N

47.9   68.5   S   O

O   O
        O
S   27.0   S   29.1   O   67.8

N
H

O
93.6   30.1   23.6

O   O   67.2   30.4

94.7   69.0

O   O

95.7

Although the origin of this variation is unclear, the correlation of bond deformation with $^1J_{CH}$ is well established.

Substitution by electronegative groups also leads to considerable increase in $^1J_{CH}$, as illustrated in Table 3.20.[45]

There can be more structure to the coupled multiplet than simple, directly bonded C–H coupling. Long-range C–H coupling ($^2J_{CH}$ and $^3J_{CH}$ can be observed directly in high-resolution ^{13}C spectra and indirectly from ^{13}C satellites in proton spectra. Some of these are illustrated in Table 3.21.[41,44a,46,47] A Karplus type dependence of $^3J_{CH}$ on the dihedral angle appears to exist.[48]

TABLE 3.18. ^{13}C CHEMICAL SHIFTS (IN PARTS PER MILLION) OF HALOMETHANES [a,8]

CH_3F	CH_3Cl	CH_3Br	CH_3I
65.7	23.9	9.0	-21.7
CH_2F_2	CH_2Cl_2	CH_2Br_2	CH_2I_2
104.4	52.9	20.4	-55.1
CHF_3	$CHCl_3$	$CHBr_3$	CHI_3
113.5	76.5	11.1	-140.9
CF_4	CCl_4	CBr_4	CI_4
117.5	95.4	-29.7	-293.5

[a]Relative to TMS; CH_4 at zero pressure = -13.2 ppm.

PROBLEM 3.7.   *Determine the structure and assign each resonance to the appropriate carbon or carbons for the compound whose ^{13}C spectrum appears in the accompanying figure (molecular formula $C_4H_{11}N$).*

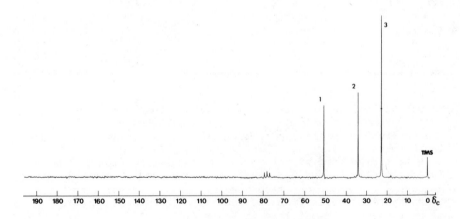

PROBLEM 3.8.   The compound whose ^{13}C spectrum appears below has a molecular formula $C_4H_8Br_2$. Deduce the structure and assign each resonance.

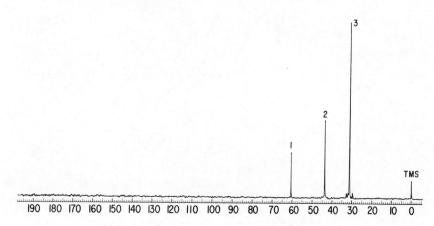

PROBLEM 3.9.   Make structure and resonance assignments for the compound whose ^{13}C spectrum appears below (molecular formula $C_6H_{15}NO$).

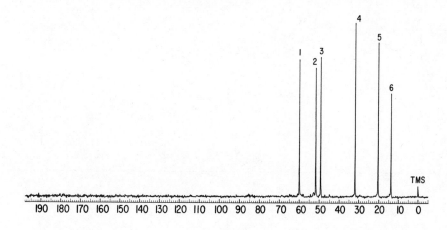

TABLE 3.19.　$^1J_{CH}$ IN ALKANES (IN HERTZ)[23f,42,44,47]

H−CH₃	H−CH₂−CH₃	(cyclohexyl)−H	(cyclobutyl)−H	(cyclopropyl)−H
125.0	124.9	123	134	161

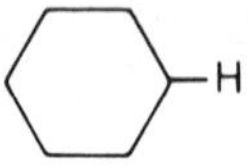

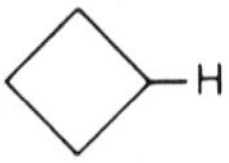

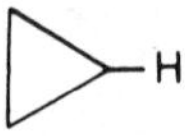

	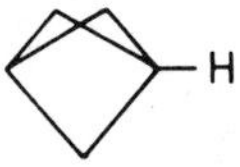		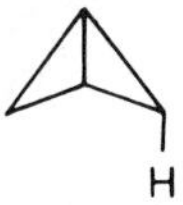	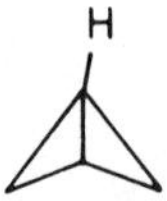
144 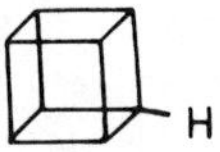	164 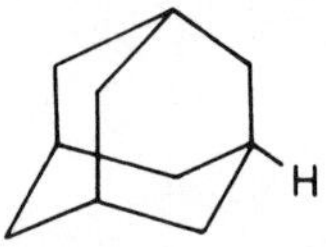	153 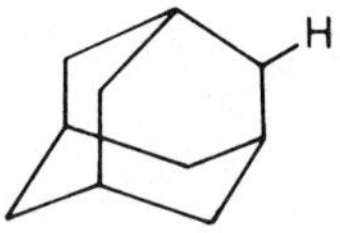	169	205

153.8	131.2	125.9

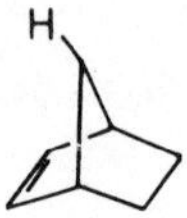

142	136.0	130.0

74

TABLE 3.20. $^1J_{CH}$ (IN HERTZ) IN SUBSTITUTED ALKANES[45]

	$^1J_{C-H}$		$^1J_{C-H}$
$H-CH_3$	125.0	$H-CH_2NO_2$	146.7
$H-CH_2COPh$	125.7	$H-CH_2F$	147.9
$H-CH_2CHO$	127.0	$H-\overset{\displaystyle OCH_3}{\underset{\displaystyle OCH_3}{CH}}$	161.8
$H-CH_2CO_2H$	130.0		
$H-CH_2C\equiv CH$	132.0		
$H-\overset{\displaystyle CO_2H}{\underset{\displaystyle CO_2H}{CH}}$	132.0	$H-\overset{\displaystyle NO_2}{\underset{\displaystyle NO_2}{CH}}$	169.4
$H-CH_2NH_2$	133.0		
$H-CH_2CN$	136.1	$H-\overset{\displaystyle F}{\underset{\displaystyle F}{CH}}$	184.5
$H-\overset{\displaystyle N(CH_3)_2}{\underset{\displaystyle N(CH_3)_2}{CH}}$	136.6	$H-\overset{\displaystyle OCH_3}{\underset{\displaystyle OCH_3}{C}}-OCH_3$	186.0
$H-CH_2OCH_3$	140.0		
$H-CH_2OH$	141.0	$H-\overset{\displaystyle F}{\underset{\displaystyle F}{CF}}$	239.1
$H-CH_2OPh$	143.0		
$H-\overset{\displaystyle CN}{\underset{\displaystyle CN}{CH}}$	145.2		137.4
			149.5
			134.6
			146.5

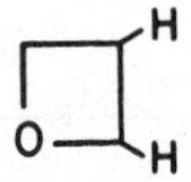

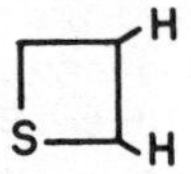

75

TABLE 3.21.   LONG-RANGE C-H COUPLING CONSTANTS (IN HERTZ) FOR
ALKANES[41,45b,46,47]

COMPOUND	TYPE	J
H_3**C**-CH$_3$	^2J	-4.5
ClC**H**$_2$-C**H**$_2$Cl	^2J	-3.4
CH$_3$C**H**$_2$CH$_3$	^2J	-4.3
CH$_3$C**H**$_2$-**C**H$_3$	^2J	1.9
CH$_3$C**H**-**C**H$_3$ $\mid$ OH	^2J	0.7
CH$_3$**C**H-**C**H$_3$ $\mid$ X	^2J	-4.0 to 4.7
CH$_3$CH-**C**H$_3$ $\mid$ X	^3J	4.6 to 5.8

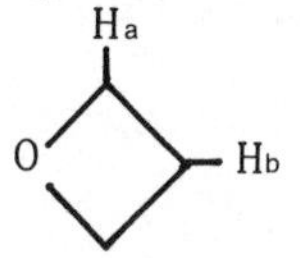

	^2J_a	-2.9
	^3J_a	+2.9
	^2J_b	-4.4

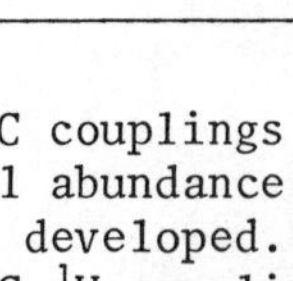

	^2J_a	-4.1
	^2J_b	-3.7
	^3J_b	+4.2

Although C-C couplings are considerably more difficult to obtain in natural abundance than are C-H values, a substantial body of data has developed. Trends analogous to those exhibited by ^{1}H-^{1}H and ^{13}C-^{1}H couplings have been elucidated. Selected values are given in Table 3.22.[41,46a,b,49,51] Of special interest is the disparity between the $^1J_{CH}$ values of the two bicyclobutanes. The negative sign is, however, theoretically predicted.[51d,e]

H_3C-CH_3 J 34.6

H_3C-CH_2OH J 37.7

$\triangleright C-CH_3$ J 44.0

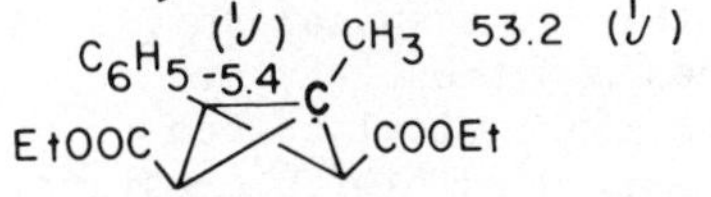

C_6H_5 (J) -5.4 CH_3 53.2 (J) EtOOC   COOEt

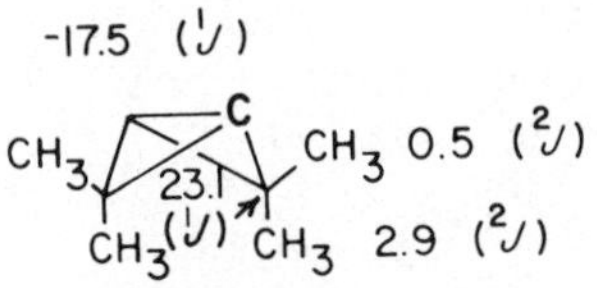

C_6H_5 CH_3 ^2J -2.4

-17.5 (J)

CH_3 C CH_3 0.5 (^2J) 23.1 (J) CH_3 CH_3 2.9 (^2J)

CH_3 CH_3 40.3 (J) CH_3 CH_3 46.9 (J)

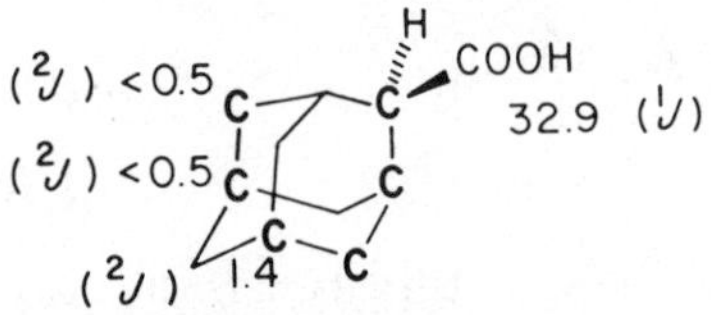

(^2J) <0.5   H   COOH   32.9 (J) (^2J) <0.5 (^2J) 1.4

HO   34.7 (J) HC 1.0 (^3J) 2.4   2.7 (^2J) (^2J) 0.7   36.0   1.9 (^2J) (J)

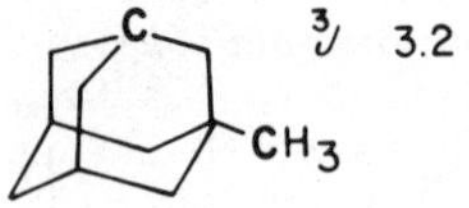

C ^3J 3.2 CH_3

C CH_3 ^3J 3.4

C—C O J 29.5

C—C S J 31.5

Br C—C J 29.9

Br J 26.8

CH_3 36.1 (J) (^2J) 8.1   29.1 (J)

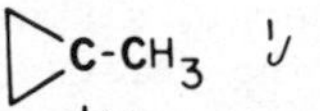

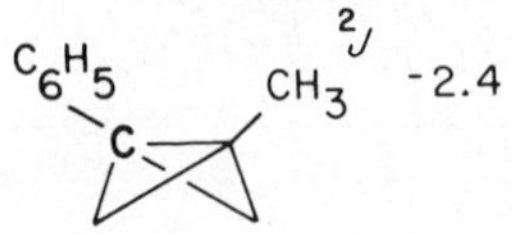

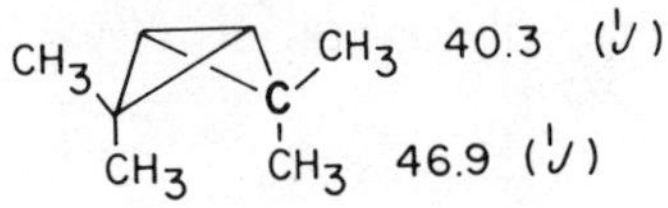

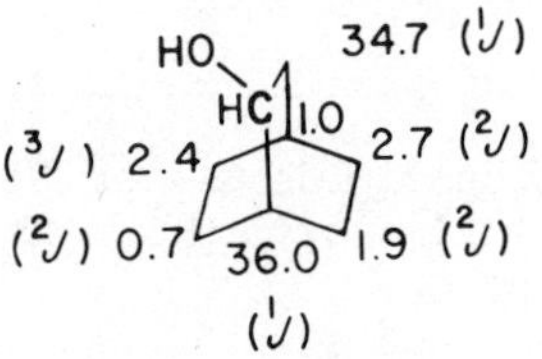

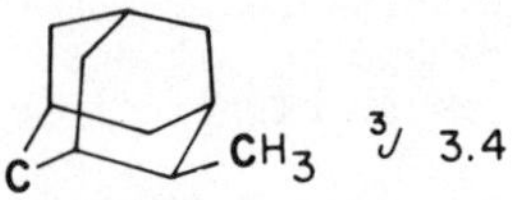

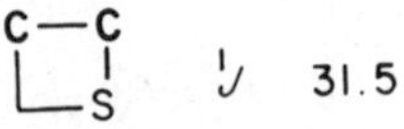

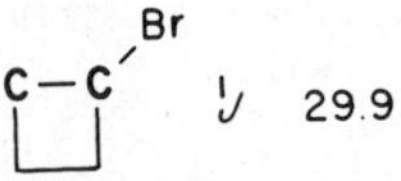

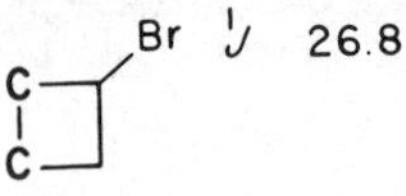

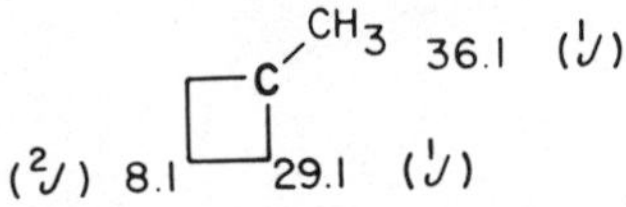

ALKENES AND DERIVATIVES

As a substituent, where conjugation is unimportant, alkene linkages have an effect not unlike an alkyl group. The lone sp^3-carbon in cycloheptatriene, for example, appears at 28.8 ppm, whereas the carbons of cycloheptane absorb at 28.5 ppm.

sp^2-Carbons (alkene and aromatic) appear considerably deshielded relative to alkanes. The chemical-shift range for substituted alkenes is 80 to 145 ppm relative to TMS. In general, alkene shifts appear 100 ± 20 ppm downfield relative to the corresponding alkane resonance. Table 3.23 gives the ^{13}C chemical shifts of all carbons for a number of simple alkenes and those of the corresponding alkanes. As expected, a terminal vinyl carbon appears at higher field than an alkyl-substituted olefinic carbon.

TABLE 3.23.   SIMPLE ALKENES AND ALKANES[1,52]

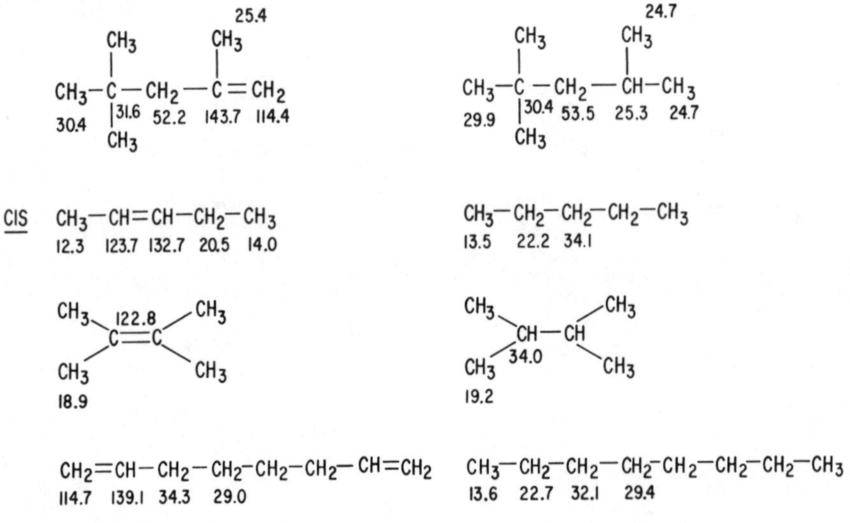

Methyl substitution at C-1 in a 1-alkene deshields C-1 by 3 to 10 ppm but shields C-2 by 4 to 9 ppm. Methyl substitution at an α-carbon causes the expected deshielding of C-1.

Dorman, Jautelat, and Roberts[52a] have analyzed some 50 alkenes and developed an empirical method whereby the carbon chemical shifts of acyclic alkenes can be calculated. On the basis of substituent parameters α, β, γ, α´, β´, and γ´ plus several correction terms, the chemical shift for any alkene carbon can be calculated with ethylene (123.3 ppm) as a reference. Table 3.24 lists the appropriate parameters.

The chemical shift for carbon 3 of 2-pentene can be calculated as follows:

$$\overset{\alpha'}{} \quad \overset{*}{} \quad \overset{\alpha}{} \quad \overset{\beta}{}$$
$$CH_3CH=CH-CH_2CH_3$$

$$\delta^{13}C = (123.3) + 1 \cdot \alpha + 1 \cdot \beta + 1 \cdot \alpha'$$
$$= 123.3 + 10.6 + 7.2 + -7.9$$
$$= 133.2 \text{ ppm}$$

If the alkene is *cis*, the *cis* correction factor is added. The calculated shift for carbon-3 of *cis*-2-pentene is thus 132.1 ppm (experimental: *cis*, 132.7 ppm; *trans*, 133.3 ppm).

Although additional correction terms are used in the original work,[52a] the seven parameters in Table 3.24 provide the investigator with a satisfactory tool for most circumstances. The correlation works less well for alicyclic alkenes, for which carbons are counted more than once where appropriate. With polyenes, alkene carbons as substituents are counted as if they were alkane carbons. In general, alkenes display substituent effects similar to those of alkanes.[53]

TABLE 3.24.   ALKENE SHIFT PARAMETERS (IN PARTS PER MILLION)[52]

α	10.6
β	7.2
γ	-1.5
α'	-7.9
β'	-1.8
γ'	1.5
cis	-1.1

PROBLEM 3.10.   *Calculate the alkene and alkane chemical shifts for the following compounds. Compare your results with the experimental values as given in Tables 3.23 and 3.25.*

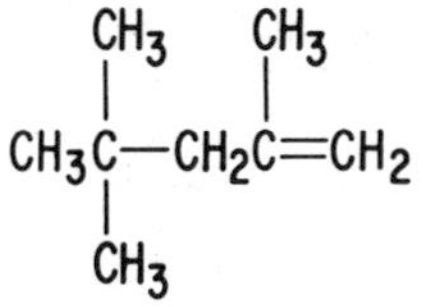

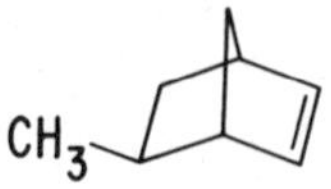

Cyclics.   The introduction of a double bond into a cyclic system is expected to cause more general changes throughout the molecule. Table 3.25[20] gives the ^{13}C chemical shifts for a number of norbornenes.

TABLE 3.25. ^{13}C CHEMICAL SHIFTS (IN PARTS PER MILLION) OF NORBORNENES[20]

	C-1	C-2	C-3	C-4	C-5	C-6	C-7	CH₃
Norbornane	36.8	30.1	30.1	36.8	30.1	30.1	38.7	
Norbornene	42.2	135.5	135.5	42.2	25.5	25.5	48.8	
1-Methyl norbornene	49.9	140.0	135.8	43.3	28.0	32.6	55.0	18.0
exo-5-Methyl norbornene	42.7	137.2	136.2	48.7	33.0	35.0	45.0	21.7
endo-5-Methyl norbornenes	43.6	137.2	132.5	47.8	33.0	34.2	50.5	19.5
syn-7-Methyl norbornene	47.8	132.4	132.4	47.8	25.9	25.9	54.7	12.5
anti-7-Methyl norbornene	46.0	137.8	137.8	46.0	21.8	21.8	53.3	14.4
Norbornadiene	50.9	143.4	143.4	50.9	143.4	143.4	75.4	

PROBLEM 3.11.   *Compare the chemical shifts of Table 3.25 with the corresponding alkanes given in Tables 3.7 and 3.8. Which carbons exhibit steric compression shifts? What is the effect of the introduction of a double bond on the other norbornyl carbons?*

Table 3.26[54] gives the ^{13}C chemical shifts for cycloheptatriene and related compounds. Carbons 3 and 4, the internal olefinic carbons, are the most deshielded in cycloheptatriene.

Methyl substitution at C-1 deshields C-1 and C-7 and shields C-2 and C-3.

TABLE 3.26. ^{13}C CHEMICAL SHIFTS (IN PARTS PER MILLION) OF CYCLOHEPTATRIENE AND NORCARADIENE COMPOUNDS[54]

	C-1	C-2	C-3	C-7	C-8	C-9	C-10
	120.4	126.8	131.0	28.1			
	130.6	122.2	128.8	40.1	24.6		
	125.9	127.7	137.2	26.6	130.3	130.8	
	123.0	124.6	130.8	49.4	31.1	27.3	
	37.7	129.0	119.2	32.3	15.7		19.7

The chemical shifts of the alkene carbon in (strained) cyclopropene[55] and cyclobutene[52a] are:

108.7 ppm   137.2 ppm

Monosubstituted Alkenes.   With substituted alkenes the situation becomes more complex.   Table 3.27[56,57] summarizes ^{13}C chemical shifts of the vinyl carbons for several monosubstituted alkenes. Substitution of electronegative groups at C-1 deshields that nucleus substantially. For bromine and iodine, the "heavy-atom" upfield shift is observed on substitution.

For alkyl vinyl ethers, the resonance form

$$CH_2^- - CH = O^+ - R$$

is a useful rationalization of the observed shifts.[53] It has been suggested that the degree of this resonance contributor is controlled mainly by the steric bulk of R; the shift at C-2 decreases as a linear function of the bulkiness of R [a plot of δ_β vs. E_s Taft values gives a straight line].[56b]

Table 3.28[58a] lists the shifts for some *cis*- and *trans*-disubstituted ethylenes and Table 3.29[59] lists shifts for polyhalo-substituted ethylenes. Carbon-13 nmr behavior of polychloropolycyclic olefins has been reported.[58b-58d] Shifts for *cis*- and *trans*-dialkyl-substituted olefins are quite close. Detailed use of proton-proton and/or proton-carbon long-range couplings would be required to distinguish isomers. The dialkyl esters of maleic acid show a considerable shift difference between *cis* and *trans* isomers, in the opposite sense from the diiodoethylenes. In a series of solvents the carbonyl carbons of diethylmaleate display a marked solvent effect, but the alkene carbons show no appreciable variation.

The resonances for alkene carbons appear in the same region as those for aromatic compounds. Since both are classes of sp^2-carbons, this is not surprising (aromatics are covered in Chapter 4). Proton nmr is useful in confirming the presence of either type of group. Partial ^1H decoupling in a cmr experiment can confirm the presence of a terminal vinyl carbon. Specific proton decoupling is often useful for elucidating specific structural features.

In the case of styrene, both alkene and aromatic carbons are present. Table 3.30[60,61] gives the ^{13}C chemical shifts for alkene

carbons in substituted styrenes; *para* substitution causes minimal change at the C-1 carbon but does affect the shift of C-2. Resonance forms of the type

help rationalize this behavior. Chemical shifts of substituted styrenes have been used to elucidate mechanisms for transmission of electronic effects.[61] In α,β-unsaturated ketones and aldehydes behavior similar to that of simple alkenes is observed as illustrated in Table 3.31.[62a] Shifts are at somewhat lower field, particularly for the C-2 (β) carbon. The resonance forms

are useful rationalizations.

TABLE 3.27. ^{13}C SHIFTS (IN PARTS PER MILLION) OF MONOSUBSTITUTED ALKENES[56,57]

SUBSTITUENT, X	C-1	C-2
	C-2 C-I	
	CH_2=CH—X	
I	85.3	130.4
Br	115.5	122.0
Cl	126.0	117.3
CO_2Et	129.7	130.4
CH_2Br	133.1	117.6
CH_2OEt	135.7	114.6
$SO_2CH=CH_2$	137.7	131.3
$OCOCH_3$	141.6	96.3
OCH_3	153.2	84.1
$O-CH_2-CH{\begin{smallmatrix}CH_3\\CH_3\end{smallmatrix}}$	152.9	85.0
$O-CH{\begin{smallmatrix}CH_3\\CH_3\end{smallmatrix}}$	151.4	87.5
$O-\overset{CH_3}{\underset{CH_3}{C}}-CH_3$	146.8	90.2

Loots et al.[62b] have extended this qualitative picture to a semiquantitative estimate of relative charge densities at each vinyl carbon. In their approach four sets of chemical shifts are required: the α,β-unsaturated ketone, the corresponding alkene, the saturated ketone, and the corresponding alkane. Values for changes in π-electron density are obtained that are intuitively reasonable and that in some cases compare favorably with theoretical calculations. However, the method is based on the assumption of a relationship between ^{13}C chemical shift and electron density that is not wholly justified.

TABLE 3.28.　^{13}C CHEMICAL SHIFTS (IN PARTS PER MILLION) OF *cis*- AND *trans*-DISUBSTITUTED ETHYLENES, XHC=CHX[58a]

SUBSTITUENT X	δ_{cis}	δ_{trans}	Δ
I	96.5	79.4	-17.1
Br	116.4	109.4	-7.0
Cl	121.3	119.4	-1.9
CN	120.8	120.2	-0.6
CH_2CH_3	131.2	131.3	0.1
CH_3	123.3	124.5	1.2
CO_2CH_3	128.7	132.4	3.7
$CO_2C_2H_5$	130.5	134.1	3.8
$CO_2C_4H_9$	129.0	132.8	3.8

PROBLEM 3.12.　　*Assign each resonance to the appropriate carbon or carbons.*

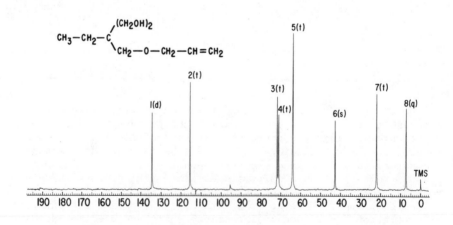

TABLE 3.29. ^{13}C CHEMICAL SHIFTS (IN PARTS PER MILLION) OF POLYHALOALKENES[59]

COMPOUND	CARBON	SHIFT
C_2H_4	--	123.3
C_2H_3Cl	CH_2 CHCl	117.2 126.1
gem-$C_2H_2Cl_2$	CH_2 CCl_2	113.3 127.1
C_2HCl_3	CHCl CCl_3	117.6 125.1
C_2Cl_4	--	121.3
C_2H_3Br	CH_2 CHBr	122.4 114.7
gem-$C_2H_2Br_2$	CH_2 CBr_2	127.2 97.0
C_2HBr_3	CHBr CBr_2	112.4 95.0
C_2Br_4	--	93.7

Allenes.   Allenes form a unique class of compounds because of the extremely low field shift of the central allenic carbon (200 to 220 ppm). Table 3.32 presents representative data for a number of substituted allenes. For a given alkyl substituent, there is a linear relationship between the number of substituents and the chemical shift of the central carbon (C-2). Considered as an additive property, a methyl group shields that carbon by 3.3 ppm, an ethyl group 4.8 ppm, and a sec-alkyl group 7 ppm. Carbons C-1 and C-3 are shielded by some 30 ppm relative to corresponding ethylene carbons but otherwise display similar substituent effects.[63e] Strain in cyclic allenes appears to have little effect.[63d]

TABLE 3.30. ^{13}C CHEMICAL SHIFTS OF SUBSTITUTED STYRENES[60,61]

X	C-1	C-2
H	136.7	113.2
p-OCH$_3$	136.5	111.5
p-NMe$_2$	137.3	108.7
p-Br	136.2	115.1
p-Me	137.5	114.0
p-NO$_2$	135.9	119.0
m-NO$_2$	134.9	116.6
o-Me	136.4	114.6
o-Br	137.5	117.2
1-Me	143.1	112.4
1-Et	150.0	110.9

Coupling.   Coupling information for alkenes is particularly useful.   Table 3.33 presents several examples of one-bond and long-
range C-H couplings. sp^2-Carbons have $^1J_{CH}$ values about 30 Hz
larger than do sp^3-carbons; as with alkanes, bond angle distortion
and substitution by electronegative groups increase the value of
$^1J_{CH}$.   The large value for cyclopropene is consistent with substantial sp hybridization for the vinyl carbons.[54]
    Values of $^2J_{CH}$ are generally small and vary in an irregular
manner.[64]   Their behavior parallels that of $^2J_{HH}$; for instance,
they become more positive when electronegative substituents are
introduced.
    Values of $^3J_{CH}$ depend on geometry and are generally larger
for trans- than for cis-coupled nuclei.[48]   A Karplus type of relationship has been calculated.[65]   These values have been

TABLE 3.31.   ALKENE CHEMICAL SHIFTS (IN PARTS PER MILLION) IN α,β-UNSATURATED KETONES[62]

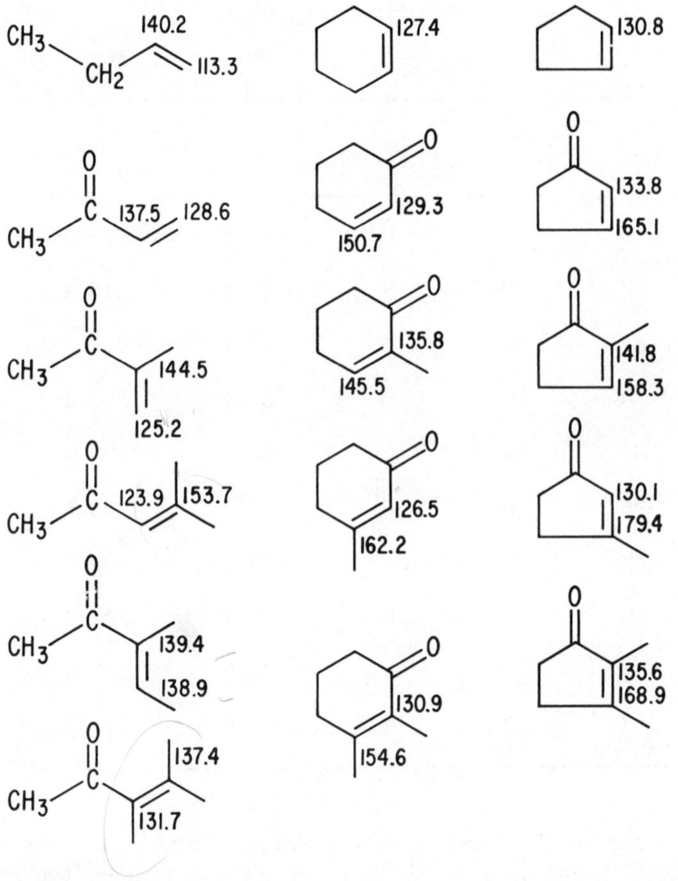

CH$_3$—CH$_2$ / 140.2, 113.3

127.4

130.8

CH$_3$—C(=O)— 137.5, 128.6

(cyclohexenone) 129.3, 150.7, O

(cyclopentenone) O, 133.8, 165.1

CH$_3$—C(=O)— 144.5, 125.2

135.8, 145.5, O

141.8, 158.3, O

CH$_3$—C(=O)— 123.9, 153.7

126.5, 162.2, O

130.1, 179.4, O

CH$_3$—C(=O)— 139.4, 138.9

130.9, 154.6, O

135.6, 168.9, O

CH$_3$—C(=O)— 137.4, 131.7

surveyed.[66,67] Values of J_{CH} for propene[68] summarize many of the trends for simple alkenes:

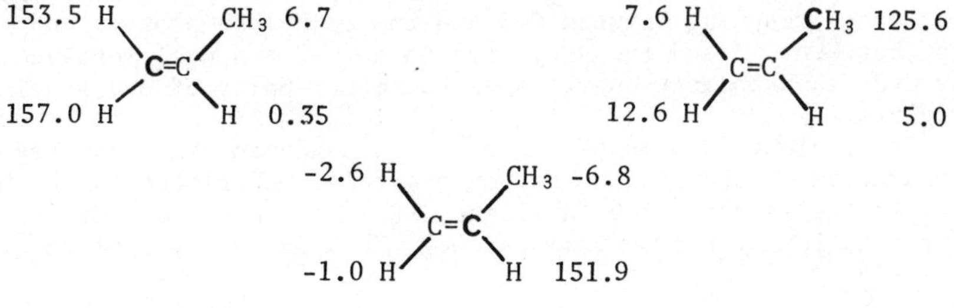

153.5 H   CH$_3$ 6.7   7.6 H   CH$_3$ 125.6
         C=C   C=C
157.0 H   H 0.35   12.6 H   H   5.0

-2.6 H   CH$_3$ -6.8
         C=C
-1.0 H   H 151.9

TABLE 3.32. ^{13}C CHEMICAL SHIFTS (IN PARTS PER MILLION) OF ALLENES[63]

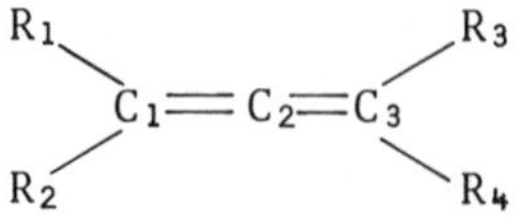

$$R_1 \quad\quad\quad R_3$$
$$\diagdown \quad\quad\quad \diagup$$
$$C_1 = C_2 = C_3$$
$$\diagup \quad\quad\quad \diagdown$$
$$R_2 \quad\quad\quad R_4$$

R_1	R_2	R_3	R_4	C_1	C_2	C_3
H	H	H	H	74.8	213.5	74.8
Me	H	H	H	84.4	210.4	74.1
Me	Me	H	H	93.4	207.3	72.1
Me	H	Me	H	85.4	207.1	85.4
Me	Me	Me	Me	92.6	200.2	92.6
Me	SMe	H	H	99.9	203.6	80.1
Ph	Ph	Ph	Ph	113.6	209.5	113.6
OMe	H	H	H	123.1	202.0	90.3
Br	H	H	H	72.7	207.6	83.8
CN	H	H	H	80.5	218.7	67.2

Thiete sulfone[69] shows a well-resolved example of short- and long-range C-H coupling in alkenes. Figure 3.3 shows the proton-coupled ^{13}C nmr spectrum for the three carbons of thiete sulfone along with shift and coupling data. The methylene carbon (C-4) is split into a C-H triplet consisting of pairs of doublets arising from long-range coupling with the two vinyl hydrogens. The long-range coupling between C-3 and the methylene protons and that between C-3 and the vinyl proton on C-2 are well resolved. For C-2, the apparent quartets are actually pairs of overlapping triplets.

In ethylene $^1J_{CC}$ is 67.6 Hz,[49] a value in accord with larger s character in the bonding carbon orbitals. Consistent with this view is the larger value in allene, 98.7 Hz, which reflects sp hybridization at the central carbon. In a manner similar to $^1J_{CH}$,

TABLE 3.33. ^{13}C-^{1}H COUPLING (IN HERTZ) IN SUBSTITUTED ALKENES[45,55,58a,59,66]

	$^1J_{C-H}$	$^2J_{C-H}$	$^3J_{C-H}$		$^1J_{C-H}$	$^2J_{C-H}$
$H_2C=CH_2$	156.4	(−)2.4		$Cl(H_\alpha)C=C(H_\beta)(H_\gamma)$	198	β(+)7.5 γ(−)7.9
$H_2C=C=CH_2$	167.8	3.9	7.7	$Cl(H_\alpha)C=C(H_\beta)(H_\gamma)$	162	α(+)6.9
1,4-dithiin (S,S ring)	179.8	8.1		$(H)(Cl)C=C(Cl)(H)$		(+)0.8
$H_2C=CHF$	159.1			$(H)(Cl)C=C(H)(Cl)$		(+)16.0
$H_2C=CHF$	200.2			$(CH_3)_2C=CH=CH$	221	
$(H)(I)C=C(I)(H)$	194.2	(−)1.4		cyclopropene $=CH$	228.2	
$(I)(H)C=C(I)(H)$	187.9	11.0		cyclobutene $=CH$	170	
methylenecyclohexane ($=CH_cH_f$)		(H_c)6.4 (H_f)11.4		$H_\alpha C(H_\beta)=C(H_\gamma)-C(=O)H$		α=162.3 β=156.6
				$H_\alpha C(H_\beta)=C(H_\gamma)-C(=O)H$		γ=162.3

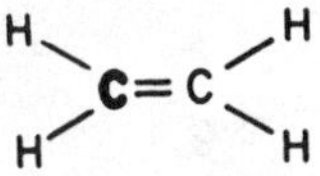

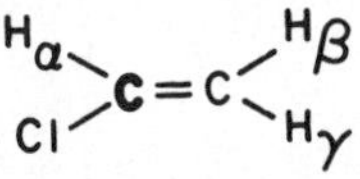

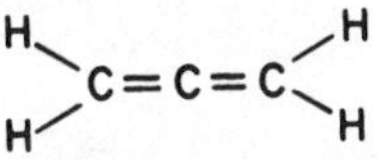

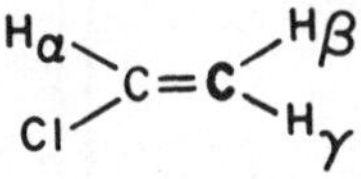

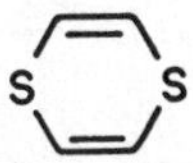

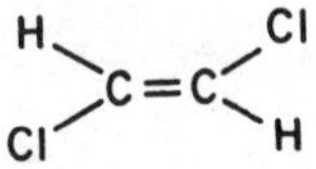

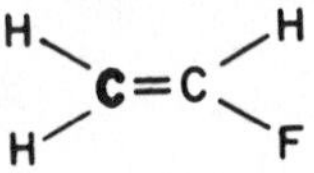

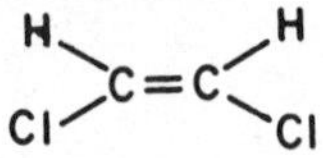

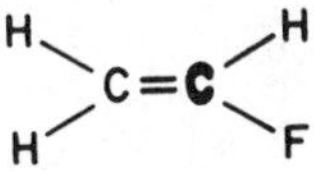

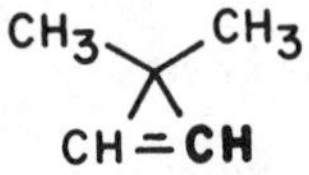

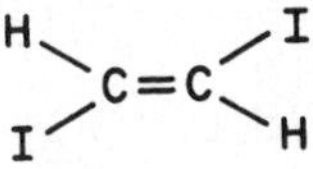

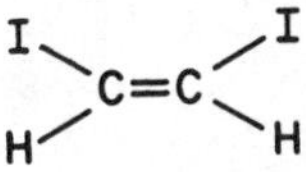

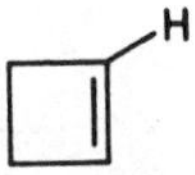

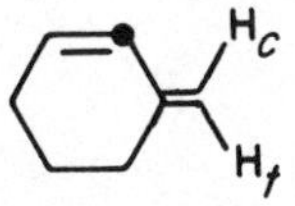

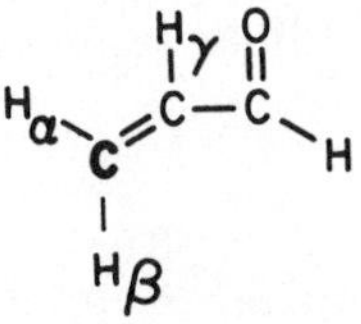

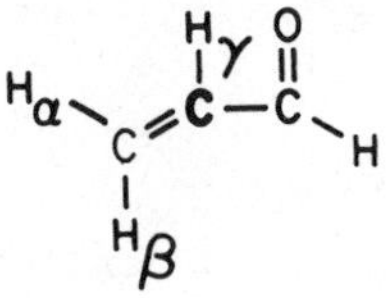

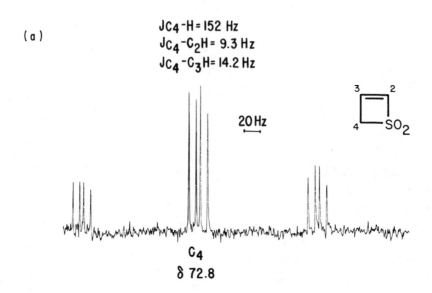

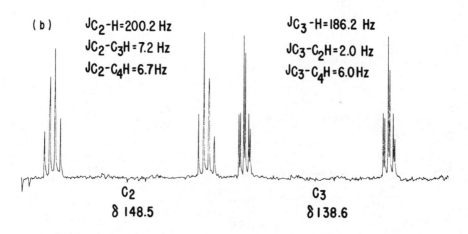

FIGURE 3.3.   Proton-coupled ^{13}C spectrum of thiete sulfone:
(a) methylene carbon region; (b) vinyl carbons.

$^1J_{CC}$ increases with substitution by electronegative elements. Interestingly, alkyl substitution also increases $^1J_{CC}$, as the following examples illustrate:

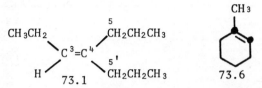

In z-4-propyl-3-heptene $^1J_{C_4C_5}$ and $^1J_{C_4C_5'}$ are not equal, having values of 41.6 and 43.0 Hz, respectively. In butadiene $^1J_{C_2C_3}$ is 53.7 Hz.[70] Available data illustrate that two-bond C-C couplings, for a variety of bonding situations in alkenes, are less than 2.5 Hz and hence are limited in usefulness.

Only one $^3J_{CC}$ value in an alkene--butadiene--has been reported, where $^3J_{CC}$ = 9.05 Hz.[70]

ALKYNES

The alkyne linkage as a substituent shields the attached carbon by 5 to 15 ppm relative to the corresponding alkane:

$$CH_3-CH_2-C \equiv C-CH_2-CH_3$$
$$8.9 \quad 10.7 \quad 82.0$$

$$CH_3-CH_2-CH_2-CH_2-CH_2-CH_3$$
$$13.7 \quad 22.7 \quad 31.7$$

Acetylenic carbons appear over the range 65 to 90 ppm for single carbon-containing compounds. Substituted alkyne carbons are deshielded relative to terminal alkyne carbons. Table 3.34[31,72] lists the ^{13}C chemical shifts for the alkyne carbons of a number of substituted acetylenes. For alkoxy ethynes, the resonance structure

$$H-\overset{-}{C}=\overset{+}{C}=OEt$$

is a useful rationalization of the results. For ethynyl phosphines or silanes, the opposite resonance contributor

$$H-\overset{+}{C}=\overset{-}{C}=P \qquad H-\overset{+}{C}=\overset{-}{C}=Si(R_3)$$

appears operative. For thioethynes, polarization of charge within the triple bond appears to occur.

TABLE 3.34. ^{13}C CHEMICAL SHIFTS (IN PARTS PER MILLION) OF SUBSTITUTED ALKYNES[71,72]

	α	β
H–C≡C–H	71.9	
$\overset{\beta}{H}$–C≡$\overset{\alpha}{C}$–C$_4$H$_9$	83.0	66.0
CH$_3$CH$_2$–$\overset{\beta}{C}$≡$\overset{\alpha}{C}$–CH$_2$–CH$_3$	82.0	82.0
$\overset{\beta}{H}$–C≡$\overset{\alpha}{C}$–C$_6$H$_5$	83.3	77.7
CH$_3\overset{\beta}{C}$≡$\overset{\alpha}{C}$–C$_6$H$_5$	85.7	85.7
C$_6$H$_5\overset{\beta}{C}$≡$\overset{\alpha}{C}$–C$_6$H$_5$	89.9	89.9
$\overset{\beta}{H}$–C≡$\overset{\alpha}{C}$–O–CH$_2$–CH$_3$	89.4	23.2
CH$_3$–CH$_2$–$\overset{\beta}{C}$≡$\overset{\alpha}{C}$–O–CH$_2$–CH$_3$	88.1	36.0
CH$_3$–$\overset{\beta}{C}$≡$\overset{\alpha}{C}$–O–CH$_3$	88.4	28.0
$\overset{\beta}{H}$–C≡$\overset{\alpha}{C}$–S–CH$_2$–CH$_3$	72.6	81.4
CH$_3$–CH$_2\overset{\beta}{C}$≡$\overset{\alpha}{C}$–S–CH$_3$	67.3	92.7
$\overset{\beta}{H}$–C≡$\overset{\alpha}{C}$–P(sBu)$_2$	83.2	92.7
(CH$_3$)$_3$Si–$\overset{\beta}{C}$≡$\overset{\alpha}{C}$–C$_6$H$_5$	104.4	92.5

Substituent parameters for alkyne shifts have been derived from a substantial body of data for acyclic alkynes (Table 3.35).[73,74] In addition, substituent parameters for calculating the effect of the triple bond on alkyl groups have been determined.

Alkyne chemical shifts have been reported for propargyl alcohols,[75] diynes, diynones, and polyynones[76] and for enynes.[77]

<u>Coupling.</u> In accord with sp hybridization at carbon, $^1J_{CH}$ is ~250 Hz (Table 3.36) and values increase slightly with electronegativity.[45] In contrast to alkanes and alkenes, $^2J_{CH}$ for alkynes can be rather large. From limited available data, $^3J_{CH}$ may display a dihedral-angle dependence. Representative ^{13}C-^{13}C couplings are:[49]

H–C≡C–H	CH$_3$C≡CH	CH$_3$–C≡C–COOCH$_3$
171.5 Hz	11.8 Hz	20.3 Hz

PROBLEM 3.13.   Determine the structure of this C_6H_{10} compound and make ^{13}C spectral assignments (TMS at S=0; CDCl$_3$ at S=77).

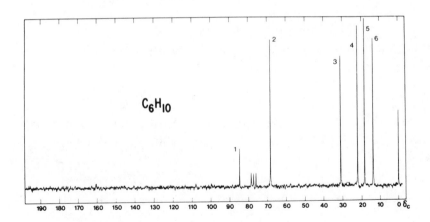

TABLE 3.35.   ALKYNE SUBSTITUENT PARAMETERS[a] (IN PARTS PER MILLION)[74]

$$\overset{\delta'}{C} - \overset{\gamma'}{C} - \overset{\beta'}{C} - \overset{\alpha'}{C} - C \equiv C - \overset{\alpha}{C} - \overset{\beta}{C} - \overset{\gamma}{C} - \overset{\delta}{C}$$

$\alpha = 6.93$	$\alpha' = -5.69$
$\beta = 4.75$	$\beta' = 2.32$
$\gamma = -0.13$	$\gamma' = -1.31$
$\delta = 0.51$	$\delta' = 0.56$

[a] Base = 72.8 ppm.

TABLE 3.36.   C-H COUPLING CONSTANTS (IN HERTZ) FOR ALKYNES[47,77,78]

	$^1J_{CH}$	$^2J_{CH}$	$^3J_{CH}$
$\overset{1}{H}-\overset{2}{C}\equiv C-H$	249	49.3	
$\overset{3}{H_3}C-\overset{2}{C}\equiv \overset{1}{C}-\overset{1}{H}$		$50.8(C_2-H_1)$	$-10.6(C_1-H_3)$
(see structure below)	$251.7(C_4-H_4)$	$50.2(C_3-H_4)$	$4.0(C_4-H_3)$
		$2.0(C_3-H_3)$	$16.3(C_3-H_1)$
			$9.5(C_3-H_2)$

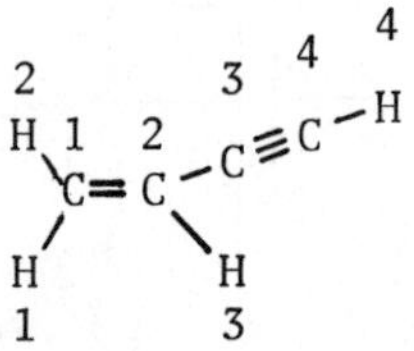

Structure for third entry:

$$\underset{\underset{1}{H}}{\overset{\overset{2}{H}}{C}} = \underset{\underset{3}{H}}{\overset{3}{C}} - \overset{4}{C} \equiv \overset{4}{C} - \overset{4}{H}$$

SPECIAL TOPICS

Deuterium Isotope Effects. The use of deuterium substitution for
assignment of resonances in proton nmr spectroscopy has its parallel in ^{13}C nmr spectroscopy. Carbons bonded to deuterium
generally display no NOE and are split by residual C-D coupling.
Lines are frequently broadened because of effects of the deuterium
quadrupole on the ^{13}C T_2s. The combination of all these factors
results in resonance lines for deuterated carbons that are substantially less intense than those of protonated ones. Hence
comparison of spectra of specifically deuterated materials with
the corresponding spectra of the protio compounds allows ready
assignment of resonances. Alternatively, if resonance lines have
already been assigned, reduction in line intensity made under
appropriate pulse conditions identifies a site of deuteration;
this may be useful, in studies of reacting compounds for example.[79]

It is well known that deuterium substitution changes resonance
positions of the substituted carbons by small but measurable
amounts. This "isotope shift" is almost always a shielding effect,
for theoretical reasons that have been described.[80] Examination
of Table 2.1 indicates the magnitudes that may be expected. The
effect is ~0.2 ppm per deuterium atom for the substituted α-
carbon and ~0.1 ppm for the β-carbon.[81a] Effects further down a
hydrocarbon chain[82a] and at more remote positions in cyclode-
canone[32b] have been reported. No effects on alkene or alkyne
resonances have been reported, but changes of the same magnitude
can be expected.

The increasing use of ^{13}C enrichment in biosynthetic studies
by ^{13}C nmr has not been hampered by the potential isotope effect
of ^{13}C on ^{13}C, and no shielding effects have been reported. An
interesting observation is the influence of ^{34}S on the resonance
position of CS_2; an isotope effect of 0.009 ppm is discernable![84]

No deuterium isotope effects on C-H or C-C couplings have
been reported. According to the respective magnetogyric ratios
$J_{CH} = 6.55 J_{CD}$, J_{CD} is related to J_{CH}.

Solvent Effects. Except for special-purpose studies in which
effects of strongly or specifically interacting solvents were explored, solvents for determining ^{13}C spectra of solids are generally chosen on the basis of convenience and sample solubility.
For the bulk of chemical-shift and coupling-constant measurements,
especially at the earlier stages of ^{13}C nmr spectroscopy, solvent-
induced changes were comparable to experimental error or else did
not affect principal conclusions. With increases in the precision
of experimental measurements, and particularly with the ability to
detect resonances in very dilute solutions, an awareness of
possible solvent effects has become important.

Of significance in this text is the report that TMS, generally accepted as a primary standard for reporting ^{13}C chemical shifts, is subject to solvent effects.[85] A variation in resonance position of almost 4 ppm was recorded in 38 solvents, and even in those commonly used for routine ^{13}C spectra, a range of ~1 ppm is displayed. Interestingly, cyclohexane was found to be much less sensitive to solvent ($\Delta\delta < 1$ ppm) and is recommended as a measurement standard when very precise values are to be reported for measurements in more than one solvent.[85]

Mantsch and Smith[86] reported chemical shifts of cholesterol, uridine, and pyridine in a variety of solvents. In this work, actual crossing of resonance positions occurred, which could have led to incorrect assignment of resonances.

Alkyl halides undergo large solvent-induced shifts.[87] A range of 5.2 ppm is exhibited by C-1 of ethyl iodide in going from cyclohexane to *N,N*-dimethylformamide. The effect diminishes with branching at the substituted carbon, disappearing in *tert*-butyl iodide. Hydrogen bonding can produce marked changes in resonance positions and in C-H coupling constants.[38] Other solvent effects are discussed in connection with aromatics (Chapter 4) and specific functional groups (Chapter 5).

REFERENCES

1. L. P. Lindeman and J. Q. Adams, *Anal. Chem.*, **43**, 1245 (1971).
2. (a) D. M. Grant and E. G. Paul, *J. Am. Chem. Soc.*, **86**, 2984 (1964); (b) L. J. M. van de Ven and J. W. de Haan, *J. Chem. Soc., Chem. Commun.*, **1978**, 94.
3. For examples of utilization of these techniques, see M. Yamazaki, T. Takeuchi, and K. Matsushita, *Kogyo Kagaku Zasshi*, **74**, 656 (1971).
4. B. V. Cheney and D. M. Grant, *J. Am. Chem. Soc.*, **89**, 5319 (1967).
5. R. Ditchfield and P. D. Ellis, in *Topics in Carbon-13 NMR Spectroscopy*, Vol. 1, G. C. Levy, Ed., Wiley, New York, 1974, p. 1.
6. (a) J. C. MacDonald, *J. Magn. Resonance*, **34**, 207 (1979); (b) H. Beierbeck, J. K. Saunders, and J. W. Apsimon, *Can. J. Chem.*, **55**, 2813 (1977).
7. (a) K. Seidman and G. E. Maciel, *J. Am. Chem. Soc.*, **99**, 659 (1977) and references cited therein; (b) J. G. Batchelor, *ibid.*, **97**, 3410 (1975); (c) G. E. Maciel, in *Topics in Carbon-13 NMR Spectroscopy*, Vol. 1, G. C. Levy, Ed., Wiley, New York, 1974.
8. E. L. Eliel and K. H. Pietrusiewicz, in *Topics in Carbon-13 NMR Spectroscopy*, Vol. 3, G. C. Levy, Ed., Wiley-Interscience, New York, 1979, Chapter 3; E. L. Eliel, V. S. Rao, and K. M. Pietrusiewicz, *Org. Magn. Resonance*, **12**, 461 (1979).

9. S. H. Grover, J. P. Guthrie, J. B. Stothers, and C. T. Tan, *J. Magn. Resonance*, **10**, 227 (1973).

10. J. B. Stothers, C. T. Tan, and K. C. Teo, *J. Magn. Resonance*, **20**, 570 (1975).

11. (a) J. B. Stothers, C. T. Tan, and K. C. Teo, *Can. J. Chem.*, **54**, 1211 (1976); (b) G. Mann, E. Kleinpeter, and H. Werner, *Org. Magn. Resonance*, **11**, 561 (1978).

12. J. G. Batchelor, *J. Magn. Resonance*, **18**, 212 (1975).

13. H. J. Schneider and E. F. Weigand, *J. Am. Chem. Soc.*, **99**, 8362 (1977).

14. T. P. Forrest and J. G. K. Webb, *Org. Magn. Resonance*, **12**, 371 (1979).

15. W. A. Ayer, L. M. Browne, S. Fung, and J. B. Stothers, *Org. Magn. Resonance*, **11**, 73 (1978).

16. D. K. Dalling and D. M. Grant, *J. Am. Chem. Soc.*, **94**, 5318 (1972); A. S. Perlin and H. J. Koch, *Can. J. Chem.*, **48**, 2639 (1970).

17. F. A. L. Anet, C. H. Bradley, and G. W. Buchanan, *J. Am. Chem. Soc.*, **93**, 258 (1971).

18. M. Christl, H. J. Reich, and J. D. Roberts, *ibid.*, **93**, 3463 (1971).

19. (a) J. J. Burke and P. C. Lauterbur, *ibid.*, **86**, 1870 (1964); (b) K. M. Creceley, R. W. Creceley, and J. H. Goldstein, *J. Phys. Chem.*, **74**, 2680 (1970); (c) G. E. Maciel and G. B. Savitzky, *ibid.*, **69**, 3925 (1965).

20. J. B. Grutzner, M. Jautelat, J. B. Dence, R. A. Smith, and J. D. Roberts, *J. Amer. Chem. Soc.*, **92**, 7107 (1970).

21. (a) G. E. Maciel and H. C. Dorn, *J. Am. Chem. Soc.*, **93**, 1968 (1971); (b) G. S. Poindexter and P. J. Kropp, *J. Org. Chem.*, **41**, 1215 (1976).

22. (a) R. R. Perkins and R. E. Pincock, *Org. Magn. Resonance*, **8**, 165 (1976); (b) H. Duddeck, *ibid.*, **9**, 528 (1977).

23. (a) H. Brouwer, J. B. Stothers, and C. T. Tan, *Org. Magn. Resonance*, **9**, 360 (1977); (b) M. Christl, *Chem. Ber.*, **108**, 2781 (1975); (c) E. Kleinpeter, H. Kühn, and M. Mühlstädt, *Org. Magn. Resonance*, **8**, 261 (1976); (d) J. B. Stothers and C. T. Tan, *Can. J. Chem.*, **55**, 841 (1977); M. Christl and W. Buchner, *Org. Magn. Resonance*, **11**, 461 (1978); (e) M. Christl and R. Herbert, *ibid.*, **12**, 150 (1979); (f) E. W. Della, P. T. Hine, and H. K. Patney, *J. Org. Chem.*, **42**, 2940 (1977).

24. K. B. Wiberg, D. E. Barth, and W. E. Pratt, *J. Am. Chem. Soc.*, **99**, 4286 (1977).

25. J. I. Kroschwitz, M. Winokur, H. J. Reich, and J. D. Roberts, *ibid.*, **91**, 5927 (1969).

26. J. D. Roberts, F. J. Weigert, J. I. Kroschwitz, and H. J. Reich, *ibid.*, **92**, 1338 (1970).

27. A. Ejchart, *Org. Magn. Resonance*, **9**, 351 (1977).

28. See p. 56 and ref. 11a.

29. G. C. Levy and R. A. Komoroski, *J. Am. Chem. Soc.*, **96**, 678 (1974).
30. (a) J. E. Sarneski, H. L. Surprenant, F. K. Molen, and C. N. Reilly, *Anal. Chem.*, **47**, 2116 (1975); (b) H. Eggert and C. Djerassi, *J. Am. Chem. Soc.*, **95**, 3710 (1973).
31. (a) W. O. Crain, Jr., W. C. Wildman, and J. D. Roberts, *J. Am. Chem. Soc.*, **93**, 990 (1971); (b) P. R. Srinivasan and R. L. Lichter, *Org. Magn. Resonance*, **8**, 198 (1976).
32. (a) E. L. Eliel and F. W. Vierhapper, *J. Org. Chem.*, **41**, 199 (1976); (b) *ibid.*, **42**, 51 (1977); (c) H. Booth and D. V. Griffiths, *J. Chem. Soc.*, *Perkin Transact II*, 111 (1975); (d) H. Booth, D. V. Griffiths, and M. L. Jozefowicz, *ibid.*, 752 (1976).
33. F. W. Vierhapper and E. L. Eliel, *J. Org. Chem.*, **44**, 1081 (1979).
34. R. Hagen and J. D. Roberts, *J. Am. Chem. Soc.*, **91**, 4504 (1969).
35. D. L. Rabenstein and T. L. Sayer, *J. Magn. Resonance*, **24**, 27 (1976).
36. (a) S. F. Nelsen and G. R. Wiseman, *J. Am. Chem. Soc.*, **98**, 3281 (1976); (b) G. Ellis and R. C. Jones, *J. Chem. Soc.*, *Perkin Transact.*, *II*, 437 (1972).
37. K. Pihlaja and T. Nurmi, *Finn. Chem. Lett.*, 141 (1977).
38. K. Jankowski and W. T. Raynes, *J. Chem. Research (S)*, 66 (1977) and references cited therein.
39. R. L. Lichter and J. D. Roberts, *J. Phys. Chem.*, **74**, 912 (1970).
40. J. H. Goldstein, V. S. Watts, and L. S. Rattet, *Prog. Nucl. Magn. Resonance*, **8**, 103 (1971).
41. J. Kowalewski, *Prog. Nucl. Magn. Resonance Spectrosc.*, **11**, Part 1 (1977).
42. K. B. Wiberg, G. M. Lampman, R. P. Ciula, D. S. Conner, P. Schertler, and J. Lavanish, *Tetrahedron*, **21**, 2749 (1965).
43. K. L. Servis, W. P. Weber, and A. K. Willard, *J. Chem. Phys.*, **74**, 3960 (1970).
44. R. Aydin and H. Günther, *Z. Naturforsch.*, **34**b, 528 (1979).
45. (a) G. E. Maciel, J. W. McIver, Jr., N. S. Ostlund, and J. A. Pople, *J. Am. Chem. Soc.*, **92**, 1, 11 (1970); (b) J. Jokisaari, J. Kuonanoja, and A.-M. Häkkinen, *Z. Naturforsch.*, **33**a, 7 (1978).
46. (a) H. Finkelmeier and W. Lüttke, *J. Am. Chem. Soc.*, **100**, 626 (1978); (b) M. Barfield, J. C. Marshall, E. D. Canada and M. R. Willcott, III, *ibid.* **100**, 7075 (1978); (c) R. E. Wasylishen, K. Chum, and J. Bukata, *Org. Magn. Resonance*, **9**, 473 (1977).
47. K. Wüthrich, S. Meiboom, and L. C. Snyder, *J. Chem. Phys.*, **52**, 230 (1970).
48. P. E. Hansen, J. Feeney, and G. C. K. Roberts, *J. Magn. Resonance*, **17**, 249 (1975) and references cited therein.
49. R. E. Wasylishen, *Annu. Rep. NMR Spectrosc.*, **7**, 250ff (1977) and references cited therein.

50. (a) F. W. Weigert and J. D. Roberts, *J. Am. Chem. Soc.*, **94**, 6021 (1972); (b) S. Berger and K. P. Zeller, *J. Chem. Soc., Chem. Commun.*, 649 (1976); (c) S. Berger, *J. Org. Chem.*, **43**, 209 (1978); (d) M. Pomerantz, R. Fink, and G. A. Gray, *J. Am. Chem. Soc.*, **98**, 291 (1976).

51. (a) M. Barfield, S. A. Conn, J. L. Marshall, and D. E. Miller, *J. Am. Chem. Soc.*, **98**, 6253 (1976); (b) J. Jokisaari, *Org. Magn. Resonance*, **11**, 157 (1978); (c) M. Stöcker and M. Klessinger, *ibid.* **12**, 107 (1979); (d) M. D. Newton and J. M. Schulman, *J. Am. Chem. Soc.*, **96**, 6295 (1974); (e) M. D. Newton, J. M. Schulman, and M. M. Manus, *ibid.*, **96**, 17 (1974).

52. (a) D. E. Dorman, M. Jautelat, and J. D. Roberts, *J. Org. Chem.*, **36**, 2757 (1971); (b) R. A. Friedel and H. L. Retcofsky, *J. Am. Chem. Soc.*, **85**, 1300 (1963); (c) P. A. Couperus, A. D. H. Clague, and J. P. C. M. van Dongen, *Org. Magn. Resonance*, **8**, 426 (1976); (d) J. W. de Haan, L. J. M. van de Ven, A. R. N. Wilson, A. E. van der Hout-Lodder, C. Altona, and D. H. Faber, *ibid.*, **8**, 477 (1976).

53. (a) A. C. Rojas and J. K. Crandall, *J. Org. Chem.*, **40**, 2225 (1975); (b) E. Taskinen, *Tetrahedron*, **34**, 353 (1978); (c) *J. Org. Chem.*, **43**, 2776 (1978).

54. H. Günther and T. Keller, *Chem. Ber.*, **106**, 1863 (1973).

55. H. Günther and H. Seel, *Org. Magn. Resonance*, **8**, 299 (1976).

56. (a) G. E. Maciel, *J. Phys. Chem.*, **69**, 1947 (1965); (b) K. Hatada, K. Nagata, and H. Yuki, *Bull. Chem. Soc. Jap.*, **43**, 3195, 3267 (1970).

57. G. Miyajima, K. Takahashi, and K. Nishimoto, *Org. Magn. Resonance*, **6**, 413 (1974).

58. (a) G. E. Maciel, P. D. Ellis, J. J. Natterstad, and G. B. Savitsky, *J. Magn. Resonance*, **1**, 589 (1969); (b) V. Mark and E. D. Weil, *J. Org. Chem.*, **36**, 676 (1971); (c) R. M. Smith, R. West, and V. Mark, *J. Am. Chem. Soc.*, **93**, 3621 (1971); (d) A. Padwa, J. Masaracchia, and V. Mark, *Tetrahedron Lett.*, **1971**, 3161.

59. G. Miyajima and K. Takahashi, *J. Phys. Chem.*, **75**, 331, 3766 (1971).

60. D. S. Khami and J. B. Stothers, *Can. J. Chem.*, **43**, 510 (1965).

61. (a) W. F. Reynolds, I. R. Peat, M. H. Freedman, and J. R. Lyerla, *Can. J. Chem.*, **51**, 1857 (1973); (b) G. K. Hamer, I. R. Peat, and W. F. Reynolds, *ibid.*, **51**, 915, 2596 (1973).

62. (a) D. H. Marr and J. B. Stothers, *Can. J. Chem.*, **43**, 596 (1976); (b) M. H. Loots, L. R. Weingarten, and R. H. Levin, *J. Am. Chem. Soc.*, **98**, 4571 (1976).

63. (a) R. Steur, J. van Dongen, M. DeBie, and W. Drenth, *Tetrahedron Lett.*, **1971**, 3307; (b) W. Runge, W. Kosbahn, and J. Winkler, *Ber. Bunsenges. Physik. Chem.*, **79**, 381 (1975); (c) W. Runge and J. Firl, *ibid.*, **79**, 913 (1975); (d) C. Charrier, D. E. Dorman, and J. D. Roberts, *J. Org. Chem.*, **38**, 2644 (1973); (e) J. K. Crandall and S. A. Sojka, *J. Am. Chem. Soc.*, **94**, 5084 (1972).

64. J. L. Marshall, D. E. Miiller, S. A. Conn, R. Seinwell, and A. M. Ihrig, *Acc. Chem. Res.*, **7**, 333 (1974).

65. R. Wasylishen and T. Schaefer, *Can. J. Chem.*, **50**, 2710 (1972).

66. (a) R. S. Butler, J. M. Kead, Jr., and J. H. Goldstein, *J. Molec. Spectrosc.*, **35**, 83 (1970); (b) K. M. Crecely, R. W. Crecely, and J. H. Goldstein, *ibid.*, **37**, 252 (1971); (c) K. Bachmann and W. von Philipsborn, *Org. Magn. Resonance*, **8**, 648 (1976); (d) N. J. Koole and M. H. A. deBie, *J. Magn. Resonance*, **23**, 9 (1976).

67. (a) C. A. Kingsbury, D. Draney, A. Sopchik, W. Rissler, and D. Durham, *J. Org. Chem.*, **41**, 3863 (1976); (b) U. Vögeli and W. von Phillipsborn, *Org. Magn. Resonance*, **7**, 617 (1975).

68. R. V. Dubs and W. von Philipsborn, *Org. Magn. Resonance*, **12**, 326 (1979).

69. G. C. Levy and D. C. Dittmer, *Org. Magn. Resonance*, **4**, 107 (1972).

70. G. Becher, W. Lüttke, and G. Schrumpf, *Angew. Chem.*, **12**, 339 (1973).

71. D. Rosenberg, J. W. de Haan, and W. Drenth, *Rec. Trav. Chim.*, **87**, 1387 (1968); D. Rosenberg and W. Drenth, *Tetrahedron*, **27**, 3893 (1971).

72. D. M. White and G. C. Levy, *Macromolecules*, **5**, 526 (1972).

73. D. E. Dorman, M. Jautelat, and J. D. Roberts, *J. Org. Chem.*, **38**, 1026 (1973).

74. W. Höbold, R. Radeglia, and D. Klose, *J. Prakt. Chem.*, **318**, 519 (1976).

75. M. T. W. Hearn, *Tetrahedron*, **32**, 115 (1976).

76. (a) M. T. W. Hearn, *J. Magn. Resonance*, **19**, 401 (1975); (b) F. Bohlmann and M. Brehm, *Org. Magn. Resonance*, **12**, 535 (1979).

77. (a) M. T. W. Hearn, *J. Magn. Resonance*, **22**, 521 (1976); (b) J. Kowalewski, M. Granberg, F. Karlsson, and R. Vestin, *ibid.*, **21**, 331 (1976).

78. J. L. Marshall, D. E. Miiller, H. C. Dorn, and G. E. Maciel, *J. Am. Chem. Soc.*, **97**, 460 (1975).

79. J. B. Stothers, C. T. Tan, A. Nickon, F. Huang, R. Sridhar, and R. Weglein, *J. Am. Chem. Soc.*, **94**, 8581 (1972).

80. H. Batiz-Hernandez and R. A. Bernheim, *Progr. NMR Spectrosc.*, **3**, 63 (1967).

81. (a) R. A. Bell, C. L. Chan, and B. G. Sayer, *J. Chem. Soc., Chem. Commun.*, **1972**, 67; (b) D. Doddrell and I. Burfitt, *Aust. J. Chem.*, **25**, 2239 (1972).

82. (a) A. P. Tulloch and M. Mazurek, *J. Chem. Soc., Chem. Commun.*, **1973**, 692; (b) F. W. Wehrli, D. Heremic, M. L. Mihailovic, and S. Milosavljenc, *ibid.*, **1978**, 302.

83. A. G. McInnes, J. A. Walter, J. L. C. Wright, and L. C. Vining, *Topics in Carbon-13 NMR Spectrosc.*, Vol. 2, G. C. Levy, Ed., Wiley-Interscience, New York, 1976, p. 125.

84. S. A. Linde and H. J. Jakobsen, *J. Magn. Resonance*, **17**, 411 (1975).

85. M. Bacon and G. E. Maciel, *J. Am. Chem. Soc.*, **95**, 2413 (1973).
86. H. H. Mantsch and I. C. P. Smith, *Can. J. Chem.*, **51**, 1384 (1973).
87. A. Marker, D. Doddrell, and V. Riggs, *J. Chem. Soc., Chem. Commun.*, **1972**, 724.

Chapter 4

Aromatics

Carbon-13 spectra of aromatic carbons assumed particular interest to organic chemists after the initial report[1] of carbon chemical-shift dependence on substituent polarity.[2] The carbons of benzene are found at 128.5 ppm, whereas ^{13}C chemical shifts of substituted aromatics appear over the wide range of 110 to 170 ppm (see Table 4.3 below). In 1961 Spiesecke and Schneider[3] provided unequivocal proof for the assignment of aromatic chemical shifts; since then the cmr literature on aromatics has been extensive.[2]

SIMPLE AROMATICS

The relationship between aromatic ^{13}C chemical shifts and the electronic nature of aromatics and aromatic-like systems was noted early; the theoretical treatment of aromatic ^{13}C chemical shifts has been discussed extensively.[3-5] As presented in Chapter 2, the total shielding constant σ for a particular nucleus is approximated by a sum of three terms:[6]

$$\sigma = \sigma_d + \sigma_p + \sigma'$$

where σ_d is the diamagnetic contribution, σ_p is the paramagnetic contribution, and σ' is the contribution from neighboring-group nuclei. The paramagnetic term σ_p has been shown to be dominant for most ^{13}C chemical shifts,[3,4,6] although σ_d has been suggested to be important, especially in interpretation of small changes.[7] The principal factors affecting σ_p are charge polarization, variation in bond order, and average excitation energy. Early work attempted to establish simple correlations between ^{13}C chemical shifts and local electron densities, but it is apparent that *no general relationship* of this type need exist.[2,4-13] However, when changes in bond order and/or the average excitation energy (which in general is not constant[14]) for related compounds do not vary much or vary in a mutually compensating way, a relationship with electron density may be apparent. Additional difficulties in seeking these correlations may arise from the approximations often used for calculating electron densities, even in methods considered highly exact.[15,16] It has been shown for the special case of benzene carbons *para* to a substituent that CNDO/2 charge densities correlate with ^{13}C shifts (Figure 4.1[3c]).

102

The use of ^{13}C chemical shifts as a probe in understanding the electronic nature of aromatic and aromatic-like systems has been further aided by use of Hammett σ and related constants[3a,c] and, more recently, by dual-substituent parameter (DSP) treatments.[17] Substituent chemical shifts (SCSs) in benzene derivatives often may be correlated by a general DSP equation of the type

$$\delta(SCS) = \kappa(P_1 + \lambda P_2)$$

where P_1 and P_2 are any pair of substituent reactivity parameters, for example, σ_I and σ_R or $\mathscr{F}$ and $\mathscr{R}$.[18] As expected, the most successful correlations have been obtained for DSP treatments with *para* carbon chemical shifts in monosubstituted benzenes.[17b] Deviations observed with *meta* and, especially, *ortho* ^{13}C shifts are often rationalized through various combinations of steric, π-inductive, σ-inductive, and field effects.[19] When systems are carefully chosen, however, single-parameter correlations may be discerned, and these may be used to derive or verify reactivity parameters.[3c,20] In the latter case *para* carbon shifts of substituted benzenes[3c] and the β-carbon of benzylidene malononitriles[20] have been used (Figure 4.2,[3c] shown for the σ^+ parameter; σ_R values used in DSP treatments can be found in ref 17b.).

RING-CURRENT EFFECTS

Although proton shielding is markedly influenced by magnetic-field-induced circulation of π electrons in a closed, conjugated system (aromatic ring currents), the same influences (~5 ppm) are less significant in determining ^{13}C resonance positions.[21] Günther and Schmickler[22] have summarized many of the arguments and limitations for ring-current effects on ^{13}C shifts. A definitive experimental determination of these influences on ^{13}C shifts was provided by DuVernet and Boekelheide.[23] They measured proton and ^{13}C chemical shifts for the R groups in **1** and **2**. Both have essentially the same geometry, but **1**, a [14] annulene, has a ring current whereas **2** does not.

R = CH_3, CH_3CH_2,

$CH_3CH_2CH_2$

1   **2**

Thus, the differences in the measured shifts for proton and carbon are attributable to the effects of the ring current in **1**. This,

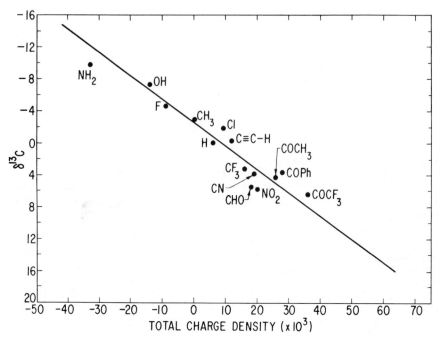

FIGURE 4.1.[3c]   Plot of total charge density (CNDO/2) versus
                    chemical shift (referenced to benzene) for the
                    para position of substituted benzenes.

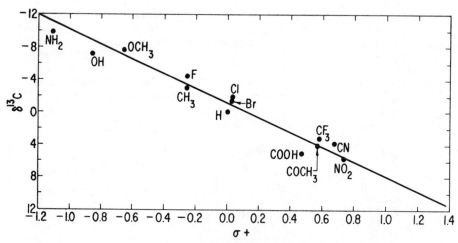

FIGURE 4.2.[3c]   Plot of ^{13}C chemical shifts at the para
                    positions for monosubstituted benzenes versus
                    σ^+. Chemical shifts relative to benzene.

in turn, depends principally on the position and the distance of the affected nucleus relative to the plane of delocalization. DuVernet and Boekelheide found that the magnitude of the ring-current effect was the same for both ^{13}C and ^{1}H, in accord with theoretical predictions of Johnson and Bovey.[24]

SOLVENT EFFECTS

There is no reason to believe that σ^+ values for a series of substituents should remain constant in strongly interacting solvent media. In various solvents an individual substituent may behave quite differently; this should be reflected in the nature of the electronic effect of the substituent on the attached aromatic ring. For example, some groups would be expected to interact strongly in hydrogen-bonding media.

One compound that would be expected to show large changes in behavior depending on solvent is acetophenone. For example, on addition of trifluoroacetic acid, the *para* carbon of acetophenone is additionally deshielded by ~2.5 ppm up to a 5:1 mole ratio of acid:acetophenone, where the effect reaches a plateau.[3c] A similar effect has been observed for methoxy. Clearly, *para* carbons of these compounds are sensitive to solvent. The chemical-shift plateau corresponds to completion of the "primary" solvation of the functional group. At this plateau "infinite dilution" is simulated insofar as the electronic perturbation of the substituent and the aromatic ring is concerned. From the shape of such dilution curves, it should be possible to gain insight into the solvation behavior of functional groups.[3c]

For a substituent such as methyl, on the other hand, little solvent-solute interaction would be expected in most solvent systems. Indeed, the *para* ^{13}C chemical shift in toluene remains constant (± 0.2 ppm) for a wide range of solvents from carbon tetrachloride to methanesulfonic acid. Table 4.1 summarizes the range of chemical shifts displayed by benzene and substituted benzenes in solvents of widely varying polarity. Since the range is so small for benzene, the large chemical-shift changes observed for the remaining substituted aromatics in Table 4.1 are real measures of solvent-solute interaction. The solvent-induced shift at the *para* carbon appears to arise primarily from a change in the electronic environment at that carbon. The large (up to several parts per million) shift differences preclude overriding contributions from solvent anisotropies arising from local ordering of solvent molecules.

These solvent effects have been used to derive solvent-dependent σ^+ values, which may be used to predict the relative reactivity of aromatic systems with strongly solvated functional

TABLE 4.1.   RANGE OF *para* ^{13}C CHEMICAL SHIFTS (IN PARTS PER MILLION) FOR BENZENE AND MONOSUBSTITUTED BENZENES IN VARIOUS SOLVENTS[3c]

SUBSTITUENT	$\Delta\delta^a$	SUBSTITUENT	$\Delta\delta^a$
H[b]	-0.1 to +0.3	$COCH_3$	4.0 to 7.2
CH_3	-2.9 to -3.2	NO_2	5.7 to 7.5
CHO	5.4 to 8.6	OCH_3	-8.0 to -5.7

[a] Chemical shift with respect to internal benzene, solute concentration 10 mole%. Positive numbers indicate deshielding relative to standard.
[b] Relative to 10 mole% benzene in CCl_4.

groups in reaction media of interest.[3c] In addition, solution studies of the ^{13}C chemical shift *para* to an individual functional group provides insight into strong solute-solvent effects with various media. For example, nitrobenzene ($pK_{BH}+ = -12.33$) appears to interact (Table 4.1) more strongly with trifluoroacetic acid (Hammett acidity, $H_0 = -3.3$) than with the stronger methanesulfonic acid ($H_0 \approx -8$). A DSP analysis carried out on the effects of aprotic solvents on *para* carbon shifts suggests that the effect depends mainly on the inductive rather than the resonance nature of the substituent.[25] Enhanced substituent polarity in the more polar solvents leading to increased polarization of the π system was suggested to explain these results.

The procedures just described can be applied in cases of protonation or ionization as well. The σ^+ value for the protonated acetyl group (estimated from the *para* chemical shift for acetophenone in trifluoromethanesulfonic acid, 16.4 ppm) is 1.9. Table 4.2 shows the solvent dependence of the ^{13}C nmr spectrum of aniline. The spectrum of aniline on protonation in methanesulfonic acid changes significantly when compared with that of aniline in carbon tetrachloride. The *ortho* and *para* carbons are deshielded considerably and the *meta* position to a lesser extent (8.1, 11.4, and 1.0 ppm, respectively). On the other hand, the *ipso* carbon is strongly shielded by -17.6 ppm. The qualitative features of these changes are also apparent in acetic acid where aniline is largely protonated. The CNDO/2 calculated changes in charge density at each carbon agree quite well with the "observed" changes in charge density derived from ^{13}C chemical-shift/charge-density correlations.[3c] Similar results

TABLE 4.2.   SOLVENT DEPENDENCE OF THE ^{13}C CHEMICAL SHIFTS OF ANILINE[a],[3c]

SOLVENT	C_1	ortho	meta	para
CCl$_4$	+18.0	-13.2	+0.9	- 9.7
CH$_3$COOH	+ 5.5	- 6.0	+1.4	- 1.1
CH$_3$SO$_3$H	+ 0.4	- 5.1	+1.9	+ 1.7
DMSO-d$_6$	+20.7	-14.3	+0.5	-12.5
Acetone-d$_6$	+20.1	-13.8	+0.6	-11.5

[a] Parts per million relative to internal benzene.

have been reported for substituted anilines.[26]

It is clear that chemical interactions between aromatic substituents and various solvent systems can be probed by the large solvent-induced chemical shifts found for ring carbons. Another application for these large shifts and also for smaller generalized solvent-induced shifts is separation of accidentally coincident carbon resonances. This is analogous to the use of solvent-induced shifts in proton nmr. The ^{13}C spectrum of 3-bromobiphenyl (Figure 4.3[27]) is illustrative. With cyclohexane as solvent, three carbons appear as a singlet in a 1000-Hz expansion (at 25.2 MHz, see Figure 4.3a). In dimethyl sulfoxide (DMSO) as solvent the three carbons are completely resolved, as shown in Figure 4.3b. The resonance assignments given in Figure 4.3 are based jointly on protonated ring carbon spin-lattice relaxation times (T_1s) and chemical shifts.

STERIC SHIFTS

In substituted benzenes adjacent substituents may experience "steric" shifts in addition to the usual electronic effects. For methylbenzenes, substituent interactions in ortho di- and polysubstituted materials shield adjacent methyl carbons.[28] The methyl group in ortho-xylene is shielded by 1.9 ppm compared with that in the meta compound, for example. Adjacent substituents need not both be alkyl, as is seen in Figure 4.4[29] which presents the spectrum of a mixture of three isomers of dianisylmethane. Although the methoxy carbons at 54.8 ppm are not clearly resolved, the three kinds of methylene carbon are separated by 5.1 ppm. The major isomer, the para-para, constitutes 70% of the mixture; its carbons are easily picked out of the spectrum. The spectrum of the ortho-para isomer (B), 25%

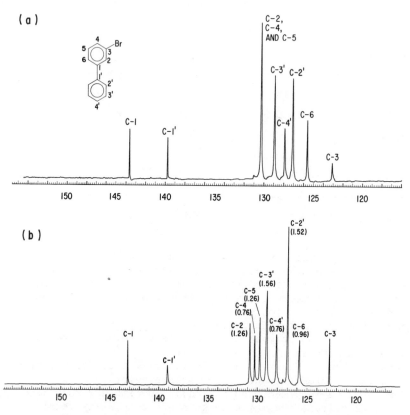

FIGURE 4.3[27]   *Spectrum of 3-bromobiphenyl in (a) cyclohexane as solvent and (b) DMSO as solvent.*

of the mixture, is superimposed; the carbons of the *ortho* portion are visible in the spectrum. The methylene carbon of the *ortho-para* isomer is 5.1 ppm upfield from that of the *para-para*. The methylene carbon of the *ortho-ortho* isomer (*C*), 5% of the mixture, is barely visible an additional 5.1 ppm upfield. For ditolylmethane isomers,[29] where the methoxy group is replaced by a methyl, the three kinds of methylene carbon are separated by 2.1 ppm and the two kinds of methyl carbon by 1.4 ppm.

Deshielding δ effects are also displayed by disubstituted benzenes. The ring methyl carbon in *o-tert*-butyltoluene is deshielded by 4 ppm compared with the methyl of *o*-xylene, although the methyl resonance positions in *o*-ethyl- and *o*-isopropyltoluene are not affected very much.[30] The data suggest a conformational dependence of the δ effect, very much in analogy to that observed in aliphatic systems (see Chapter 3).

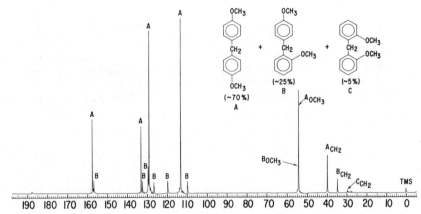

FIGURE 4.4[29]   Carbon-13 spectrum of a mixture of three isomers of dianisylmethane. Assignments are as noted.

Chemical shifts of cyclophanes may also be interpretable in terms of steric constraints. The bridging methylene carbons of [2.2]metacyclophanes bearing alkyl groups *ortho* to the bridge are shielded relative to unstrained models, whereas the inner aryl carbons are substantially deshielded. Chemical shifts for parent cyclophanes are given with the structures.[21a,31] The rings of these compounds exhibit substantial distortion from planarity.

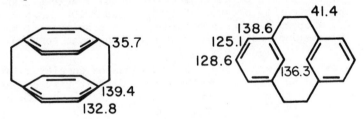

(chemical shifts in ppm)

ASSIGNMENT OF AROMATIC ^{13}C SPECTRA

Substituent Effects.   Despite the susceptibility of aromatic ^{13}C chemical shifts to the influences discussed previously, sub-stituent effects are remarkably additive. Table 4.3[2,3c,32] pre-sents the ^{13}C chemical shifts for a number of substituted benzenes (relative to benzene as a standard = 128.5 ppm). For a substituent such as methoxy, the *ortho* and *para* positions are shielded and the carbon of attachment (*ipso* carbon) is deshielded but the *meta* position is not much different from benzene. Chemical shifts for polysubstituted aromatics are for the most part predictable on the basis of the additivity of the chemical

shifts of the corresponding monosubstituted benzenes. The shifts
shown in Table 4.3 represent approximate substituent parameters.

The ^{13}C spectrum of bisphenol-A is shown in Figure 4.5. The
two kinds of aliphatic carbon are easily distinguished by chemical
shift and relative peak heights. There are four different kinds
of aromatic carbon present. The shift for C-1 is nearly identical

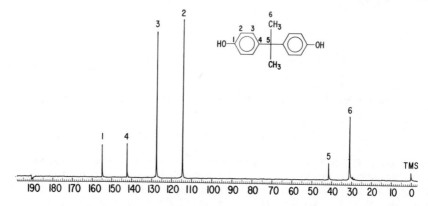

FIGURE 4.5.   Spectrum of bisphenol-A, assignments are as noted.

to that for the carbon of attachment in phenol. Both C-1 and C-4
are of lower intensity because of longer relaxation times and
lower NOEs, whereas C-3 has a chemical shift near that of benzene,
and C-2 is shielded, as expected for a carbon ortho to an oxygen
function.

In a similar manner, the ^{13}C resonances of complex chlorinated
aromatics could be assigned on the basis of additivity of phenyl
and chlorine substituents,[33] even though the total chemical-shift
range spanned only 16 ppm in one study.[33a]

Spectral Assignments from Relaxation Measurements.   Carbon-13
spectra of substituted benzenes often may be interpreted
unambiguously by use of chemical-shift parameters (e.g., Table
4.3) and off-resonance ^1H-decoupled spectral information. Sometimes further data are required to complete spectral assignments.
The spin-lattice relaxation times (T_1) of protonated ring carbons
in substituted benzenes yield information about molecular
symmetry that can be combined with chemical-shift data (see
Chapter 8).

The T_1 values for the protonated ring carbons of m-nitrobiphenyl given in Figure 4.6 facilitate the assignment of
resonances.[27] Preferred rotation around molecular axes collinear

TABLE 4.3. ^{13}C SUBSTITUENT EFFECTS OF MONO-SUBSTITUTED BENZENES^2

| SUBSTITUENT | POSITION | | | | REFERENCE |
	C-1	*ortho*	*meta*	*para*	
Br	- 5.5	+ 3.4	+1.7	- 1.6	
CF$_3$	+ 2.6	- 3.3	-0.3	+ 3.2	
CH$_3$	+ 8.9	+ 0.7	-0.1	- 2.9	
CN	-15.4	+ 3.6	+0.6	+ 3.9	
C≡C-H	- 6.1	+ 3.8	+0.4	- 0.2	
COCF$_3$	- 5.6	+ 1.8	+0.7	+ 6.7	
COCH$_3$	+ 9.1	+ 0.1	0.0	+ 4.2	
COCl	+ 4.6	+ 2.4	0.0	+ 6.2	
CHO	+ 8.6	+ 1.3	+0.6	+ 5.5	a
COOH	+ 2.1	+ 1.5	0.0	+ 5.1	
COPh	+ 9.4	+ 1.7	-0.2	+ 3.6	
Cl	+ 6.2	+ 0.4	+1.3	- 1.9	
F	+34.8	-12.9	+1.4	- 4.5	
H	0.0	-	-	-	
NCO	+ 5.7	- 3.6	+1.2	- 2.8	
NH$_2$	+18.0	-13.3	+0.9	- 9.8	
NH$_3$+,^{h}	+ 0.1	- 5.8	+2.2	+ 2.2	
NO$_2$	+20.0	- 4.8	+0.9	+ 5.8	
OCH$_3$	+31.4	-14.4	+1.0	- 7.7	
OH	+26.9	-12.7	+1.4	- 7.3	
Ph	+13.1	- 1.1	+0.4	- 1.2	
SH	+ 2.3	+ 1.1	+1.1	- 3.1	
SCH$_3$	+10.2	- 1.8	+0.4	- 3.6	b
SO$_2$NH$_2$	+15.3	- 2.9	+0.4	+ 3.3	
O$^-$	+39.6	- 8.2	+1.9	-13.6	
OPh	+29.2	- 9.4	+1.6	- 5.1	
OAc	+23.0	- 6.4	+1.3	- 2.3	

111

TABLE 4.3. Continued.

SUBSTITUENT	C-1	*ortho*	*meta*	*para*	REFERENCE
NMe$_2$	+22.6	-15.6	+1.0	-11.5	
NEt$_2$	+19.9	-15.3	+1.4	-12.2	
NHAc[d]	+11.1	- 9.9	+0.2	- 5.6	
CH$_2$OH	+12.3	- 1.4	-1.4	- 1.4	
I[f]	-32.0	+10.2	+2.9	+ 1.0	c
SiMe$_3$	+13.4	+ 4.4	-1.1	- 1.1	
SnMe$_3$	+13.1	+ 7.2	-0.4	- 0.4	
CH=CH$_2$	+ 9.5	- 2.0	+0.2	- 0.5	
CO$_2$Me[e]	+ 1.3	- 0.5	-0.5	+ 3.5	
COCl	+ 5.8	+ 2.6	+1.2	+ 7.4	
CHO	+ 9.0	+ 1.2	+1.2	+ 6.0	
COEt[e]	+ 7.6	- 1.5	-1.5	+ 2.4	
COi-Pr[e]	+ 7.4	- 0.5	-0.5	+ 4.0	
COt-Bu[e]	+ 9.4	- 1.1	-1.1	+ 1.7	
Li	-43.2	-12.7	+2.4	+ 3.1	g
MgBr	-35.8	-11.4	+2.7	+ 4.0	g
▽	+15.1	- 3.3	-0.6	- 3.6	b,g
HN▽	+12.2	- 2.9	-0.3	- 1.8	b
O▽	+ 9.1	- 0.1	-3.0	- 0.5	b
PMe$_2$	+13.6	+ 1.6	-0.6	- 1.0	b

[a]Parts per million relative to internal benzene standard; positive shifts deshielding. Solute concentration, 10% in CCl$_4$.
[b]Parts per million relative to benzene. Data obtained relative to internal TMS and converted using δ_C=128.5 for benzene. Solute concentration ~10% in CDCl$_3$.
[c]Neat liquids examined unless noted otherwise. Data converted using δ_C=128.7 for benzene.
[d]In *N,N*-dimethylformamide.
[e]Original data obtained relative to CS$_2$. Conversion δ_C=192.8 was used for CS$_2$.
[f]See Footnote [e]. Conversion δ_C=193.7 was used for CS$_2$.
[g]Solvent unspecified.
[h]CF$_3$CO$_2$H solvent.

PROBLEM 4.1.   *Determine the structure of this C_7H_8O compound. Assign all resonances (TMS at 0.0 ppm).*

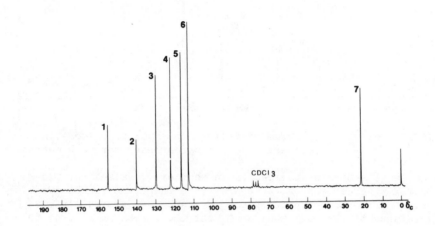

PROBLEM 4.2.   *Identify the compound whose ^{13}C spectrum appears here. Assign each resonance to the appropriate carbon or carbons. The molecular formula is $C_{14}H_9Cl_5$.*

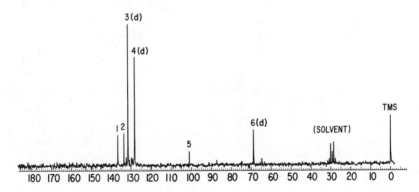

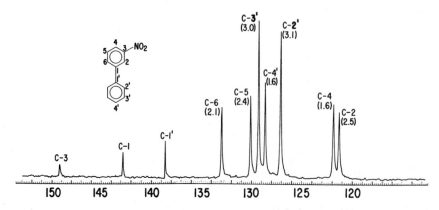

FIGURE 4.6.[27]   *Spectrum of m-nitrobiphenyl; T_1 values (in sec) are noted in parentheses for each carbon.*

with large substituents results in shorter T_1 values for protonated ring carbons on those axes (see Chapter 8). The two
carbons *ortho* to the nitro group, C-2 and C-4, are easily distinguished by the much shorter T_1 for the carbon *para* (C-4) to
the large phenyl substituent. Similarly, T_1 for C-4', the
carbon *para* to the *m*-nitrophenyl group, is shorter than T_1 for
C-2' and C-3'. The T_1 value for C-6 is somewhat shorter than
T_1 for C-2 and C-5 because it is *para* to the nitro group.

PROBLEM 4.3.   *Spectra a and b are of isomers of compounds with molecular formula $C_{18}H_{14}$. Identify each isomer.*

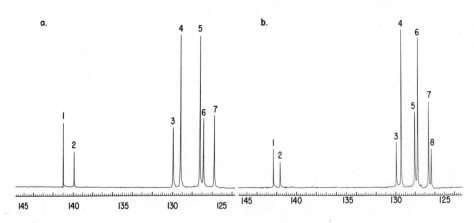

PROBLEM 4.4.   *Spectrum of m-nitrobenzoic acid (in acetone-d_6); T_1 values are noted for the peaks representing protonated carbons. Assign each resonance to the appropriate carbon.*

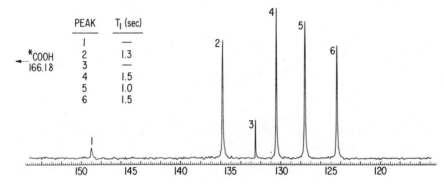

POLYNUCLEAR ARENES

Polycyclic aromatic hydrocarbons have been of considerable interest because of the ease with which their π-electron systems can be treated theoretically. It is, thus, not surprising that a number of papers dealing with the correlation of theoretical parameters and ^{13}C chemical shifts have appeared.[2] Much of the work has been summarized comprehensively by Hansen.[34]

Table 4.4 presents the chemical shifts of several related polynuclear aromatics.[34,39,42] In general, the nonprotonated quaternary carbons are at lowest field. It has been found that nonalternate hydrocarbons[35] have a larger chemical-shift range (14 to 22 ppm) than do alternate systems[4,5] such as naphthalene (8 ppm), phenanthrene (9.2 ppm), and pyrene (6.4 ppm). It has been suggested that this difference in cmr behavior might offer experimental differentiation between these two classes of polynuclear aromatics. Figure 4.7[35] shows the spectrum of fluoranthene. In fluoranthene the quaternary carbons are deshielded whereas all the protonated carbons are shielded relative to benzene. Carbon 15 is the lowest in intensity, as might be expected for carbons so remote from protons. Data for acepleiadylene,[36] helicenes,[37] and bifluorenyl dimers[38] have been reported.

From correlations of ^{13}C chemical shifts with theoretically calculated charge densities and other parameters it has been shown that the carbon shifts are not simply related to π-electronic charge but that both π and σ charge variations play a

TABLE 4.4. ^{13}C CHEMICAL SHIFTS (IN PARTS PER MILLION) OF POLYNUCLEAR AROMATICS[34,39,42]

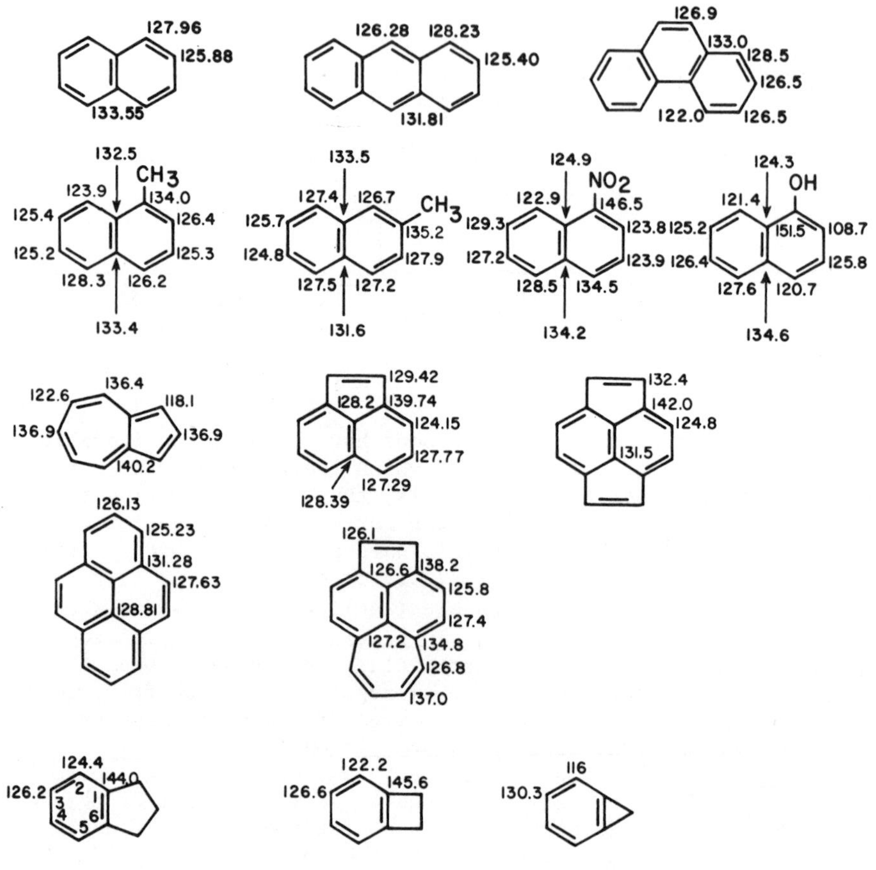

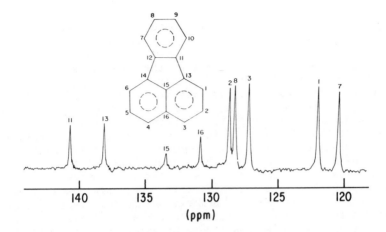

FIGURE 4.7.[35]   Proton-decoupled ^{13}C magnetic resonance
spectrum of fluoranthene.

role in alternate and nonalternate hydrocarbons.[35] A minor role
for ring currents has also been suggested.

Substituent effects in naphthalenes and anthracenes parallel
those found in substituted benzenes, and SCS values are reasonably
additive.[34] Exceptions arise with compounds having *ortho* or
peri (1,8) interactions in naphthalenes. It should be noted
that insofar as ^{13}C shifts reflect electron delocalization,
effects of conjugating substituents (e.g., OH, NH$_2$, and carbonyls)
in naphthalenes are not transmitted appreciably to the unsubsti-
tuted ring. This is in accord with simple considerations of
resonance structures. A DSP treatment also supports this view.[40]
As in aliphatic compounds, assignments of resonances can be
facilitated by deuterium substitution and T_1 measurements.[41]

Benzocycloalkenes (Table 4.4) also display characteristic
changes in the chemical shifts of the aromatic carbons. Carbons
2 and 5 become progressively shielded as ring size of the cyclo-
alkene decreases.[42]

HETEROAROMATICS

Heterocyclic aromatic compounds have been of particular
interest in cmr studies. Numerous reports have been made considering particularly the question of experimental chemical shift
correlation with theoretical quantities.

Five-Membered Heterocycles.   Carbon-13 spectra of heterocyclic
aromatics have many of the qualitative features of substituted

alkenes, yet the effect of the heteroatom is not as marked as for alkenes. The chemical shifts of the α- and β-carbons of furan differ by 33 ppm, with the α-carbon deshielded whereas the β-carbon is shielded. For thiophene, the ^{13}C shifts of the α- and β-carbons are nearly identical.[43] This is in analogy with sulfur-substituted simple alkenes. In pyrrole the effect of nitrogen is to shield both the α- (~10 ppm) and β- (~20 ppm) carbons relative to benzene. Pyrrole ^{13}C resonances vary considerably more with solvent than do those of furan or thiophene.[43] Additional nitrogen substitution in a five-membered ring results in a deshielding at the α-carbon with minimal change at the β-carbon. Selenium deshields both C-2 and C-3 relative to sulfur in thiophene, but the chemical-shift differences are comparable. Table 4.5[43-50] gives the ^{13}C chemical shifts for several five-membered heterocycles. Substituted materials are also shown. Introduction of a polar substituent at the 2-position results in a large effect at C-3 and C-5 as well as at the carbon of attachment. Expectations based on simple canonical forms appear to be operative. For a 3-substituent, on the other hand, the 2,3-double bond behaves much like an isolated double bond.

Reasonable correlation of chemical shifts with theoretically obtained parameters has been found by a number of workers.[43,51-55]

Six-Membered Heterocycles.   When one refers to six-membered heterocyclic aromatics, pyridines are perhaps the first to come to mind. Pyridine itself is one of the most intensively studied molecules, even from the early days of nmr spectroscopy, probably because it serves as a stringent test for nmr theories. Retcofsky and Friedel[56] have presented extensive data on 2-, 3-, and 4-substituted pyridines. Table 4.6 presents representative data.

The three carbons of pyridine are deshielded compared with those of the five-membered pyrrole and behave more like those of substituted benzenes. Substituent effects are comparable to those in substituted benzenes. Carbons *para* to substituents in 2-substituted pyridines correlate in a linear fashion with σ_p. Carbons of attachment in both 2- and 3-substituted pyridines correlate in a linear fashion with those of the corresponding monosubstituted benzenes.[57] As in benzene, there is little *meta* effect. Ring carbons are deshielded (~10 ppm) by each α-nitrogen and by γ-nitrogens (~3 ppm each) but are shielded (~-3 ppm) by β nitrogens as illustrated in Table 4.7.[58-63]

Table 4.7 also includes ^{13}C chemical shifts for pyridinium ion,[61] in which C_α is *shielded*, C_γ deshielded, and C_β relatively unchanged compared to pyridine. Shielding of the α-carbon on protonation is a common phenomenon for azabenzenes and has been attributed to both changes in N-C bond order[63a] and in the

OXYGEN

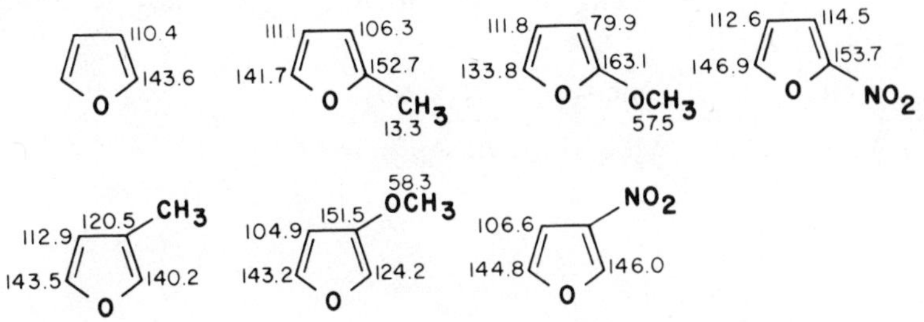

110.4 143.6

111.1 106.3 141.7 152.7 CH₃ 13.3

111.8 79.9 133.8 163.1 O OCH₃ 57.5

112.6 114.5 146.9 153.7 O NO₂

112.9 120.5 CH₃ 143.5 140.2 O

104.9 151.5 58.3 OCH₃ 143.2 124.2 O

106.6 NO₂ 144.8 146.0 O

NITROGEN

108.2 118.5 N H

108.0 121.6 N 35.4 CH₃

108.1 105.9 116.7 127.2 N CH₃ H 12.4

105.8 134.6 N N H

105.4 139.0 129.8 N N 38.2 CH₃

122.8 N 136.3 N H

129.6 N 120.3 138.3 N 32.6 CH₃

130.6 N NH N

134.7 N N–CH₃ 41.5 N

HN N 147.9

143.3 N–N N H

144.2 N–N N 33.7 CH₃

153.4 N N–CH₃ N 38.8

134.3 N 125.5 N N 35.7 CH₃

30.7 CH₃ N N 144.1 N

144.7 N 36.0 CH₃–N 152.6 N

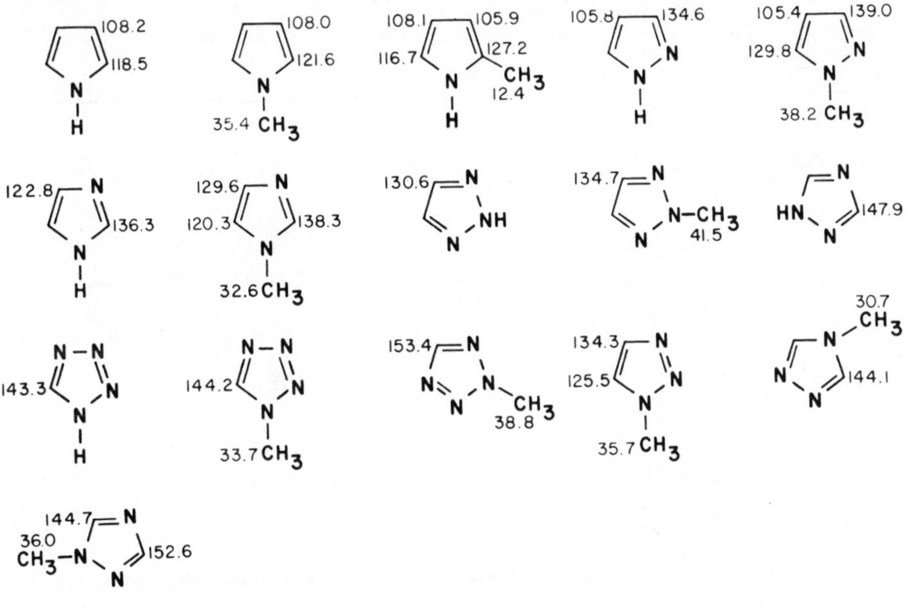

TABLE 4.5. Continued.

SULFUR

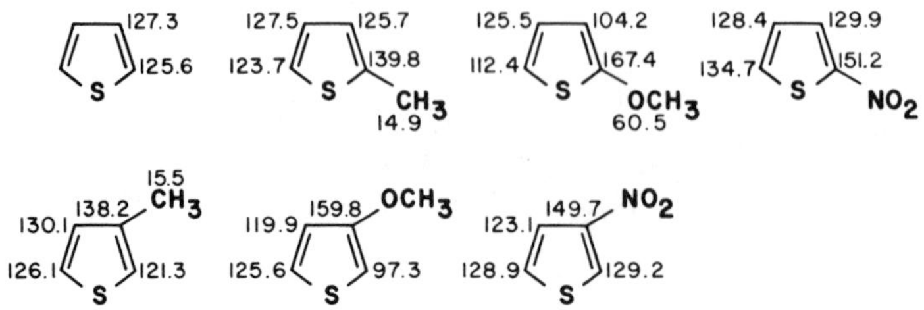

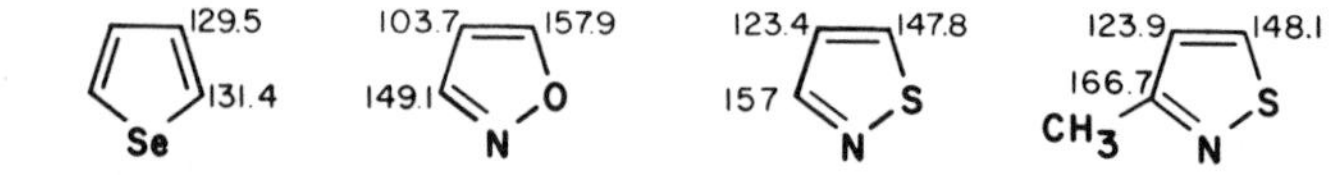

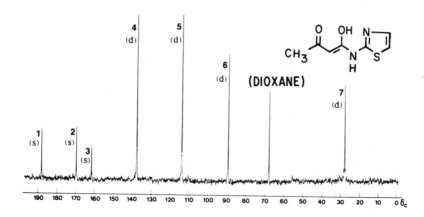

127.3
125.6
S

127.5 125.7
123.7 139.8
S CH₃
14.9

125.5 104.2
112.4 167.4
S OCH₃
60.5

128.4 129.9
134.7 151.2
S NO₂

15.5
CH₃
138.2
130.1
126.1 121.3
S

159.8 OCH₃
119.9
125.6 97.3
S

149.7 NO₂
123.1
128.9 129.2
S

129.5
131.4
Se

103.7 157.9
149.1
N O

123.4 147.8
157
N S

123.9 148.1
166.7
CH₃
N S

PROBLEM 4.5.   Assign the ^{13}C spectrum of N-(2-thiazolyl)-acetoacetamide.

190 180 170 160 150 140 130 120 110 100 90 80 70 60 50 40 30 20 10 0 δ_c

1 (s)   2 (s)   3 (s)   4 (d)   5 (d)   6 (d)   (DIOXANE)   7 (d)

CH₃ — C(=O) — CH = C(OH) — NH — (2-thiazolyl)

TABLE 4.6. ^{13}C CHEMICAL SHIFTS OF SUBSTITUTED PYRIDINES[56],[57b]

POSITION OF SUBSTITUENT	SUBSTITUENT	C-2	C-3	C-4	C-5	C-6
2	H	150.6	124.5	136.4	124.5	150.6
	OCH$_3$	164.8	111.4	138.2	116.9	148.0
	CHO	153.3	121.8	137.7	128.3	150.5
	Br	143.2	128.9	139.7	123.9	151.0
	CN	134.3	129.2	138.4	127.8	151.7
	Me	159.9	123.4	137.2	122.0	149.8
	F	164.6	110.5	141.8	122.1	148.3
3	CHO	152.2	132.3	136.4	125.0	155.2
	Br	151.4	121.6	139.2	125.5	148.7
	CN	153.7	111.2	140.5	124.7	153.7
	Me	150.5	133.8	137.3	124.1	147.8
	F	138.1	159.6	122.4	124.4	145.7
4	CHO	151.5	122.8	141.9	122.8	151.5
	Br	151.8	127.8	133.4	127.8	151.8
	CN	151.5	126.7	120.9	126.7	151.5
	Me	150.1	126.7	147.7	126.7	150.1
	F	152.5	111.8	168.7	111.8	152.5

average excitation energy.[63b]   The corresponding shielding on *N*-methylation is countered by an opposing β effect, so that the net change is small (see Table 4.7)[61]:

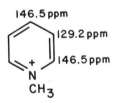

Formation of pyridine *N*-oxide yields a different situation. On the one hand, oxygen in a β position to carbon 2 is expected to rather markedly deshield that carbon. On the other hand, the formal negative charge is extensively delocalized to C-2 and C-4, which are correspondingly shielded. This effect is enhanced by generation of a positively charged nitrogen. The chemical shifts probably reflect a balance among these effects.[64]   As shown,

TABLE 4.7. ^{13}C CHEMICAL SHIFTS (IN PARTS PER MILLION) OF SIX-MEMBERED NITROGEN HETEROCYCLES[58-63]

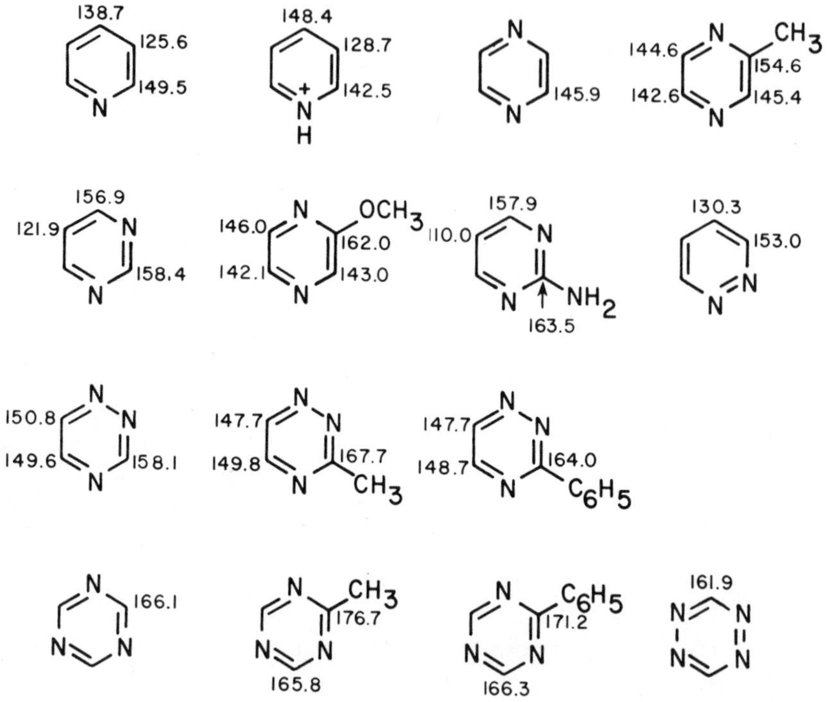

protonation converts the oxide to an N-hydroxypyridinum ion, in which the most marked change occurs at C-4.[64a]

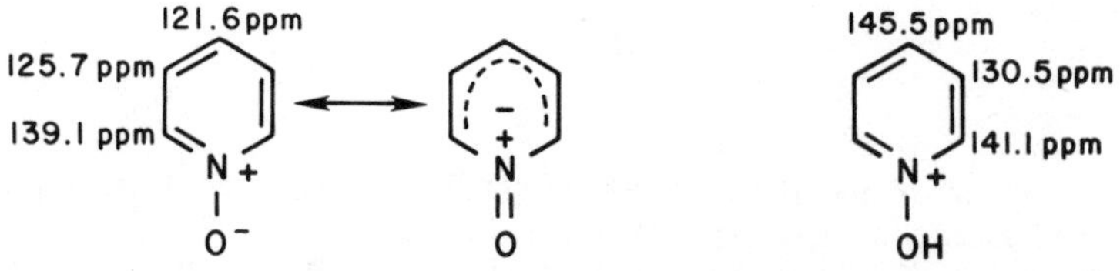

The effects of heteroatoms other than nitrogen are demonstrated in Table 4.8.[65,66] As with pyridine, it is the α-carbon that is influenced most strongly in the neutral compounds. The

TABLE 4.8. ^{13}C CHEMICAL SHIFTS (IN PARTS PER MILLION) OF SIX-MEMBERED OXYGEN AND GROUP V HETEROCYCLES[65],[66]

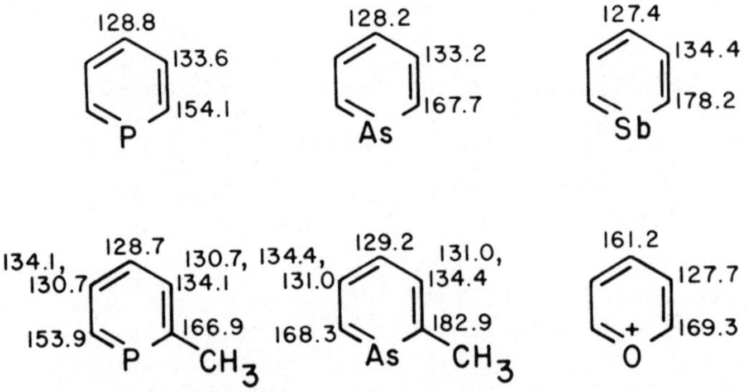

progressive deshielding from phosphorus to antimony is not pre-dicted from current theories of chemical shifts.

Polycyclic Heteroaromatics.   Indole is perhaps the simplest of the polycyclic heteroaromatics. Compared with pyrrole (Table 4.5), the α-carbon (C-2) is deshielded and the β-carbon (C-3) is

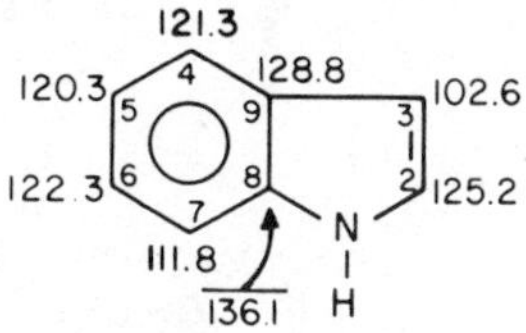

shielded. The effects of methyl substitution, especially on the benzene moeity, have been clarified.[67] For 1,2-dimethylindole, a combination of ortho-para aromatic directional effects, steric, and α,β-alkene substituent effects appears operative.

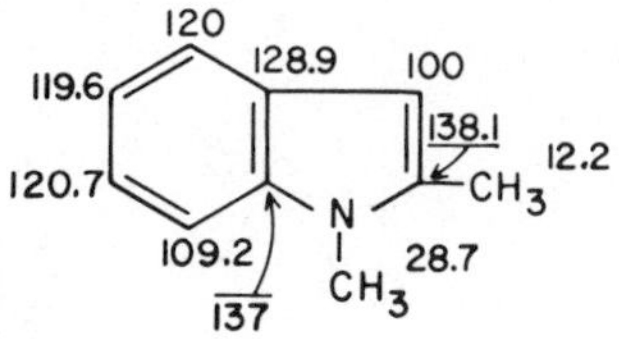

Chemical shifts of benzofuran[68] and benzothiophene[69] may be compared with the corresponding nonannelated compounds (Table 4.5). Other polycyclic five-membered heterocycles have been reported,[70-76] some of which are given in Table 4.9.[70-74]

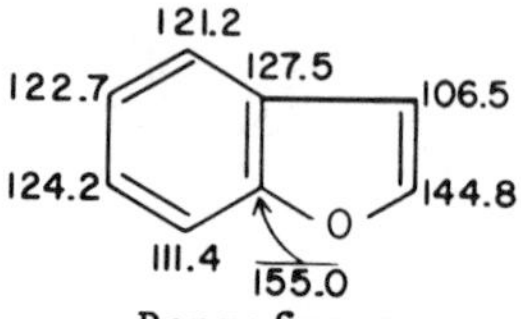

Benzofuran

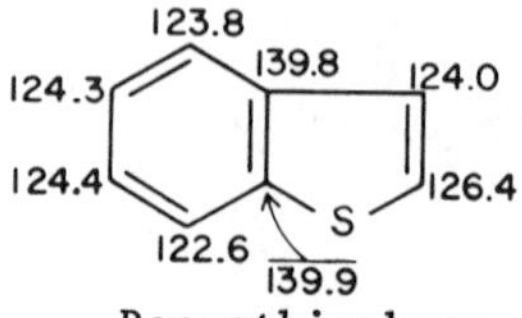

Benzothiophene

Quinoline and isoquinoline are the simplest polycyclic six-membered heteroaromatics.[77a] These and other representative examples are given in Table 4.10.[77-80] Other classes of compounds studied include azafluorenes,[81] pyrazolopyrimidines,[82] phenazines,[83] phenothiazines,[84] and diazanaphthalenes.[85]

TABLE 4.9.   [13]C CHEMICAL SHIFTS (IN PARTS PER MILLION) OF POLYCYCLIC FIVE-MEMBERED HETEROCYCLES[70-74]

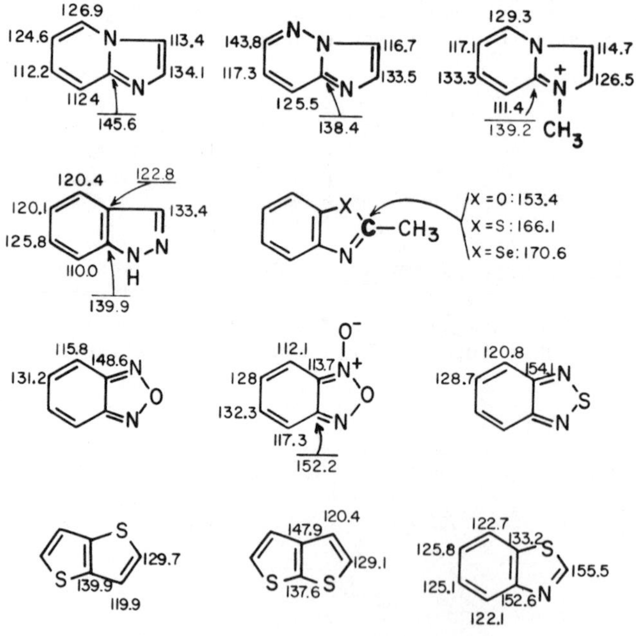

TABLE 4.10. ^{13}C CHEMICAL SHIFTS (IN PARTS PER MILLION) OF POLYCYCLIC SIX-MEMBERED HETEROAROMATICS[8],[77-80]

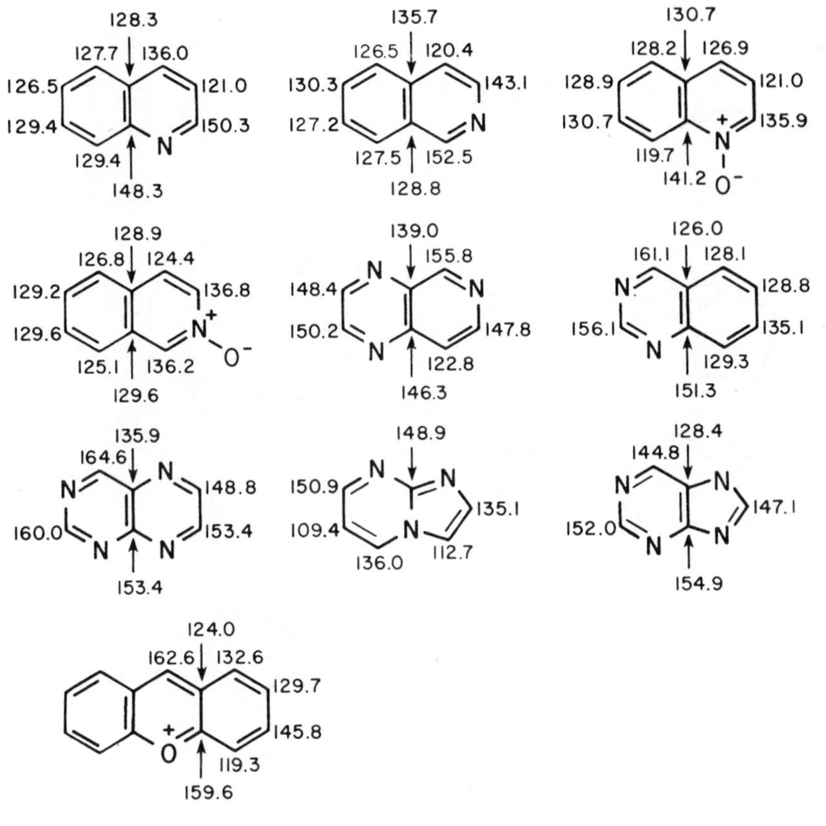

128.3
127.7 | 136.0
126.5 ⟨ ⟩ 121.0
129.4 ⟨ ⟩ 150.3
129.4 N
148.3

135.7
126.5 | 120.4
130.3 ⟨ ⟩ 143.1
127.2 ⟨ ⟩ N
127.5 | 152.5
128.8

130.7
128.2 | 126.9
128.9 ⟨ ⟩ 121.0
130.7 ⟨ ⟩ + 135.9
119.7 N
141.2 O$^-$

128.9
126.8 | 124.4
129.2 ⟨ ⟩ 136.8
129.6 ⟨ ⟩ N$^+$
125.1 | 136.2 O$^-$
129.6

139.0
N | 155.8
148.4 N
150.2 N 147.8
122.8
146.3

126.0
161.1 | 128.1
N 128.8
156.1 N 135.1
N 129.3
151.3

135.9
164.6 |
N N 148.8
160.0 N N 153.4
153.4

148.9
N N
150.9 135.1
109.4 N
136.0 112.7

128.4
144.8 |
N N
152.0 N N 147.1
154.9

124.0
162.6 | 132.6
129.7
+ 145.8
O 119.3
159.6

COUPLING CONSTANTS

Any discussion of ^{13}C couplings in aromatic systems properly begins with benzene. Günther *et al.*[86] have obtained very precise values for all C-H couplings in benzene by measurements on deuterium-decoupled pentadeuteriobenzene. The values are $^1J_{CH}=157.65$ Hz, $^2J_{CH}=1.16$ Hz, $^3J_{CH}=7.63$ Hz, $^4J_{CH}=-1.22$ Hz (experimental error ±0.09 Hz). The $^1J_{CH}$ and $^3J_{CH}$ values compare favorably with corresponding values in alkenes (Table 3.33); $^1J_{CH}$ also is consistent with the trend displayed by alkene values as a function of ring size (Table 4.11); the decrease may be rationalized in terms of decreasing s character in the external

TABLE 4.11. ^{13}C-H COUPLING CONSTANTS (IN HERTZ) IN CYCLIC
ALKENES AND AROMATIC COMPOUNDS[86]

	$^1J_{CH}$		$^1J_{CH}$
 (cyclopropene, H)	228.2	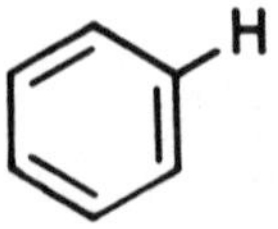 (benzene, H)	157.7
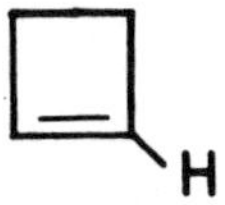 (cyclobutene, H)	170	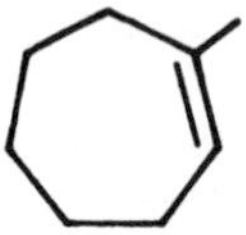 (cycloheptene)	156.2
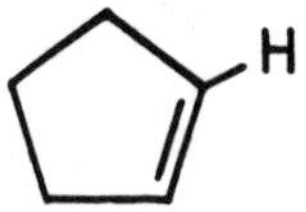 (cyclopentene, H)	161.6	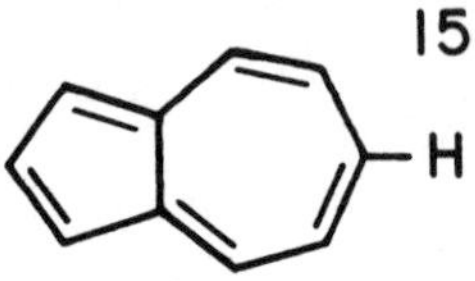 (azulene, H)	154
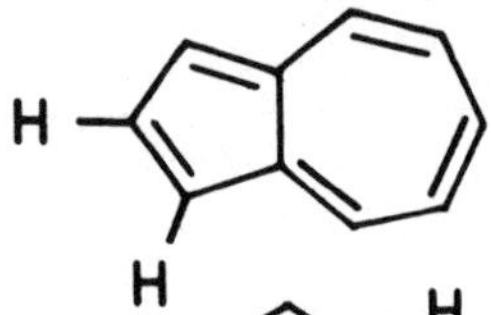 (azulene, H, H)	164, 179	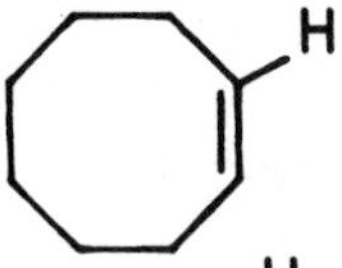 (cyclooctene, H)	156.0
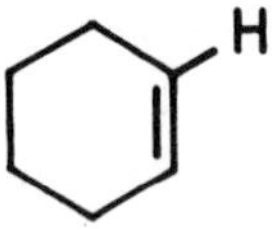 (cyclohexene, H)	158.4	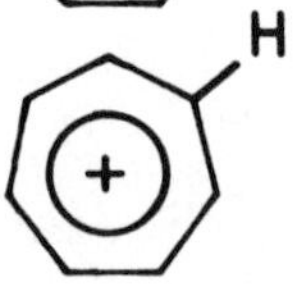 (tropylium +, H)	166.8 ($^3J_{CH}$ =9.99)

C-H bond. In this context the increase in $^1J_{CH}$ exhibited by
tropylium ion is exceptional and may be associated with the
presence of the positive charge.[86] Values of J_{CH} in naphthalene
allow effects of geometry to be discerned.[87]

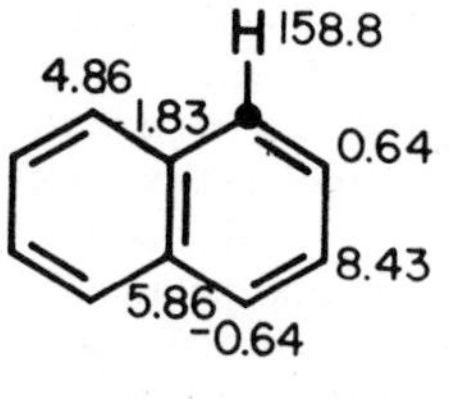

(naphthalene, left: H 158.8, 4.86, 1.83, 0.64, 8.43, 5.86, −0.64)

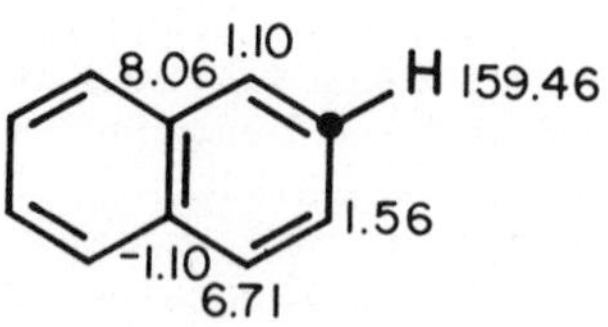

(naphthalene, right: 8.06, 1.10, H 159.46, 1.56, −1.10, 6.71)

(J_{CH} in Hz)

The effect of substituents on J_{CH} may be seen in Table 4.12, which presents values selected from a more comprehensive study.[88] These C-H values may be contrasted with those of fluorobenzene,[89] in which J_{CF} decreases with distance from fluorine.

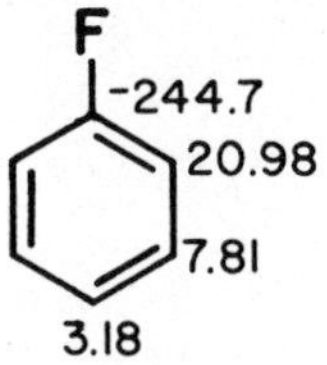

F
-244.7
20.98
7.81
3.18

(J_{CF} in Hz)

Incorporation of a heteroatom such as nitrogen into a ring increases J_{CH}, as exemplified by pyridine.[90] In analogy with carbocyclic systems, C-H couplings in quinolines also show influences of geometry.[91]

TABLE 4.12.   ^{13}C-H COUPLING CONSTANTS (IN HERTZ) IN MONO-SUBSTITUTED BENZENES[88]

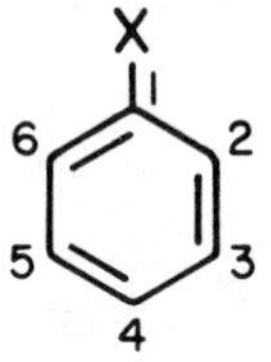

X
6 2
5 3
4

X	$^2J_{12}$	$^3J_{13}$	$^4J_{14}$	$^1J_{22}$	$^3J_{26}$	$^3J_{35}$
F	-4.89	10.95	-1.73	162.55	4.11	9.02
NO_2	-3.57	9.67	-1.75	168.12	4.45	8.18
OCH_3	-2.79	9.22	-1.51	158.52	4.80	8.73
CHO	0.29	7.19	-1.26	160.95	6.25	7.58
CH_3	0.54	7.61	-1.40	155.89	6.59	7.91
$Si(CH_3)_3$	4.19	6.34	-1.10	156.14	8.63	7.26

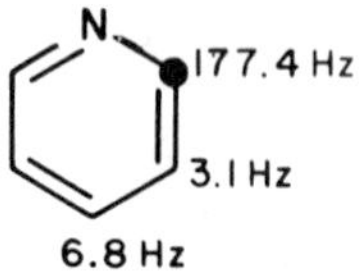

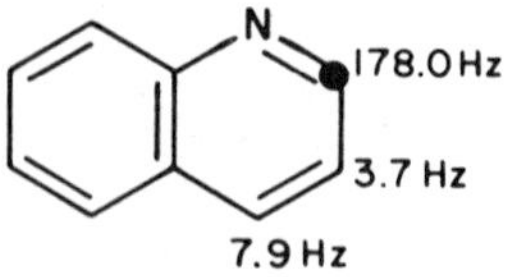

Values for methylpyridines[92] and methylquinolines have been reported.[91,93] Table 4.13 gives C-H coupling constants for some five-membered heterocycles.[94]

Carbon-carbon couplings have been determined in several classes of compounds both at the natural-abundance level of ^{13}C and with singly and doubly enriched materials. Extensive tabulations have been published.[95,96] The $^{1}J_{CC}$ values are in the range 57 to 60 Hz; these values are increased by electronegative substituents (cf. values for substituted naphthalenes and phenanthrenes[97,98]).

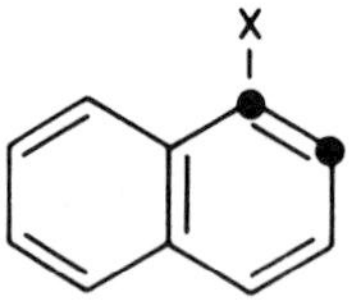

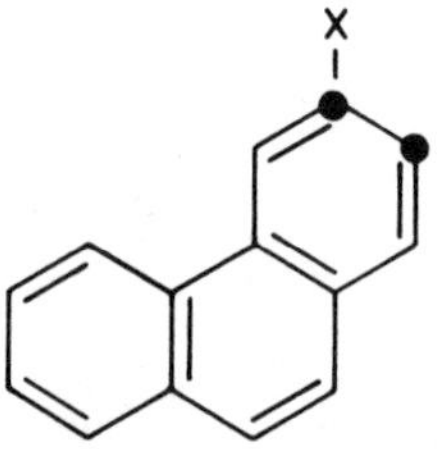

X		
H	60.3	
OH	68.8	66.8
OAc	74.0	72.0
CH$_3$		60.1
CHO		60.1
CN		62.8

An interesting change in $^{1}J_{CC}$ with ring size has been reported.[99] The unusually large value for J_{ab} of benzocyclopropene (in the following structures) has been related to a corresponding increase in s character of the benzene C-C orbital. The related decrease in J_{bc} is similarly rationalized.

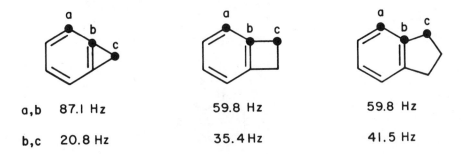

| a,b | 87.I Hz | | 59.8 Hz | | 59.8 Hz |
| b,c | 20.8 Hz | | 35.4 Hz | | 41.5 Hz |

TABLE 4.13. ^{13}C-H COUPLING CONSTANTS (IN HERTZ) FOR SOME FIVE-MEMBERED HETEROCYCLES[94]

	C_3			C_4			C_5		
	H_3	H_4	H_5	H_3	H_4	H_5	H_3	H_4	H_5
isothiazole	183.5	8.2	11.3	14.2	173.3	5.4	6.2	8.0	187.1
isoxazole	187.6	6.2	6.2	9.6	184.6	14.3	4.4	10.9	203.5
N-methylpyrazole	183.8	8.4, 5.7		9.8	175.5	9.8			~187

Carbon-carbon couplings over more than one bond are less than 10 Hz, and frequently less than 5 Hz.[95]   Three-bond couplings are larger in magnitude than are two-bond values. The latter may be either positive or negative in sign and may depend on bond order.

One-bond coupling constants between aromatic and directly attached carbons are somewhat larger than in aliphatic derivatives, reflecting the larger s character of the aromatic ring. Longer-range values are less than 4 Hz, and $^3J_{CC}$ in general is larger than $^2J_{CC}$. Illustrative values for substituted toluenes are given in Table 4.14.[96,100]

TABLE 4.14.   CARBON-CARBON COUPLINGS (IN HERTZ) FOR SUBSTI-
TUTED TOLUENES[96],[100]

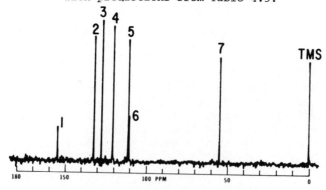

X	$^1J_{17}$	$^2J_{27}$	$^3J_{37}$	$^4J_{47}$	$^3J_{57}$	$^2J_{67}$
H	44.19	3.1	3.84	0.86		
2-NO$_2$	44.55	~1.0	1.53	0.50	3.60	2.20
2-NH$_2$	45.00	~1.0	1.64	0.62	3.87	2.47
2-CN	43.82	~0.8	2.40	0.73	3.87	2.83
4-NO$_2$	43.45	3.46	3.87	--	--	--
4-NH$_2$	45.91	3.17	4.19	< 0.5	--	--

    Carbon-carbon couplings have been reported for benzoic acid
derivatives,[101] dehydroanthracene and phthalic acid derivatives,[102] phenylethylenes,[103] and cyclopentadienones.[103]

PROBLEM 4.6.   *Determine the structure of the following com-*
                  pound (Formula: C_7H_7BrO) and complete ^{13}C
                  spectral assignments. Compare these results
                  with predictions from Table 4.3.

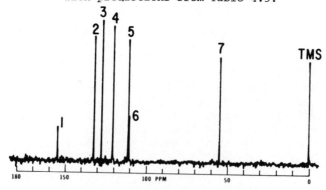

REFERENCES

1. P. C. Lauterbur, *Ann. N. Y. Acad. Sci.*, **70**, 841 (1958).
2. For reviews and critical discussions, see (a) G. L. Nelson and E. A. Williams, *Progr. Phys. Org. Chem.*, **12**, 229 (1976); (b) D. F. Ewing, *Org. Magn. Resonance*, **12**, 499 (1979). The latter article tabulates over *700* substituent chemical shifts.
3. (a) H. Spiesecke and W. G. Schneider, *J. Chem. Phys.*, **35**, 731 (1961); (b) R. Ditchfield and P. D. Ellis, *Topics in Carbon-13 NMR Spectroscopy*, Vol. 1, G. C. Levy, Ed., Wiley, New York, 1974, p. 1 and references cited therein; (c) G. L. Nelson, G. C. Levy, and J. D. Cargioli, *J. Am. Chem. Soc.*, **94**, 3090 (1972).
4. G. E. Maciel, in ref. 3b, p. 53.
5. A. B. Strong, D. Ikenberry, and D. M. Grant, *J. Magn. Resonance*, **9**, 145 (1973).
6. A. Saika and C. P. Slichter, *J. Chem. Phys.*, **22**, 26 (1954).
7. (a) J. Mason, *J. Chem. Soc., Perkin Transact. II*, **1976**, 167; (b) *J. Chem. Soc. A.*, **1971**, 1038.
8. D. M. Grant et al., *J. Am. Chem. Soc.*, **93**, 1880, 1887 (1971); **92**, 2386(1970).
9. D. T. Clark and J. W. Emsley, *Molec. Phys.*, **12**, 365 (1967).
10. J. E. Bloor and D. L. Breen, *J. Am. Chem. Soc.*, **89**, 6835 (1967).
11. A. Allerhand and E. A. Trull, *Annu. Rev. Phys. Chem.*, **21**, 317 (1970).
12. K. A. K. Ebraheem and G. A. Webb, *Org. Magn. Resonance*, **9**, 241 (1977).
13. G. E. Maciel and H. C. Dorn, *J. Magn. Resonance*, **24**, 251 (1976).
14. N. C. Baird and K. C. Teo, *J. Magn. Resonance*, **24**, 87 (1976).
15. (a) H. Günther, J. Prestien, and P. Joseph-Nathan, *Org. Magn. Resonance*, **7**, 339 (1975); (b) L. Ernst, *J. Magn. Resonance*, **21**, 241 (1976).
16. (a) For a more complete discussion, see D. W. Jones, in *Nuclear Magnetic Resonance*, Vol. 7, Specialist Periodical Report, The Chemical Society, London, 1978, pp. 31ff; (b) J. Bromilow and R. T. C. Brownlee, *Tetrahedron Lett.*, **1975**, 2113.
17. (a) C. G. Swain and E. C. Lupton, *J. Am. Chem. Soc.*, **90**, 4328 (1968); (b) J. Bromilow, R. T. C. Brownlee, D. J. Craik, V. O. Lopez, and R. W. Taft, *J. Org. Chem.*, **44**, 4766 (1979); (c) J. Bromilow and R. T. C. Brownlee, *ibid.*, 1261.
18. M. Godfrey, *J. Chem. Soc., Perkin Transact. II*, **1977**, 769.
19. (a) D. A. Dawson and W. F. Reynolds, *Can. J. Chem.*, **53**, 373 (1975); (b) W. Adcock, B. D. Gupta, and W. Kitching, *J. Org. Chem.*, **41**, 1498 (1976); (c) N. Inamoto, S. Masuda, and K. Tokumaru, *Tetrahedron Lett.*, **1976**, 3707,3711.

20. (a) T. B. Posner and C. D. Hall, *J. Chem. Soc., Perkin Transact. II*, **1976**, 729; (b) A. Cornelis, S. Lambert, and P. Laszlo, *J. Org. Chem.*, **42**, 381 (1977).

21. (a) R. H. Levin and J. D. Roberts, *Tetrahedron Lett.*, **1973**, 135; (b) H. Günther and H. Schmickler, *Angew. Chem. Internat. Ed. Engl.*, **2**, 243 (1973); (c) H. Günther, H. Schmickler, U. H. Brinker, K. Nachtkamp, J. Wassen, and E. Vogel, *ibid.*, **12**, 760 (1973); (d) R. J. Abraham, G. E. Hawkes, and K. M. Smith, *J. Chem. Soc., Perkin Transact. II*, **1974**, 627.

22. H. Günther and H. Schmickler, *Pure Appl. Chem.*, **44**, 807 (1975).

23. R. DuVernet and V. Boekelheide, *Proc. Nat. Acad. Sci. (USA)*, **71**, 296 (1974).

24. C. E. Johnson and F. A. Bovey, *J. Chem. Phys.*, **29**, 1012 (1958).

25. J. Bromilow, R. T. C. Brownlee, R. D. Topsom, and R. W. Taft, *J. Am. Chem. Soc.*, **98**, 2020 (1976).

26. C. Nagata, O. Hamada, and S. Taneka, *Nippon Kagaku Kaishu*, **1976**, 1029.

27. G. C. Levy, J. D. Cargioli, and F. A. L. Anet, *J. Am. Chem. Soc.*, **95**, 1527 (1973).

28. (a) W. R. Woolfenden and D. M. Grant, *J. Am. Chem. Soc.*, **88**, 1496 (1969); (b) L. Ernst, *Tetrahedron Lett.*, **1974**, 3079; (c) D. K. Dalling, K. H. Ladner, D. M. Grant, and W. R. Woolfenden, *J. Am. Chem. Soc.*, **99**, 7142 (1977).

29. W. E. Smith, unpublished results.

30. H. Pearson, *J. Chem. Soc., Chem. Commun.*, **1975**, 912.

31. (a) T. Takemura and T. Sato, *Can. J. Chem.*, **54**, 3412 (1976); (b) T. Sato, T. Takemura, and M. Kainosho, *J. Chem. Soc., Chem. Commun.*, **1974**, 97; (c) T. Takemura, K. Tokita, S. Kondo, and N. Mori, *Chem. Lett.*, **1977**, 865.

32. (a) B. M. Lynch, *Can. J. Chem.*, **55**, 541 (1977); (b) M. P. Simonnin, M. -J. Pouet, and F. Terrier, *J. Org. Chem.*, **43**, 855 (1978); (c) P. E. Rakita, J. P. Srebo, and L. S. Worsham, *J. Organomet. Chem.*, **104**, 27 (1976).

33. (a) N. K. Wilson, *J. Am. Chem. Soc.*, **97**, 3573 (1975); (b) U. Edlund and Å. Norstrom, *Org. Magn. Resonance*, **9**, 196 (1977).

34. P. E. Hansen, *Org. Magn. Resonance*, **12**, 109 (1979).

35. A. J. Jones, T. D. Alger, D. M. Grant, and W. M. Litchman, *J. Am. Chem. Soc.*, **92**, 2386 (1970).

36. A. J. Jones, P. D. Gardner, D. M. Grant, W. M. Litchman, and V. Boekelheide, *ibid.*, **92**, 2395 (1970).

37. N. Defay, D. Zimmerman, and R. H. Martin, *Tetrahedron Lett.*, **1971**, 1871.

38. H. A. Staab, K. S. Rao, and H. Brunner, *Chem. Ber.*, **104**, 2634 (1971).

39. B. M. Trost and W. B. Herdle, *J. Am. Chem. Soc.*, **98**, 4080 (1976).

40.   W. Kitching, M. Bullpitt, D. Gartshore, W. Adcock, T. C. Khor, D. Doddrell, and I. D. Rae, *J. Org. Chem.*, **42**, 2411 (1977).

41.   (a) W. Kitching, M. Bullpitt, and D. Doddrell, *Org. Magn. Resonance*, **6**, 289 (1974); (b) H. E. Gottlieb, *Isr. J. Chem.*, **16**, 57 (1977).

42.   (a) W. Adcock, B. D. Gupta, T. C. Khor, D. Doddrell, D. Jordan, and W. Kitching, *J. Am. Chem. Soc.*, **96**, 1595 (1974); (b) E. L. Motell, D. Lauer, and G. E. Maciel, *J. Phys. Chem.*, **77**, 1865 (1973).

43.   (a) S. Gronowitz, I. Johnson, and A. -B. Hörnfeldt, *Chemica Scripta*, **7**, 76 (1975); (b) *ibid.*, **7**, 211 (1975); (c) M. T. W. Hearn, *Aust. J. Chem.*, **29**, 107 (1976); (d) F. Fringuelli, S. Gronowitz, A. -B. Hörnfeldt, I. Johnson, and A. Taticchi, *Acta Chem. Scand. B.*, **28**, 125 (1974).

44.   J. Elguero, C. Marzin, and J. D. Roberts, *J. Org. Chem.*, **39**, 357 (1974).

45.   M. Garreau, G. J. Martin, M. C. Martin, M. Morel, and C. Paulmier, *Org. Magn. Resonance*, **6**, 648 (1974).

46.   J. Gainer, G. A. Howarth, W. Hoyle, and W. M. Roberts, *Org. Magn. Resonance*, **8**, 226 (1976).

47.   N. Plavac, I. W. J. Still, M. S. Chauhan, and D. M. McKinnon, *Can. J. Chem.*, **53**, 836 (1975).

48.   R. Faure, A. Assaf, E. -J. Vincent, and J. -P. Aune, *J. Chim. Phys.*, **75**, 727 (1978).

49.   R. E. Wasylishen, T. R. Clem, and E. D. Becker, *Can. J. Chem.*, **53**, 596 (1975).

50.   F. Fringuelli, S. Gronowitz, A. -B. Hörnfeldt, I. Johnson, and A. Taticchi, *Acta. Chem. Scand.*, **28B**, 175 (1974).

51.   B. M. Lynch, *Chem. Commun.*, **1968**, 1337.

52.   R. J. Pugmire and D. M. Grant, *J. Am. Chem. Soc.*, **90**, 2513 (1967).

53.   W. Adam, A. Grimison, and G. Rodriguez, *Tetrahedron*, **23**, 2513 (1967).

54.   H. Hartmann and R. Radeglia, *J. Prakt. Chem.*, **317**, 657 (1975).

55.   K. A. K. Ebraheem, G. A. Webb, and M. Witanowski, *Org. Magn. Resonance*, **11**, 27 (1978).

56.   (a) H. L. Retcofsky and R. A. Friedel, *J. Phys. Chem.*, **71**, 3592 (1967); (b) *ibid.*, **72**, 290 (1968).

57.   For a similar discussion for ^{19}F shifts in 5-substituted-2-fluoropyridines, see (a) C. S. Giam and J. L. Lyle, *J. Chem. Soc. B.*, **1970**, 1516; (b) R. L. Lichter and R. E. Wasylishen, *J. Am. Chem. Soc.*, **97**, 1808 (1975).

58.   C. J. Turner and G. W. H. Cheeseman, *Org. Magn. Resonance*, **6**, 663 (1974).

59.   J. Riaud, M. Th. Chenon, and N. Lumbroso-Bader, *J. Am. Chem. Soc.*, **99**, 6838 (1977).

60. (a) J. Riaud, M. Th. Chenon, and N. Lumbroso-Bader, *Tetrahedron Lett.*, **1974**, 3123; (b) C. J. Turner and G. W. H. Cheeseman, *Org. Magn. Resonance*, **8**, 357 (1976).

61. P. van de Weijer and C. Mohan, *Org. Magn. Resonance*, **9**, 53 (1977).

62. S. Braun and G. Frey, *Org. Magn. Resonance*, **7**, 194 (1975).

63. (a) R. J. Pugmire and D. M. Grant, *J. Am. Chem. Soc.*, **90**, 697 (1968); (b) W. Adam, A. Grimison, and G. Rodriguez, *J. Chem. Phys.*, **50**, 645 (1969).

64. (a) F. A. L. Anet and I. Yavari, *J. Org. Chem.*, **41**, 3589 (1976); (b) R. J. Cushley, D. Naugler, and C. Ortiz, *Can. J. Chem.*, **53**, 3419 (1975); (c) H. Günther and A. Gronenborn, *Heterocycles*, **11**, 337 (1978).

65. A. J. Ashe, III, R. R. Sharp, and J. W. Tolan, *J. Am. Chem. Soc.*, **98**, 5451 (1976).

66. A. T. Balaban and V. Wray, *Org. Magn. Resonance*, **9**, 16 (1977).

67. (a) G. W. Gribble, R. B. Nelson, J. L. Johnson and G. C. Levy, *J. Org. Chem.*, **40**, 3720 (1975); (b) R. R. Fraser, S. Passannanti, and F. Piozzi, *Can. J. Chem.*, **54**, 2915 (1976).

68. N. Platzer, J. J. Basselier, and P. Demerseman, *Bull. Soc. Chim. Fr.*, **1974**, 877.

69. P. Geneste, J. -L. Olive, S. N. Ung, M. E. A. E. Faghi, J. W. Easton, H. Beierbeck, and J. K. Saunders, *J. Org. Chem.*, **44**, 2887 (1979).

70. R. J. Pugmire, J. C. Smith, D. M. Grant, B. Stanovnik, and M. Tisler, *J. Hetercyclic Chem.*, **13**, 1057 (1976).

71. (a) J. Elguero, A. Fruchier, and M. del Carmen Pardo, *Can. J. Chem.*, **54**, 1329 (1976); (b) A. Fruchier, E. Alcade, and J. Elguero, *Org. Magn. Resonance*, **9**, 235 (1977).

72. (a) S. Florea, W. Kimpenhaus, and V. Farcasan, *Org. Magn. Resonance*, **9**, 133 (1977); (b) S. N. Sawhney and D. W. Boykin, *J. Org. Chem.*, **44**, 1136 (1974).

73. I. Yavari and F. A. L. Anet, *Org. Magn. Resonance*, **8**, 158 (1976).

74. S. Gronowitz, I. Johnson, and A. Bugge, *Acta Chem. Scand.*, **30B**, 417 (1976).

75. P. D. Clark, D. F. Ewing, and R. M. Scrowton, *Org. Magn. Resonance*, **8**, 252 (1976).

76. (a) R. Faure, J. Elguero, E. J. Vincent, and R. Lazaro, *Org. Magn. Resonance*, **11**, 617 (1978); (b) R. Faure, J. P. Galy, E. J. Vincent, and J. Elguero, *Can. J. Chem.*, **56**, 46 (1978).

77. (a) J. A. Su, E. Siew, E. V. Brown, and S. L. Smith, *Org. Magn. Resonance*, **10**, 122 (1977); (b) *ibid.*, **11**, 565 (1978).

78. (a) U. Ewers, H. Günther, and L. Jaenicke, *Angew. Chem.*, **87**, 356 (1975); (b) J. P. Geerts, A. Nagel, and H. C. van der Plas, *Org. Magn. Resonance*, **8**, 607 (1976).

79.   R. J. Pugmire, D. M. Grant, L. B. Townsend, and R. K. Robins, *J. Am. Chem. Soc.*, **95**, 2791 (1973).

80.   E. Dradi and G. Gatti, *J. Am. Chem. Soc.*, **97**, 5472 (1975).

81.   C. Mayor and C. Wentrup, *J. Am. Chem. Soc.*, **97**, 7467 (1975).

82.   R. A. Earl, R. J. Pugmire, G. R. Revankar, and L. B. Townsend, *J. Org. Chem.*, **40**, 1822 (1975).

83.   E. Breitmaier and U. Hollstein, *J. Org. Chem.*, **41**, 2104 (1976).

84.   G. Fronza, R. Modelli, G. Scapini, G. Ronsisvalle, and F. Vittorio, *J. Magn. Resonance*, **23**, 437 (1976).

85.   P. van de Weijer, D. M. W. van dem Ham, and D. Van der Meer, *Org. Magn. Resonance*, **9**, 281 (1977).

86.   H. Günther, H. Seel, and M. -E. Günther, *Org. Magn. Resonance*, **11**, 97 (1978).

87.   H. Seel, R. Aydin, and H. Günther, *Z. Naturforsch.*, **33b**, 353 (1978).

88.   L. Ernst, V. Wray, V. A. Chertokov, and N. M. Sergeyev, *J. Magn. Resonance*, **25**, 123 (1977).

89.   (a) V. A. Chertkov and N. M. Sergeyev, *J. Magn. Resonance*, **21**, (1976); (b) V. Wray and D. N. Lincoln, *ibid.*, **18**, 374 (1975); (c) L. Ernst, D. N. Lincoln, and V. Wray, *ibid.*, **21**, 115 (1976); (d) V. Wray, L. Ernst, and E. Lustig, *ibid.*, **27**, 1 (1977); (e) L. Ernst and V. Wray, *ibid.*, **28**, 373 (1977); (f) W. S. Brey, L. W. Jaques, and H. J. Jakobsen, *Org. Magn. Resonance*, **12**, 243 (1979).

90.   H. Günther, H. Seel, and H. Schmickler, *J. Magn. Resonance*, **28**, 145 (1977).

91.   S. R. Johns and R. I. Willing, *Aust. J. Chem.*, **29**, 1617 (1976).

92.   Y. Takeuchi, *Org. Magn. Resonance*, **7**, 181 (1975).

93.   P. A. Claret and A. G. Osborne, *Org. Magn. Resonance*, **8**, 147 (1976).

94.   R. E. Wasylishen and H. M. Hutton, *Can. J. Chem.*, **55**, 619 (1977).

95.   P. E. Hansen, *Org. Magn. Resonance*, **11**, 215 (1978).

96.   R. E. Wasylishen, *Annu. Rep. NMR Spectrosc.*, **7**, 245 (1977).

97.   P. E. Hansen, O. K. Poulsen, and A. Berg, *Org. Magn. Resonance*, **7**, 475 (1975).

98.   S. Berger and K. P. Zeller, *Org. Magn. Resonance*, **11**, 303 (1978).

99.   H. Günther and W. Herrig, *J. Am. Chem. Soc.*, **97**, 5594 (1975).

100. J. L. Marshall and A. M. Ihrig, *Org. Magn. Resonance*, **5**, 235 (1973).

101. J. L. Marshall, L. G. Faehl, A. M. Ihrig, and M. Barfield, *J. Am. Chem. Soc.*, **98**, 3406 (1976).

102. (a) P. E. Hansen and A. Berg, *Org. Magn. Resonance*, **8**, 591 (1976); (b) **12**, 50 (1979).

103. (a) P. E. Hansen, O. I. Poulsen, and A. Berg, *Org. Magn. Resonance*, **8**, 632 (1976); (b) *ibid.*, **12**, 43 (1979).

Chapter 5

Organic Functional Groups

One advantage of ^{13}C nmr in organic chemistry is that many of the important functional groups contain carbon and are thus directly observable. Figure 5.1 illustrates the general chemical-shift ranges for each class of functional group.

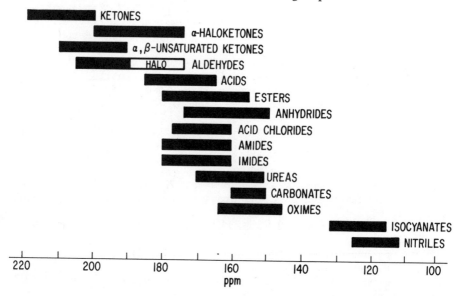

FIGURE 5.1.   Functional-group ^{13}C chemical shifts.

CARBONYLS

The carbonyl carbon has been the most investigated of all carbons by cmr. The ease of isotopic enrichment and the lack of extensive ^{13}C-^{1}H coupling to most carbonyl carbons made early cmr studies attractive.[1] Although there is considerable overlap, the various carbonyl groups have distinctive chemical shifts (Table 5.1). Some correlation with carbonyl π-bond polarity has been found.[2]

TABLE 5.1.   CARBONYL CHEMICAL SHIFTS (IN PARTS PER MILLION)

Structure	Shift	Structure	Shift
$CH_3\overset{O}{\overset{\|}{C}}CH_3$	205.2	$CH_3-\overset{O}{\overset{\|}{C}}-O-\overset{O}{\overset{\|}{C}}-CH_3$	167.3
$Ph-\overset{O}{\overset{\|}{C}}-CH_3$	196.9	$Ph-\overset{O}{\overset{\|}{C}}-O-\overset{O}{\overset{\|}{C}}-Ph$	162.8
$CH_3-\overset{O}{\overset{\|}{C}}-H$	200.5	$CH_3-\overset{O}{\overset{\|}{C}}-Cl$	170.5
$Ph-\overset{O}{\overset{\|}{C}}-H$	190.7	$Ph-\overset{O}{\overset{\|}{C}}-Cl$	168.7
$CH_3-\overset{O}{\overset{\|}{C}}-OH$	177.3	$CH_3-\overset{O}{\overset{\|}{C}}-N(CH_3)_2$	170.7
$Ph-\overset{O}{\overset{\|}{C}}OH$	173.5	$Ph-\overset{O}{\overset{\|}{C}}-N(CH_3)_2$	170.8
$CH_3-\overset{O}{\overset{\|}{C}}-OCH_3$	171.0	$CH_3-O-\overset{O}{\overset{\|}{C}}-OCH_3$	156.5
$Ph-\overset{O}{\overset{\|}{C}}-OCH_3$	167.0	$Ph-O-\overset{O}{\overset{\|}{C}}-O-Ph$	152.0

The carbonyl carbon nuclei of aliphatic ketones are the most highly deshielded of all carbonyls. Alkyl substitution at the α-carbon causes a deshielding of 2 to 3 ppm. Table 5.2 illustrates this effect for acetone.[3] An isotope shift of 0.28 ppm for the carbonyl carbon of acetone-d_6 relative to the carbonyl carbon in acetone-h_6 has been observed[4] (see also Table 2.1).

Incorporation of the keto function into a ring produces a varied effect, as illustrated in Table 5.3.[5-6] Cyclopentanones, for example, are deshielded by 5 ppm from both cyclohexanone and cyclobutanone. As with acyclic ketones, alkyl substitution at α-carbons deshields the carbonyl carbon, whereas β-substitution has little effect.

TABLE 5.2.   EFFECT OF ALKYL SUBSTITUTION ON ACETONE[3] (IN PARTS PER MILLION)

Structure	Shift	Structure	Shift
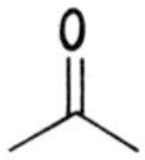 (acetone)	205.2	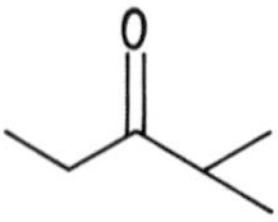 (methyl isopropyl ketone)	213.2 (8.0)
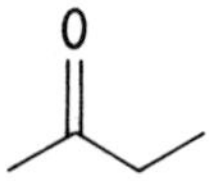 (methyl ethyl ketone)	207.7 (2.5)[a]	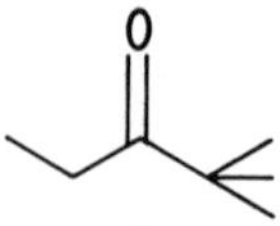 (methyl tert-butyl ketone)	214.5 (9.3)
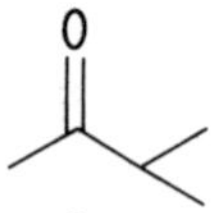 (methyl isopropyl)	210.7 (5.5)	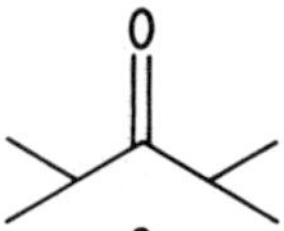 (diisopropyl ketone)	216.5 (11.3)
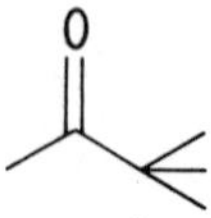 (methyl tert-butyl)	211.8 (6.6)	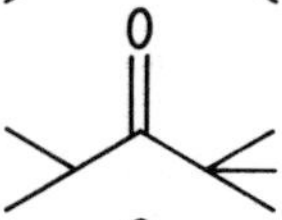	218.5· (13.3)
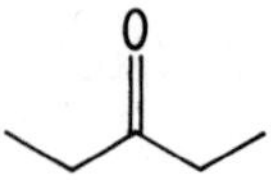 (diethyl ketone)	210.1 (4.9)	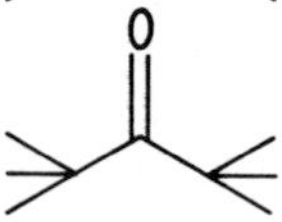	216.5 (11.3)

[a]Value in parentheses corresponds to the deshielding with respect to acetone.   Average is 2.4 ppm/CH_3.

Ring size is important not only for alicyclic carbonyls but for carbonyls adjacent to rings as well.   The sensitivity of acetyl carbonyl chemical shifts to the size of adjacent rings has been shown.   The carbonyl chemical shifts for cyclopropyl, cyclobutyl, cyclopentyl, and cyclohexyl methyl ketones are 206.8, 208.0, 207.7 and 210.2 ppm, respectively.[7]

Substituted Ketones.   It was shown previously that alkyl-substituted aliphatic ketones, whether acyclic or cyclic, have carbonyl chemical shifts in the range 200 to 220 ppm; moderate deshielding occurs with each additional α-alkyl substituent. α-Substitution by halogen or α,β-unsaturation, on the other hand, results in a marked shielding for carbonyl resonances.   Chemical shifts for some α-haloketones are shown in Table 5.4.[8-10]

Placement of an sp^2 center α to a carbonyl function causes a shielding of as much as 10 ppm (Table 5.5).   Simple α,β-unsaturated ketone carbonyl resonances can be expected in the

TABLE 5.3.   [13]C CHEMICAL SHIFTS OF CARBONYL CARBONS IN
CYCLOALKANONES[5-6]

 cyclobutanone	208.2	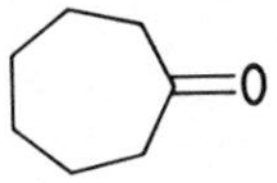 cycloheptanone	211.7
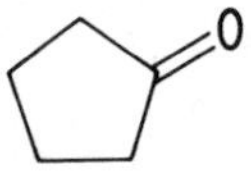 cyclopentanone	213.9	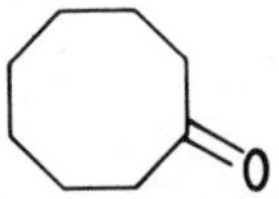 cyclooctanone	215.9
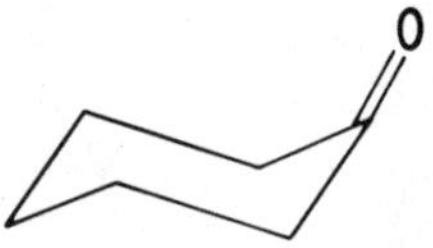	208.8	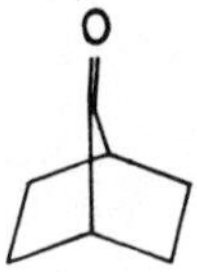	216.2
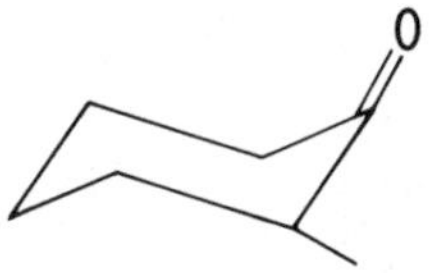	210.3	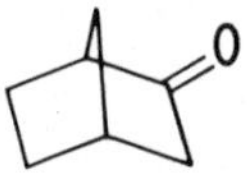	215.3
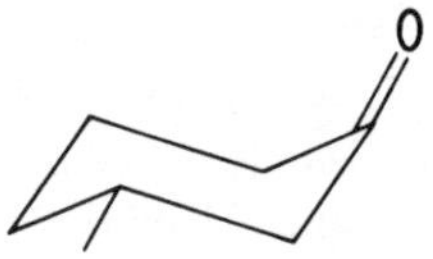	208.4	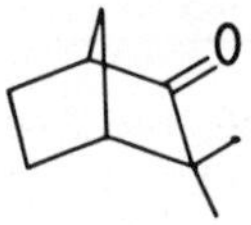	219.4
		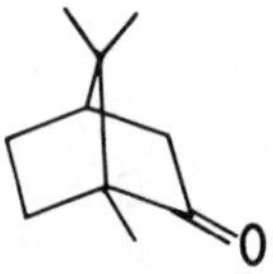	216.7

range 190 to 210 ppm. The expected deshielding resulting from
α-alkyl substitution or incorporation into five-membered rings
can also be seen in these conjugated ketones (see Table
5.5).[6,8,10-11]

Of special note is the shielded carbonyl of cyclopropenone
at 155.1 ppm.[11] This is consistent with substantial contribution
of the aromatic dipolar canonical form shown in which π-bond

TABLE 5.4.   [13]C CHEMICAL SHIFTS (IN PARTS PER MILLION) OF CARBONYL CARBONS IN α-HALOKETONES[8-10]

CH$_3$—C(=O)—CH$_3$	205.2	Cl$_3$C—C(=O)—CCl$_3$	175.9
Cl—CH$_2$—C(=O)—CH$_3$	200.7	CF$_3$—C(=O)—CH$_3$	187.7
Cl—CH$_2$—C(=O)—CH$_2$—Cl	194.9	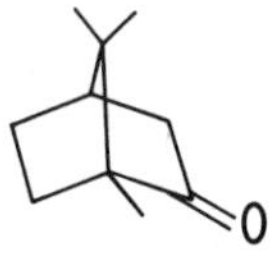	216.7
CCl$_3$—C(=O)—CH$_3$	186.3	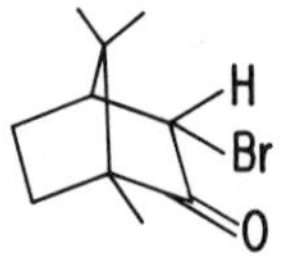	210.1

character is reduced and the carbon nucleus is shielded. The effect is enhanced by the presence of electron-donating substituents, as shown for the bis(diisopropylamino) and diethoxy compounds.[11]

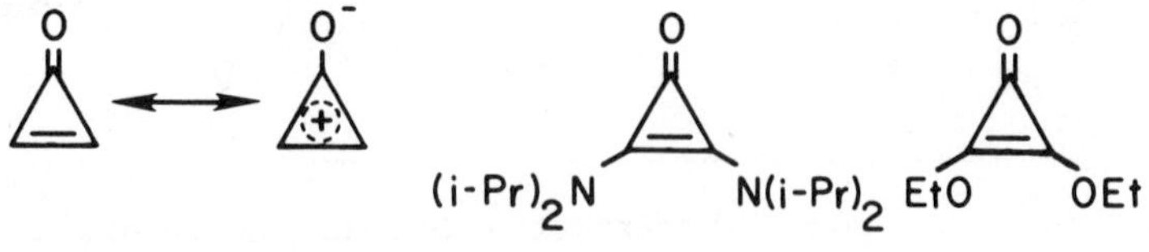

(i-Pr)$_2$N   N(i-Pr)$_2$ EtO   OEt

134.0 ppm   137.4 ppm

TABLE 5.5. [13]C CHEMICAL SHIFTS (IN PARTS PER MILLION) OF CARBONYL CARBONS IN α,β-UNSATURATED KETONES[6,8,10-11]

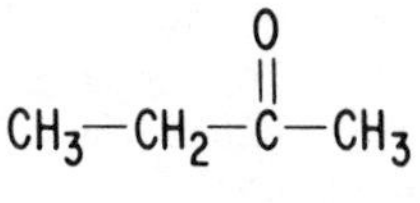	207.2	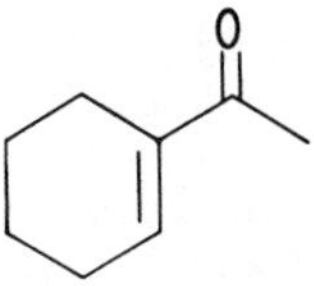	198.3
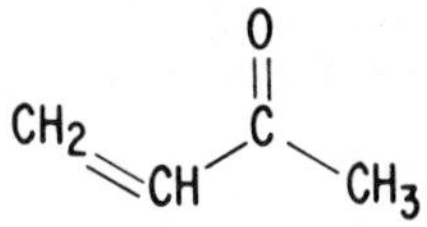	198.1	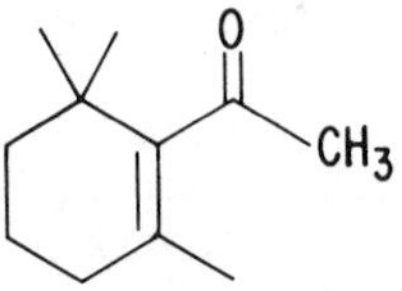	206.4
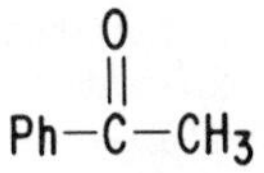	196.9	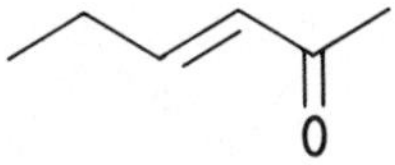	196.5
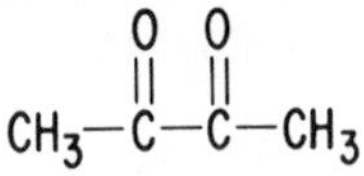	198.6	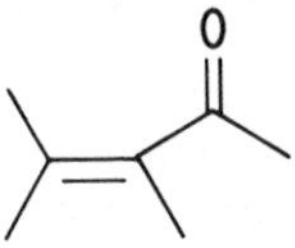	202.6
	155.1	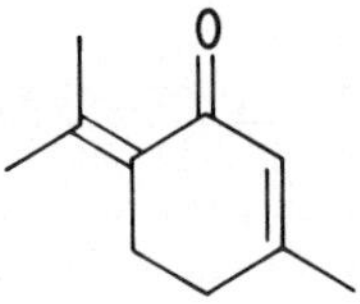	190.2
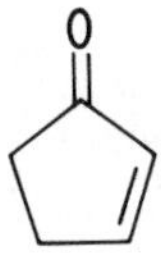	208.1	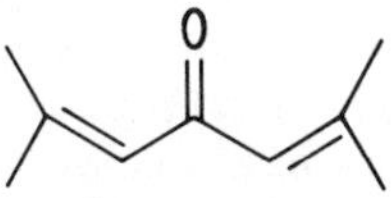	188.8
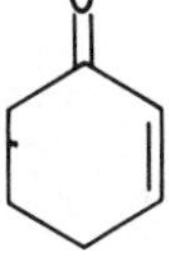	197.1		203.3
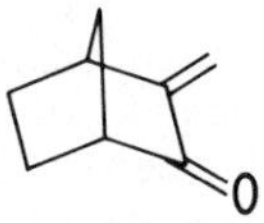	203.8		
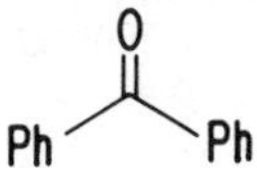	194.8		

Phenyl Ketones.   Neither *meta* nor *para* substitution appreciably affects the carbonyl chemical shift in acetophenone (range 196.4 to 198.6 ppm).[12a] By contrast, *ortho* ring substitution does result in significant change in the carbonyl resonance position; deshielding·on the order of 2 to 8 ppm is common. This is generally attributed to steric inhibition of coplanarity between the carbonyl group and the phenyl ring. For α-substituted acetophenones, the expected trends hold.[12b] Changes in chemical shifts in α-alkyl acetophenones resulting from *meta* or *para* ring substitution have been noted.[12c]

Solvent Effects.   Carbon-13 chemical shifts of ketones are sensitive to solvents, particularly solvents capable of hydrogen bonding to the carbonyl oxygen. These effects were noted for acetophenone in Chapter 4. Aliphatic ketones also exhibit strong solvent-induced shifts,[13-15] as shown for acetone in Table 5.6; the carbonyl chemical shift of ketones determined as a function of medium acidity in H_2SO_4-H_2O mixtures yields a useful titration curve for the $pK_{BH}{}^+$ (actually the Hammett acidity, H_0, at half-protonation) of the carbonyl group.[15-16]

The sensitivity of carbonyl chemical shifts to solvent or concentration changes makes it difficult to compare chemical shifts from different studies; reported differences of more than 1 ppm for the same compound under similar conditions are not uncommon. Detailed physicochemical studies have been carried out, including effects of interaction with Lewis acids.[17]

Quinones and Related Polyketones. Quinone carbonyl resonances lie in the range 170 to 190 ppm. Generally, *ortho*-quinones are shielded more than the *para* isomers are.[18,19] Table 5.7 gives representative data, along with values for cyclic oxocarbons.[20]

Aldehydes.   The organic chemistry of aldehydes is quite similar to that of ketones. Both classes of carbonyl compounds are generally considered as a unit. Aldehydic carbonyl groups exhibit most of the shift properties of the corresponding ketones but generally are shielded by 5 to 10 ppm. Aldehydic carbonyl groups are easily identified by observation of doublet signals in off-resonance [1]H-decoupled experiments. Moreover, aldehydic carbonyl resonances generally are more intense than ketone carbonyl signals at the same concentration because of higher NOE values and shorter relaxation times (both result from the directly attached proton). Table 5.8[3,8] gives several examples of aldehydic chemical shifts; chemical shifts for the corresponding methyl ketones are also given.

TABLE 5.6.   CHANGE (IN PARTS PER MILLION) IN CARBONYL CHEMICAL SHIFT FOR ACETONE AS A FUNCTION OF SOLVENT[13a]

Neat (position)	(205.2)	Chloroform	+ 2.3
Cyclohexene	-2.4	Methanol	+ 3.7
Diethyl ether	-2.0	Acetic acid	+ 6.2
Carbon tetrachloride	-1.3	Phenol	+ 8.7
Benzene	-0.8	Formic acid	+ 9.1
1,4-Dioxane	0.0	Dichloroacetic acid	+11.9
Iodomethane	0.0	Trifluoroacetic acid	+14.1
2-Propanol	+1.9	Sulfuric acid	+39.2
Acetonitrile	+2.1		

Benzaldehydes[21] exhibit only minor changes on *meta* or *para* substitution; *ortho* substitution produces much the same effect as in *ortho*-acetophenones.

Carboxylic Acids.   Carboxylic acid carbonyl resonances can be found in the range 165 to 185 ppm, as illustrated in Table 5.9.[8,22] The corresponding anions are deshielded by ∿5 ppm.[23] Solvent effects are quite marked (see Figure 5.2[24]). For acetic acid, relatively large dilution effects are observed with acetone, a smaller effect is observed with chloroform, and only small effects are seen in water or cyclohexane. The results are understandable in terms of the influences of making or breaking of hydrogen bonds between the carbonyl group of acetic acid, the OH of acetic acid, and the solvent molecules.

PROBLEM 5.1.   The following ^{13}C nmr spectrum is obtained from a compound of molecular formula $C_8H_{10}O_4$. Determine the structure and give spectral assignments (note: no off-resonance decoupling information is provided; peak intensities are misleading).

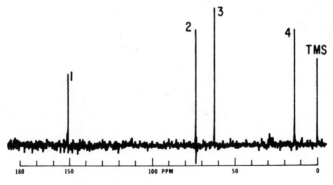

TABLE 5.7.   [13]C CHEMICAL SHIFTS (IN PARTS PER MILLION) OF
CARBONYL CARBONS IN QUINONES AND CYCLIC OXOCARBONS[18-20]

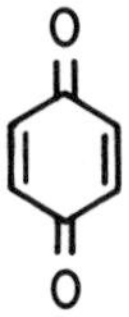

187.0

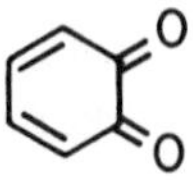

180.2

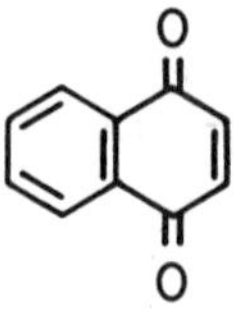

184.6

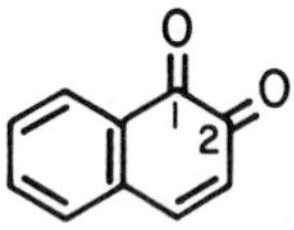

179.8 (C-1)
180.9 (C-2)

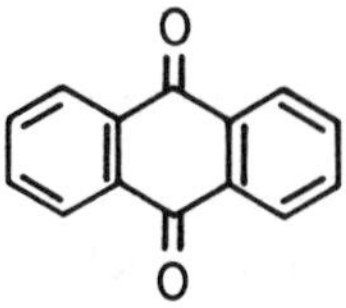

183.0

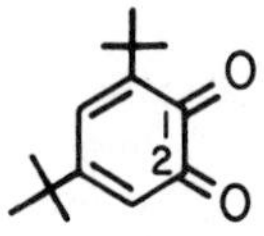

179.6 (C-1)
180.6 (C-2)

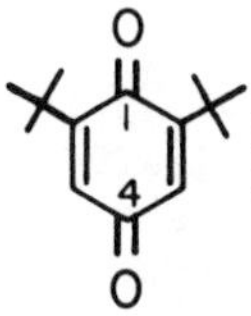

187.7 (C-1)
188.6 (C-4)

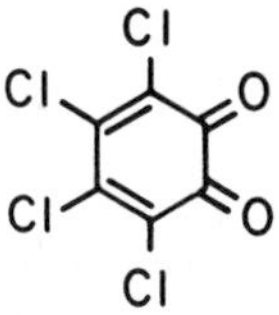

186.8

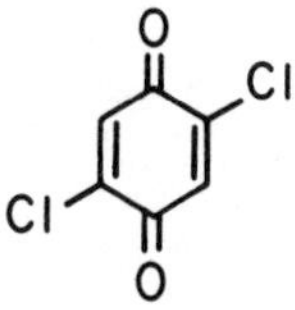

176.6

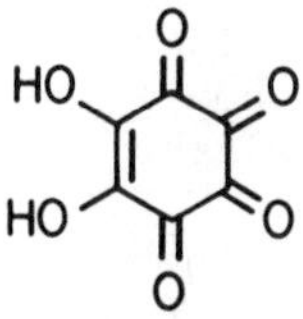

175.0, 173.9

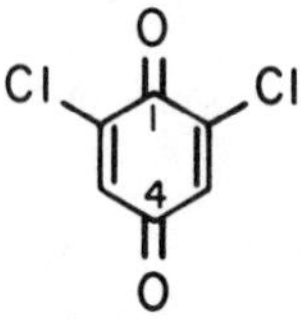

172.7 (C-1)
182.5 (C-4)

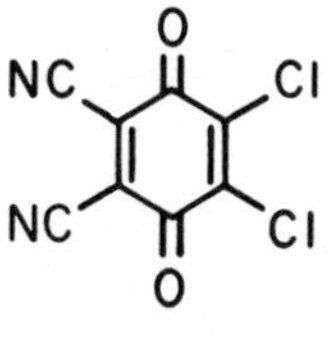

170.1

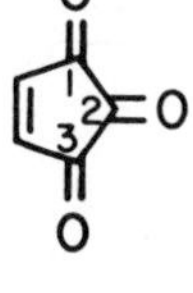

182.4 (C-1,3)
188.4 (C-2)

144

TABLE 5.8. ^{13}C CHEMICAL SHIFTS (IN PARTS PER MILLION) OF CARBONYL CARBONS IN ALDEHYDES COMPARED WITH CORRESPONDING METHYL KETONES[3],[8]

Structure	Shift	Structure	Shift
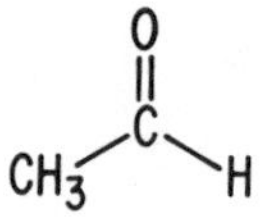 CH_3–CO–H	200.5 (205.2)	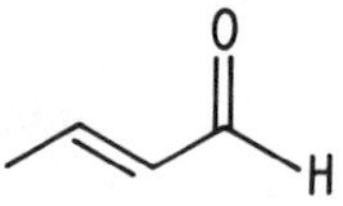 $CH_3CH{=}CH$–CO–H	192.3 (197.4)
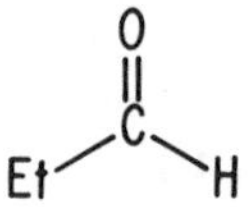 Et–CO–H	202.7 (207.7)	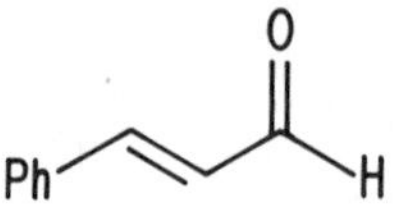 Ph–$CH{=}CH$–CO–H	192.7 (197.7)
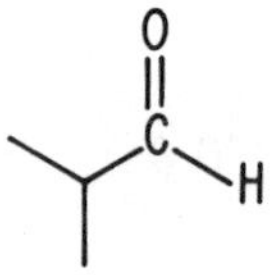 (CH_3)_2CH–CO–H	204.9 (210.7)	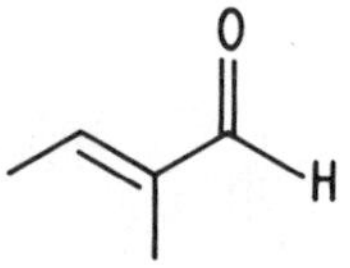 $CH_3CH{=}C(CH_3)$–CO–H	193.7 (198.9)
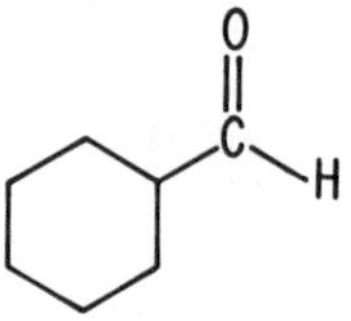 cyclohexyl–CO–H	202.7 (210.2)	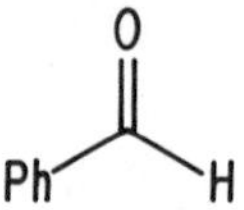 Ph–CO–H	190.7 (195.6)
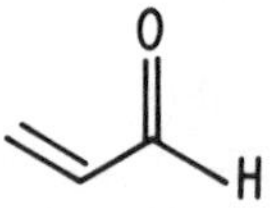 $CH_2{=}CH$–CO–H	193.3 (198.1)	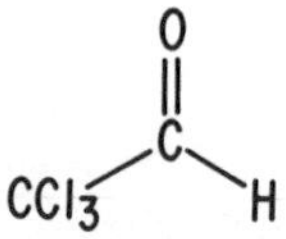 CCl_3–CO–H	175.9 (186.3)

Esters.   Esters present a more complex problem. Not only can substitution at the α-carbon affect the chemical shift of the carbonyl carbon, but the nature of R in the R'COOR linkage also may be expected to affect the chemical shift of the carbonyl carbon. Table 5.10[6],[8],[22],[25] presents the experimental data. Ester carbonyls are generally found in the range 160 to 180 ppm. Substitution at the carbon α to the carbonyl has a greater effect than variation of the alcohol residue. Without question in the latter case, unsaturation induces an upfield shift; alkyl substitution also results in a small upfield shift.

Anhydrides.   Carbonyl nuclei of acyclic anhydrides are shielded by ∿10 ppm compared with the corresponding carboxylic acids (in the region 150 to 175 ppm). The effects of ring size and substitution are as described for other carbonyl carbons (see Table 5.11).[26]

TABLE 5.9. ^{13}C CHEMICAL SHIFTS (IN PARTS PER MILLION) FOR CARBONYL CARBONS IN CARBOXYLIC ACIDS[6,8,22]

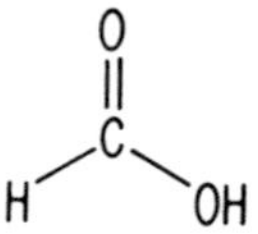

Structure	Shift	Structure	Shift
H–C(=O)–OH	166.7	Ph–C(=O)–OH	173.5
CH₃–C(=O)–OH	177.3	CH₂Cl–C(=O)–OH	174.7
(CH₃)₂CH–C(=O)–OH	184.8	CHCl₂–C(=O)–OH	170.6
cyclopentyl–C(=O)–OH	180.7	CCl₃–C(=O)–OH	168.0
CH₂=CH–C(=O)–OH	173.2	CF₃–C(=O)–OH	166.0
maleic acid (cis, COOH/COOH)	167.1	norbornane-COOH	181.4
fumaric acid (trans, HOOC/COOH)	166.6	norbornane-COOH	180.5

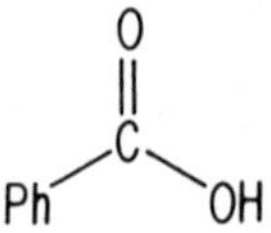

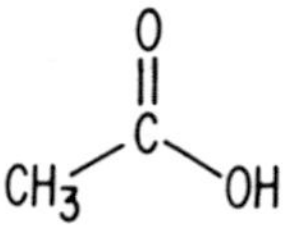

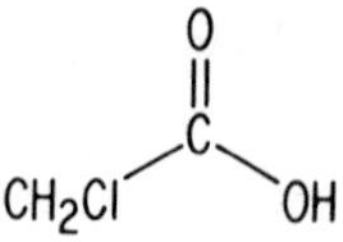

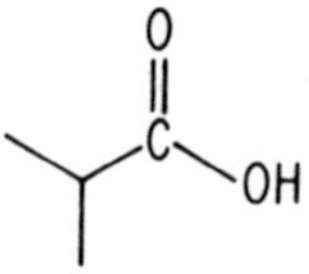

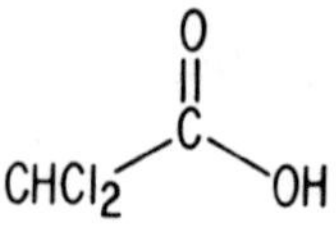

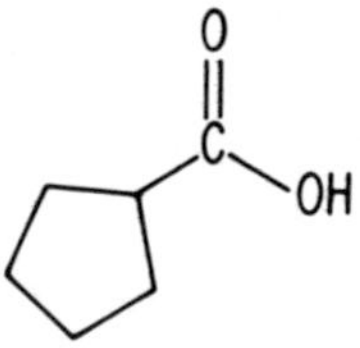

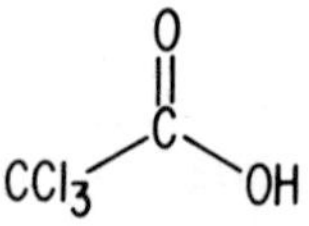

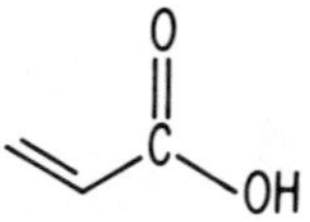

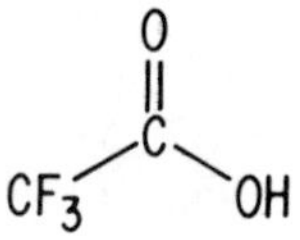

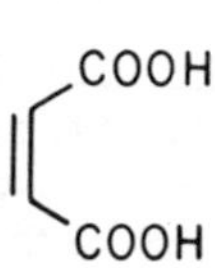

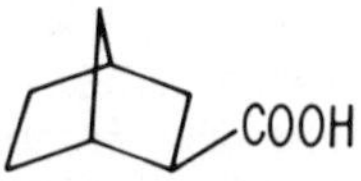

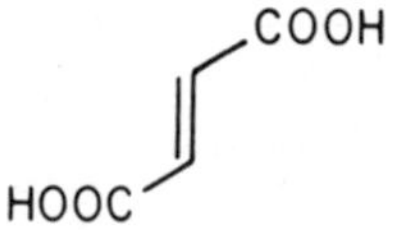

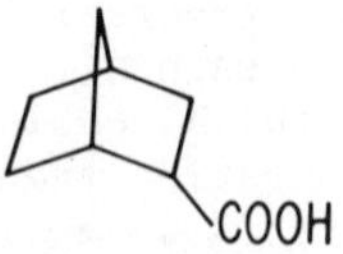

146

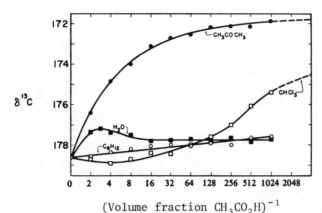

FIGURE 5.2.[24]   *Carbon-13 chemical shifts of $CH_3{}^{13}CO_2H$ versus (volume fraction of acetic acid)$^{-1}$ in the solvents acetone, water, cyclohexane, and chloroform.*

PROBLEM 5.2.   *A compound whose molecular formula is $C_8H_{12}O_2$ has the following ^{13}C spectrum. Deduce the structure and assign each carbon resonance to the appropriate carbon or carbons.*

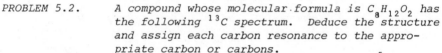

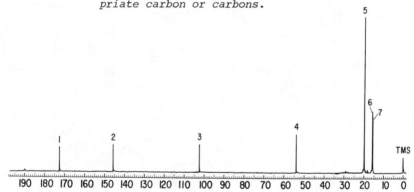

Table 5.11 presents a number of examples.[26] Effects of alkyl and 3-aryl substitution have been reported.[26b,c]

<u>Acid Halides.</u>   Acid chloride carbonyls experience a shielding of 4 to 8 ppm compared with the corresponding carboxylic acid. For acid bromides and iodides, further shieldings of some 5 and 10 ppm, respectively, are observed. Table 5.12 provides illustrative data.

TABLE 5.10. ^{13}C CHEMICAL SHIFTS (IN PARTS PER MILLION) OF CARBONYL CARBONS IN CARBOXYLIC ESTERS[6,8,22,25]

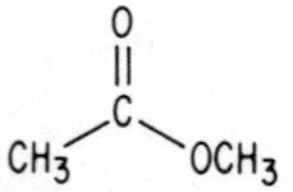

Structure	Shift	Structure	Shift	Structure	Shift
CH$_3$–CO–OCH$_3$	171.0	Ph–CO–OCH$_3$	167.0	CH$_3$–CO–CO–OEt (*)	161.6
CH$_3$–CO–OEt	170.3	Ph–CO–OEt	164.9	CH$_3$–CO–CH$_2$–CO–OEt (*)	167.7
CH$_3$–CO–O–iPr	169.8	norbornane–COOCH$_3$	175.7	CH$_3$–CO–CH$_2$–CH$_2$–CO–OEt (*)	173.9
CH$_3$–CO–O–cyclopentyl	169.9	norbornane–COOEt	174.6	β-propiolactone	171.2
CH$_3$–CO–O–CH=CH$_2$	167.7	EtOCO–CH=CH–COOEt (trans)	163.3	γ-butyrolactone	178.0
CH$_3$–CO–O–C(=CH$_2$)CH$_3$	168.9	EtOCO–CH=CH–COOEt (cis)	164.8	δ-valerolactone	175.2
CH$_2$=CH–CO–OCH$_3$	164.5	EtOCOCH$_2$–CH$_2$–COOEt	170.8	dimethyl-δ-valerolactone (CH$_3$, CH$_3$)	169.8
		H–CO–OEt	160.7		

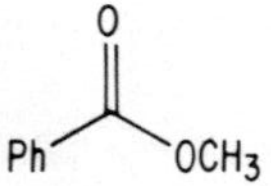

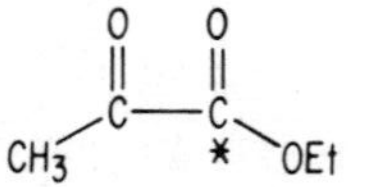

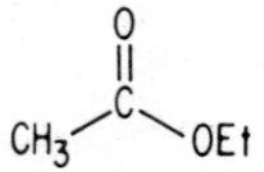

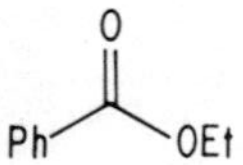

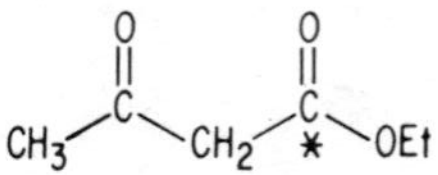

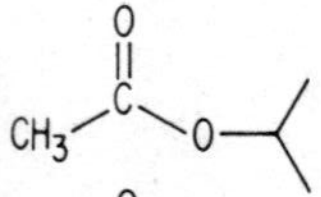

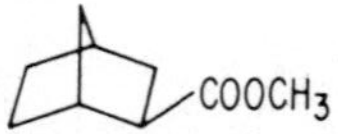

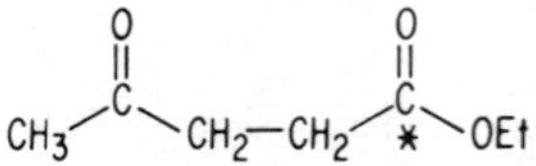

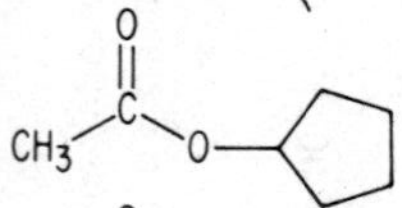

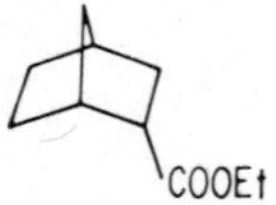

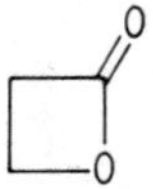

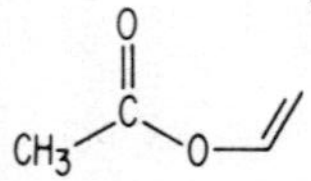

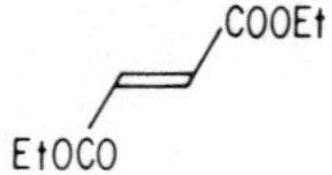

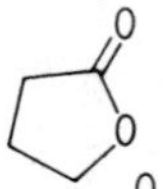

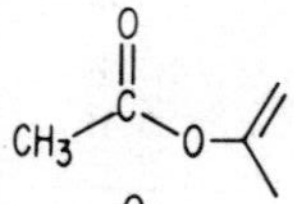

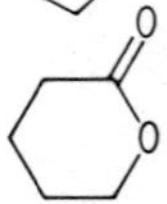

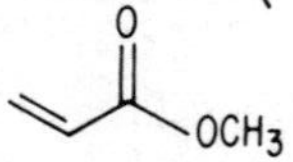

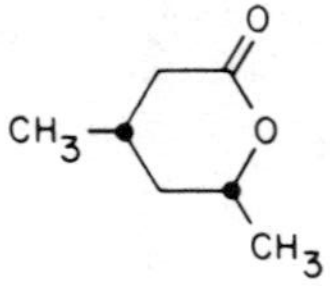

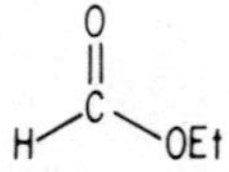

148

TABLE 5.11. ^{13}C CHEMICAL SHIFTS (IN PARTS PER MILLION) OF
CARBONYL CARBONS IN CARBOXYLIC ACID ANHYDRIDES[26]

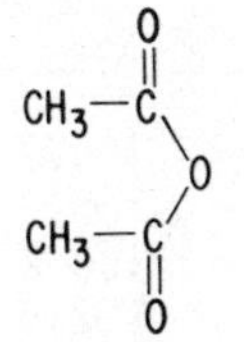	167.3	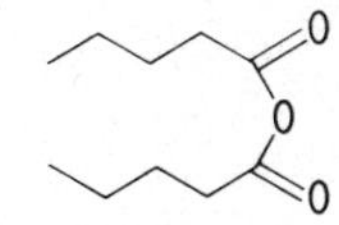	169.6	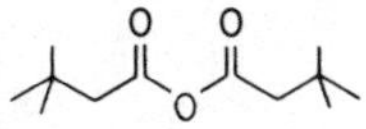	174.0
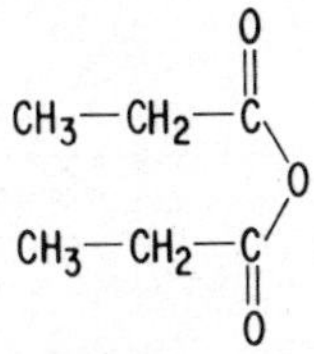	170.8	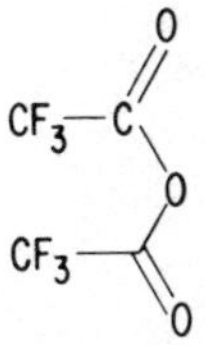	151.5	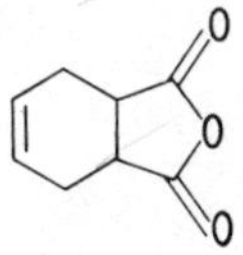	175.3
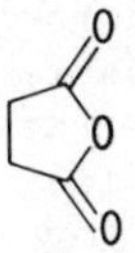	173.1	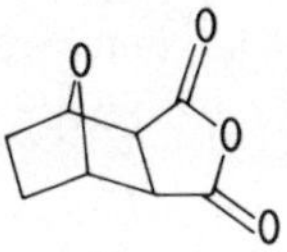	172.9	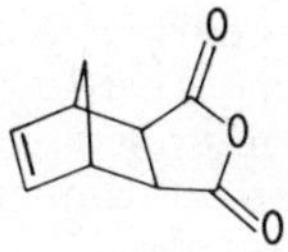	172.4
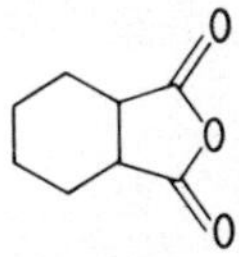	174.0	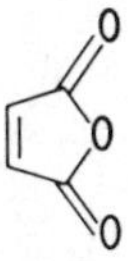	165.9	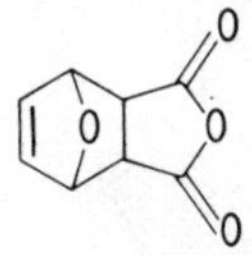	171.7
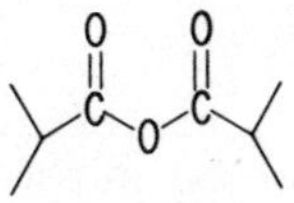	172.8	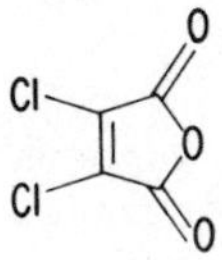	158.6	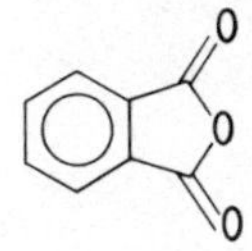	163.6
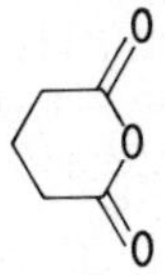	168.2	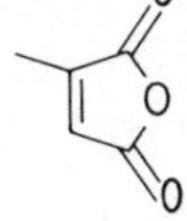	167.3 165.1	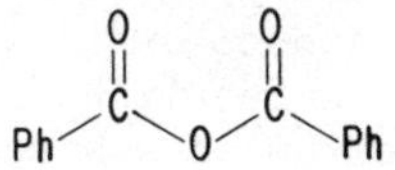	162.8

TABLE 5.12. ^{13}C CHEMICAL SHIFTS (IN PARTS PER MILLION) OF CARBONYL CARBONS IN CARBOXYLIC ACID HALIDES[8,27b,28]

CH_3COCl	170.5	PhCOCl	168.7
CH_3COBr	166.5	$ClCH_2COCl$	169.7
CH_3COI	159.8	$Cl_2CHCOCl$	166.2
CH_3CH_2COCl	174.5	Cl_3CCOCl	163.5
$(CH_2CH_2COCl)_2$	174.3		

Amides. The carbonyl resonances of amides are found in the region 160 to 180 ppm (Table 5.13).[29] The conformation of the substituent(s) on nitrogen relative to the carbonyl group has a significant effect on the carbonyl chemical shift--for example, in N-methylformamide.[29b] A difference of 3.3 ppm is observed for

the carbonyl resonances of the two isomers. This is a reflection of both steric and anisotropic differences in the environment of the two methyl groups. Comparable effects are experienced by α-amido substituent carbons. Smaller effects are noted further from the amide linkage. Figure 5.3[29b] illustrates this fact with the spectrum of N,N-di-n-butylformamide. Similar behavior has been observed for N-nitrosamines.[27]

An extensive compilation of N,N-dialkylamide chemical shifts, with N-alkyl groups ranging from 1 to 12 carbons, has been reported.[31] The carbonyl values can be classified conveniently according to the parent acid (Table 5.14).

Benzamide carbonyl shifts lie at 169 to 170 ppm and are independent of para substitution.[32] Replacement of amide protons by deuteria results in an isotope shift of the carbonyl carbon that can allow assignment of these resonances in oligopeptides.[33]

TABLE 5.13.   [13]C CHEMICAL SHIFTS (IN PARTS PER MILLION) FOR THE CARBONYL CARBONS OF AMIDES AND LACTAMS[29]

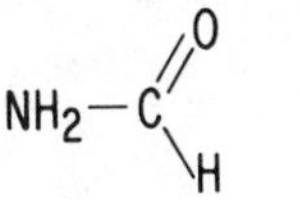

Structure	Shift	Structure	Shift
NH_2–C(=O)H	165.5	CH_3–C(=O)NH_2	172.7
CH_3(H)N–C(=O)H	163.4	CH_3–CH_2–C(=O)NH_2	177.2
H(CH_3)N–C(=O)H	166.7	nBu(nBu)N–C(=O)CH_3	169.1
CH_3(CH_3)N–C(=O)H	162.7	CH_3–C(=O)–NH–CH_3	171.6
nBu(nBu)N–C(=O)H	162.3	C_6H_5(CH_3)N–C(=O)H	162.1
caprolactam	178.6	C_6H_5–C(=O)–N(CH_3)CH_3	170.8
CH_2=C(CH_3)–C(=O)NH_2	170.3	C_6H_5–NH–C(=O)CH_3	168.3
3-pyrrolin-2-one	175.9	4-NO_2-C_6H_4–NH–C(=O)CH_3	169.2
1,4-diphenylazetidin-2-one (C6H5, N–C6H5)	166.2		

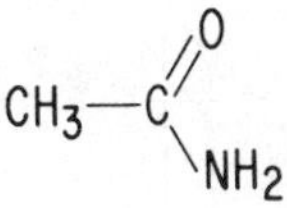

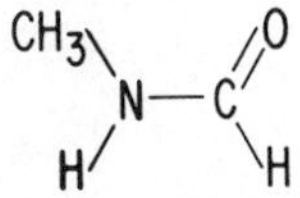

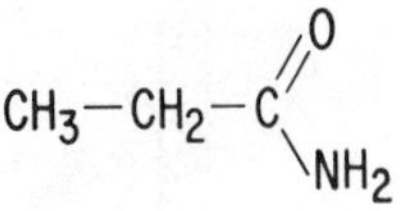

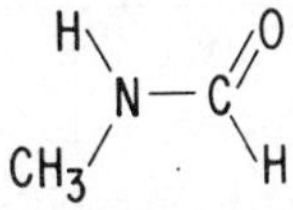

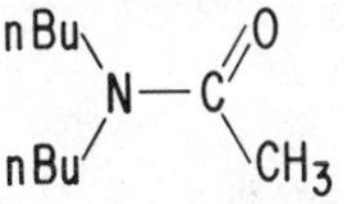

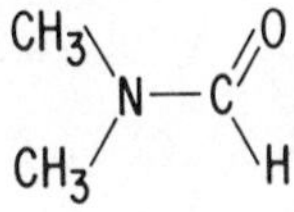

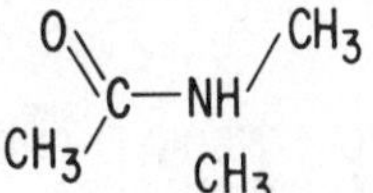

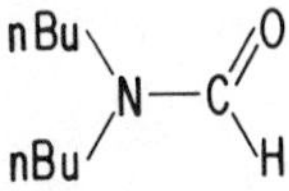

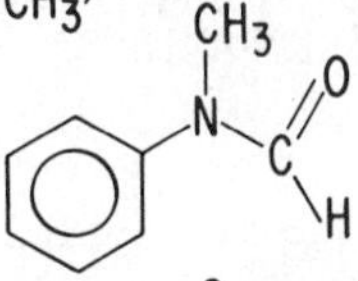

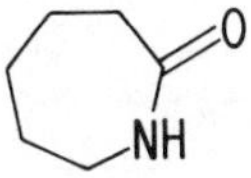

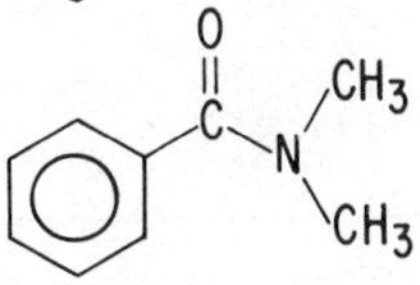

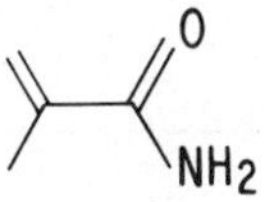

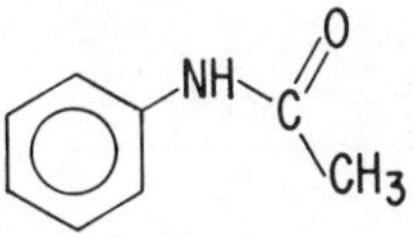

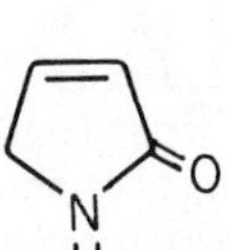

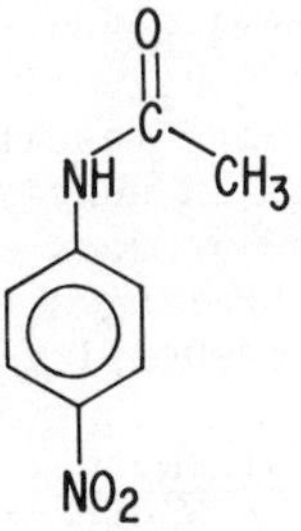

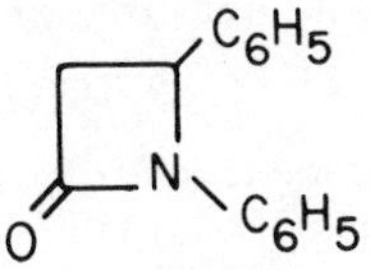

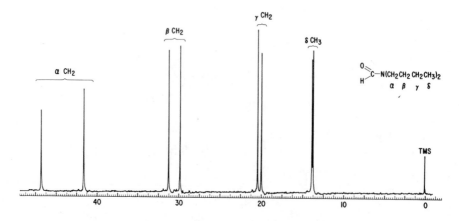

FIGURE 5.3.[29b]   *Carbon-13 nmr spectrum of N,N-di-n-butylforma-*
mide (aliphatic region shown).

TABLE 5.14. ^{13}C CARBONYL CHEMICAL SHIFT RANGES (IN PARTS PER MILLION) OF *N,N*-DIALKYLAMIDES[31]

$$R-CO-NR_2'$$

R	δ	
	CDCl$_3$	C$_6$D$_6$
H	162.2-162.8	161.6-162.2
CH$_3$	169.6-170.4	168.6-169.3
CH$_3$CH$_2$	173.2-173.5	171.6-172.4
CH$_3$CH$_2$CH$_2$	172.2-172.7	170.9-171.4

Imides.   The carbonyl resonances of imides fall in the range 160 to 180 ppm (Table 5.15).   Incorporation of the imide linkage into a five-membered ring results in the expected deshielding. *N*-Substitution results in shielding for both aliphatic and aromatic substituents, as does introduction of unsaturation.   An *N*-phenyl substituent bearing a *para* electron-withdrawing group has no effect on the carbonyl chemical shift.

Ureas, Uracils, and Carbamates.   The replacement of the alkyl group of an amide by an additional amino function to yield a urea results in a further shielding   (Table 5.16).   Carbamate

TABLE 5.15.   IMIDE CARBONYL CHEMICAL SHIFTS (IN PARTS PER MILLION)[34]

Structure	Shift	Structure	Shift
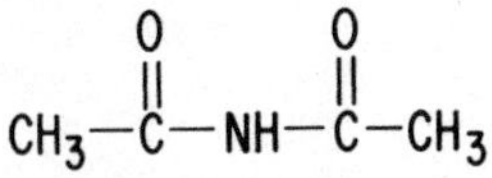 $CH_3-C(=O)-NH-C(=O)-CH_3$	171.8	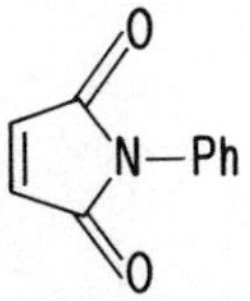 maleimide N–Ph	170.0
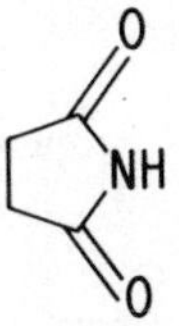 succinimide NH	179.1	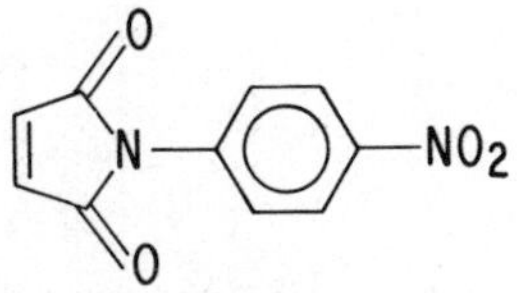 maleimide N–C₆H₄–NO₂	170.0
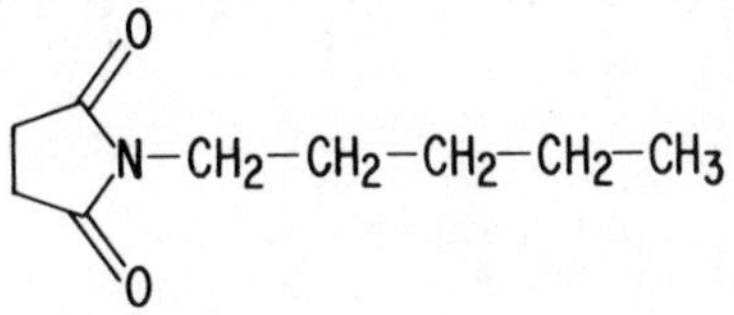 succinimide N–CH₂–CH₂–CH₂–CH₂–CH₃	177.5	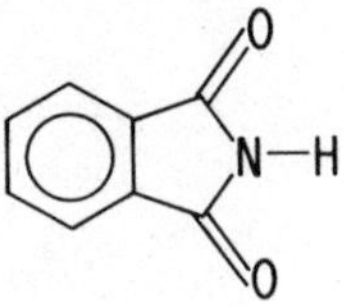 phthalimide N–H	169.7
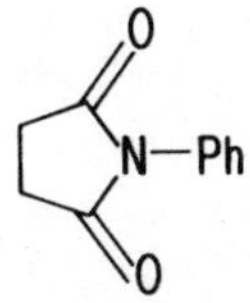 succinimide N–Ph	177.9	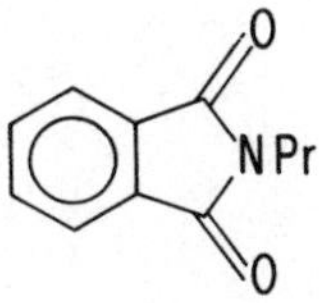 phthalimide N Pr	168.5
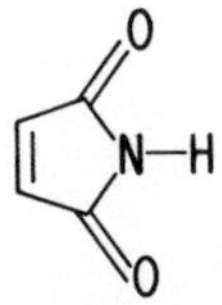 maleimide N–H	172.5	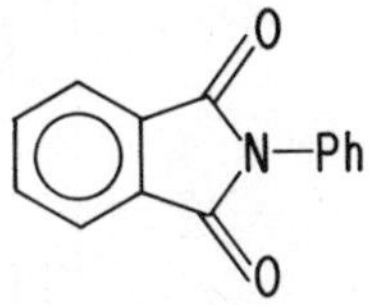 phthalimide N–Ph	167.7

linkages appear at slightly higher field (157.8 for methylethyl-carbamate and 154.0 for isopropylphenylcarbamate).[36]

In the case of uracil, thymine, and 5-halouracils, the behavior of the two different carbonyl carbons is that expected on the basis of substitution (Table 5.17).[37] A similar situation exists for barbituric acids and related compounds (Table 5.18).[38] For chiral substitution at C-5, C-4 and C-6 are magnetically nonequivalent by ∿0.3 ppm.[38]

PROBLEM 5.3.   A compound whose molecular formula is $C_6H_6N_2O$
gave the following ^{13}C spectrum. Determine the
structure and assign the six resonance lines.

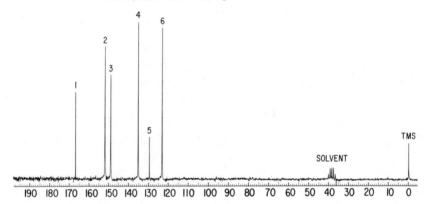

PROBLEM 5.4.   The compound whose ^{13}C spectrum follows has
a molecular formula C_8H_9NO. There are two
isomers present (1, 2, 3, 5, 8, and 10 are
resonances of the predominant isomer). Deduce
the structure and predominant isomer of the
compound and assign all resonances to the
appropriate carbon(s).

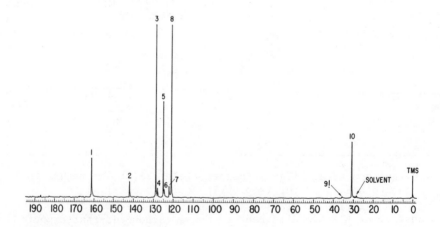

TABLE 5.16.   CARBONYL CHEMICAL SHIFTS (IN PARTS PER MILLION) OF UREAS.[35]

$H_2N-C-NH_2$ (C=O)	161.3	$PhNHCNH_2$ (C=O)	156.6
CH_3NHCNH_2 (C=O)	159.8	$PhNHCNHPh$ (C=O)	152.6
$CH_3NHCONHCH_3$	159.5	$H_2N-C-NH-C-NH_2$ (C=O, C=O)	155.5
$(CH_3)_2NCON(CH_3)_2$	164.8	biuret	

TABLE 5.17.   [13]C CHEMICAL SHIFTS (IN PARTS PER MILLION) FOR CARBONYL CARBONS IN SUBSTITUTED URACILS[37]

	X	CARBON 4	CARBON 2
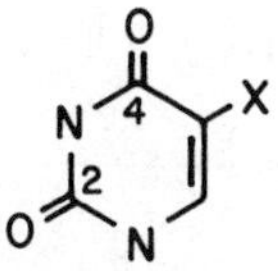	H	165.3	152.9
	CH_3	165.7	152.3
	F	158.9	150.9
	I	162.4	152.1

Carbonates.   Carbonate carbonyl resonances are quite insensitive to substitution and appear at 150 to 160 ppm (Table 5.19).[34a,36a,39]

Coupling.   Routine coupling information for carbonyl carbons can be obtained most easily for formyl carbons, the only carbonyls possessing directly attached protons.   The $^1J_{CH}$ values for formyl derivatives cover a wide range,[40a] as illustrated in Tables 5.20 and 5.21.

The $^1J_{CH}$ value for substituted formyl compounds has been found to correlate with C-H stretching frequencies in infrared spectra as well as σ_I and σ^* substituent constants.[41]   For aldehydes, it has been shown that two bond couplings, $^2J_{CH}$, between formyl hydrogens and carbons α to the carbonyl group are quite large (Table 5.21).   The $^2J_{CH}$ value between carbonyl

	R	C_2	C_4	C_6		
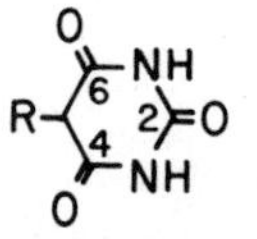 (barbituric acid, NH/NH)	H	152.2	168.5	168.5	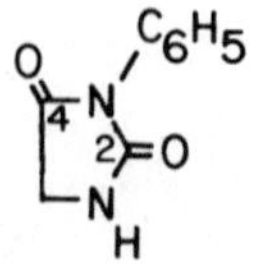 (C_6H_5 hydantoin)	157.6 (C-2)[b]
	CH_3	150.9	170.8	170.8		172.8 (C-4)
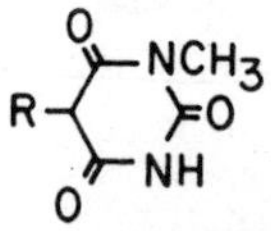 (NCH_3/NH)	H	152.2	167.3	166.9		
	CH_3	151.3	170.1	169.2		
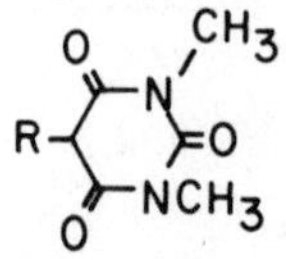 (NCH_3/NCH_3)	H	152.5	166.0	166.0		
	CH_3	151.9	169.2	169.2		

[a]$2M$ in DMSO-d_6.
[b]In morpholine; reported to be shielded by ~3 ppm in DMSO.[38f]

TABLE 5.19.   [13]C SHIFTS (IN PARTS PER MILLION) FOR CARBONYL
CARBONS OF CARBONATES[34a,36a,39]

DIMETHYL CARBONATE	156.5	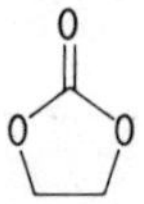 (ethylene carbonate) 156.7	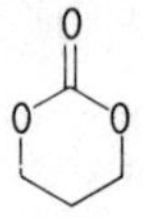 (propylene/trimethylene carbonate) 155.2
DIETHYL CARBONATE	155.8		
DIPHENYL CARBONATE	152.0	$CH_3CH_2\,\underset{CH_3}{CHO}\overset{O}{\overset{\|}{C}}-OCH_3$	155.6
DIBUTYL CARBONATE	155.5		

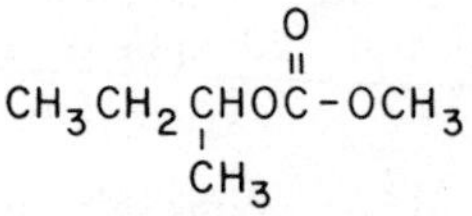

TABLE 5.20.   DIRECT ^{13}C-H COUPLING CONSTANTS (IN HERTZ) FOR FORMYL CARBONS[40]

$$H-\overset{\overset{\displaystyle O}{\|}}{C}-Ph \quad 173.7 \qquad\qquad H-\overset{\overset{\displaystyle O}{\|}}{C}-N(CH_3)_2 \quad 191.2$$

$$H-\overset{\overset{\displaystyle O}{\|}}{C}-O_{(aq.)}^{\ominus} \quad 194.8 \qquad\qquad H-\overset{\overset{\displaystyle O}{\|}}{C}-OH \quad 222.0$$

$$H-\overset{\overset{\displaystyle O}{\|}}{C}-H \quad 172.0 \qquad\qquad H-\overset{\overset{\displaystyle O}{\|}}{C}-OCH_3 \quad 226.2$$

$$H-\overset{\overset{\displaystyle O}{\|}}{C}-CH_3 \quad 172.4 \qquad\qquad H-\overset{\overset{\displaystyle O}{\|}}{C}-F \quad 267.0$$

PROBLEM 5.5[36b]   *The FT ^{13}C spectrum and structure for carnosin are given here. Assign each resonance to the appropriate carbon (answer, ref. 36b).*

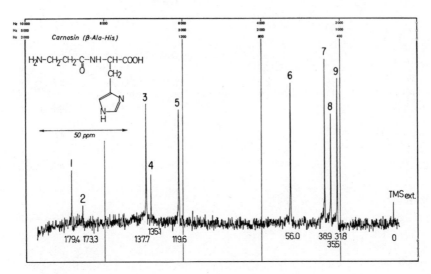

157

carbons and protons on adjacent carbons is usually 6 to 7 Hz, whereas $^1J_{CC}$ varies over a wide range (Table 5.22).[28]   Carbon-proton coupling across a carbonyl has been observed.   The long-range coupling $^3J_{CH}$ in acetone is 1.5 Hz.[42]   Long-range C-H couplings have been used for conformational analysis.[43]

TABLE 5.21.   DIRECT AND TWO-BOND ^{13}C-H COUPLINGS (IN HERTZ) FOR THE FORMYL PROTON IN SOME ALIPHATIC ALDEHYDES[28]

COMPOUND	$^1J(^{13}CH)$	$^2J(^{13}C-CH)$
$C_2H_5CH_2CHO$	170.3	24.8
$C_2H_5CHClCHO$	183.9	26.5
$C_2H_5CHBrCHO$	183.3	27.6
$C_2H_5CHICHO$	183.2	28.9
CH_3CHO	172.0	26.6
$ClCH_2CHO$		32.5
Cl_2CHCHO	198.0	35.8
Cl_3CCHO	207.0	46.3
Br_3CCHO		51.6

TABLE 5.22. ^{13}C COUPLINGS (IN HERTZ) FOR $C_AH_3C_BOX$ COMPOUNDS[40b]

SUBSTITUENT X	$^1J_{C_AC_B}$	$^2J_{C_BH_A}$
CH_2CH_3	38.4	
H	39.4	6.6
CH_3	40.1	5.9
C_6H_5	43.3	6.2
I	46.5	7.5
$O^-Na^+(aq)$	51.6	5.82
$N(CH_3)_2$	52.2	5.95
Br	54.1	7.5
Cl	56.1	7.45
OH	56.7	6.70
OC_2H_5	58.8	6.79

The $^1J_{CC}$ values in carboxylic acids vary between 54 and 59 Hz; however, two-bond values in saturated compounds do not exceed 1.8 Hz, and $^3J_{CC}$ displays a crude Karplus type of dependence on dihedral angle.[44,45a]   Similar trends are observed in unsaturated fragments.[45b]   Some of these values are summarized

in Table 5.23.   Carbon-carbon couplings have been explored for
use in peptide conformational analysis.[46]

TABLE 5.23. ^{13}C-^{13}C COUPLING CONSTANTS (IN HERTZ) FOR
CARBOXYLIC ACIDS[45]

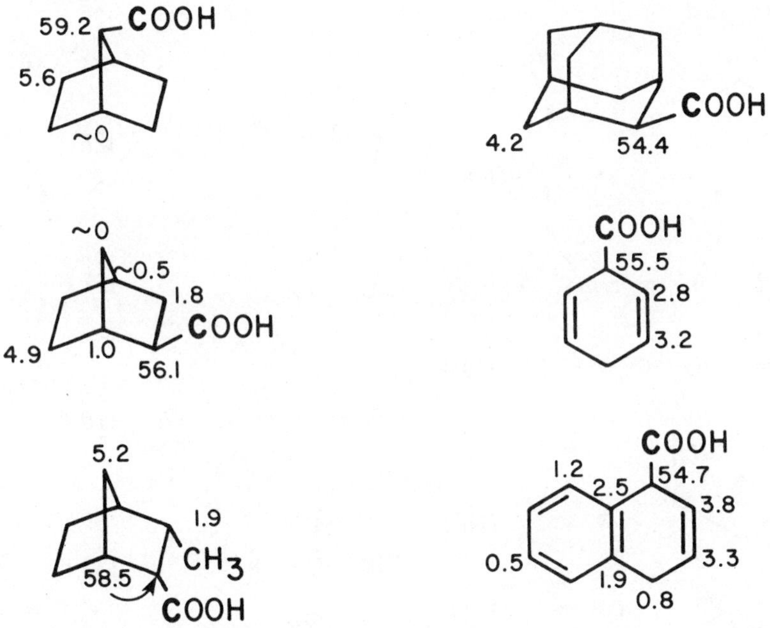

OTHER CARBON-CONTAINING FUNCTIONAL GROUPS

There are several other important carbon-containing functional groups that should be considered. Descriptions of their
cmr behavior follow.

Nitriles.   Carbon-13 chemical shifts for organonitrile carbons
appear at 112 to 126 ppm. Alkyl substitution at the α-position
induces deshielding   of perhaps 3 ppm. In general, aliphatic
nitrile carbons are not particularly sensitive to substitution
and only moderately sensitive to solvent effects. Table
5.24[6a,10a,36a,47] presents illustrative chemical-shift data.

TABLE 5.24. ^{13}C CHEMICAL SHIFTS (IN PARTS PER MILLION) OF NITRILE CARBONS[6a,10a,36a,47a]

CH_3CN	117.7	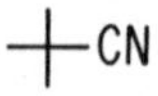 ⊢CN	125.1
 CH₂=CH—CN (allyl)	117.5	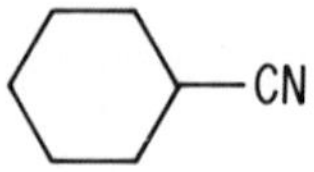 cyclohexyl—CN	121.5
 CH₂=CH—CH₂—CN	117.6	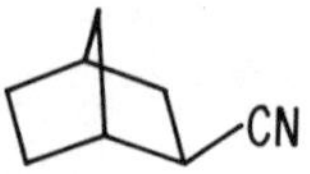 norbornyl—CN	123.4
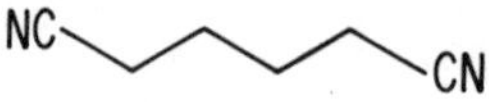 NC—(CH₂)₄—CN	119.3		
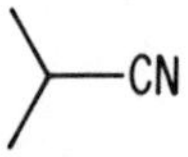 (CH₃)₂CH—CN	123.7	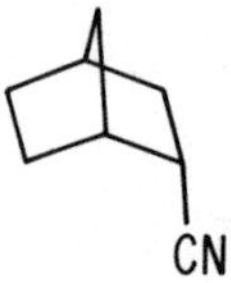 norbornyl—CN	122.6
Ph—CH₂—CN	118.0		
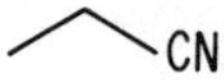 CH₃CH₂—CN	120.8	PhO—CH₂—CH₂—CN	118.4
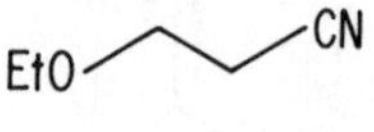 EtO—CH₂CH₂—CN	118.2	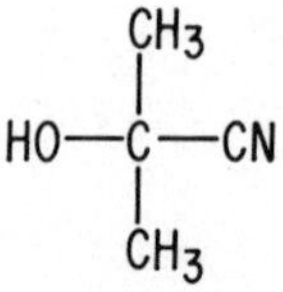 HO—C(CH₃)₂—CN	123.5
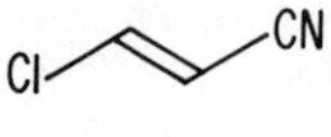 Cl—CH=CH—CN	114.4		
Ph—CN	118.7	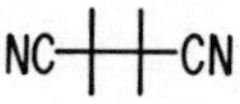 NC—⊣⊢CN	123.5
p—F—PhCN	114.1	N(CH₂CN)₃ (nitrilotriacetonitrile)	115.1

The nitrile carbon of benzonitrile is sensitive to *meta* and *para* substitution, but not in a systematic manner[48] except when solvent effects are accounted for (with polar substituents, it is necessary to use highly dilute solutions in non-polar solvents[49]). Solvent effects are not actually large. Benzonitrile absorbs at 117.6 ppm in CCl$_4$ solution but at 119.1 ppm in trifluoroacetic acid solution.[10a] Cyanide ion (aqueous KCN),[43a] on the other hand, is considerably deshielded, absorbing at 168.5 ppm.

From proton-coupled ^{13}C spectra for acetonitrile, $^2J_{CH}$ is 9.9 Hz and J_{CC} is 56.5 Hz. For isobutyronitrile these values become 9.2 and 54.8 Hz, respectively.[47b]

Isocyanides.   The chemical shifts of isocyanides are more like cyanide ion than nitriles:[50]

CH$_3$NC	157.5
CH$_3$CH$_2$NC	157.8
(CH$_3$)$_3$CNC	156.2

Oximes, Hydrazones, and other Carbonyl Derivatives.   The carbon resonances of ald- and ketoximes appear in the region 145 to 163 ppm, as illustrated in Table 5.25.[29a,51] Such resonances are some 50 ppm to higher shielding relative to the corresponding carbonyl resonances. Where substitution is asymmetric, differences in oxime carbon chemical shift are observed, depending on the conformation of the oxime N-OH. Such conformational isomerism also has a profound effect on the chemical shift of the α-carbon.

The ^{13}C spectrum of acetone oxime has three resonances, the derivatized carbonyl carbon at 154.4 ppm and the two non-equivalent CH$_3$ groups at 21.5 and 14.7 ppm. The difference between the chemical shifts of the two methyl groups, 6.8 ppm, is primarily a "steric" shift. This is clearly indicated in the ^{13}C spectrum of methylethylketone oxime (Figure 5.4),[29b] where the two oxime substituents are not sterically identical and thus the two isomers are not present in equal amounts. In methylethylketoxime the observed isomer ratio is 77:23 in favor of the isomer with the N-OH group facing the smaller methyl group.

Table 5.26 summarizes the ranges of chemical shifts displayed by imines, phenylhydrazones, imino ethers, semicarbazones, amidines, and oximes.[51b] The geometrical isomerism possible in these compounds may also be reflected in the ^{13}C spectra of these compounds.[52a,b] The imino carbon shifts of benzilidene benzylamines are reported to correlate with Hammett σ_R values.[52c]

TABLE 5.25. ^{13}C CHEMICAL SHIFTS (IN PARTS PER MILLION) OF
OXIME CARBONS[29b],[51]

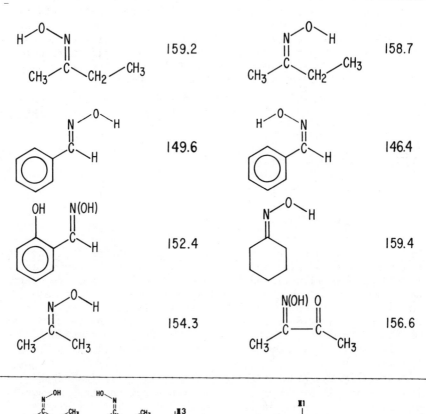

H-O-N=C(CH₃)-CH₂-CH₃	159.2
N-O-H, C(CH₃)-CH₂-CH₃	158.7
N-O-H benzaldehyde oxime	149.6
H-O-N benzaldehyde oxime	146.4
OH N(OH) salicylaldoxime	152.4
N-O-H cyclohexanone oxime	159.4
N-O-H (CH₃)₂C=N	154.3
N(OH) O, CH₃-C-C-CH₃	156.6

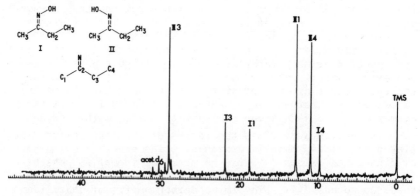

FIGURE 5.4.[29b]   *Alkyl region ^{13}C spectrum of methylethylketone*
oxime δ(I)C=N, 159.2; δ(II)C=N, 158.7 ppm.

TABLE 5.26.   RANGE OF ^{13}C CHEMICAL SHIFTS (IN PARTS PER MILLION) FOR IMINE FUNCTIONAL GROUPS[51b]

Imines	168-175 (aliphatic) 157-163 (aromatic)[a]	Oximes	154-159 (aliphatic) 148-155 (aromatic)[a]
Imino ethers	153-156 (aromatic)	Semi-carbazones	158-160 (aliphatic)
Amidines	149-154 (aliphatic) 149-151 (aromatic)	Phenyl-hydrazones	145-149 (aliphatic) 145-146 (aromatic)

[a]Includes ethylenic compounds.

Carbodiimides and other Linear Carbon Compounds.   The central sp-hybridized carbon of carbodiimides is analogous to that of allene and of ketenimines. However, the chemical shift of this carbon in the two values shown[53] is displaced considerably to higher field (see Table 3.32), even compared with the more nearly

(CH$_3$)$_2$CHN=C=NCH(CH$_3$)$_2$ 〈 〉—N=C=N—〈 〉   RR'C=C=NR"

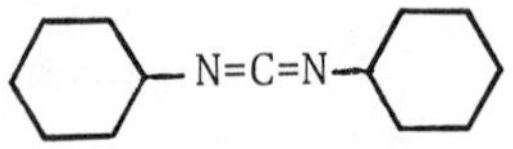

140.2 ppm   139.9 ppm   ∿190 ppm

analogous ketenimines.[54] The chemical shift resembles that of the linear CO$_2$ (δ_C = 125 ppm)[55] and of the central sp-carbon in cumulenes (δ_C = 120 to 150 ppm).[56] Strikingly, the central carbon of carbon suboxide, C$_3$O$_2$, is one of the most shielded carbon atoms known, with δ_C = -14.6 ppm.[57]

Diazo Compounds and Ketenes.   The terminal carbons of these compounds are unusually shielded, in contrast to ordinary alkenes (Table 3.27).[55,58,59] A relationship with enhanced electron density as implied by the dipolar resonance structures given has been suggested.

CH$_2$=N̅=N̅ ⟷ C̅H$_2$-N̅≡N   CH$_2$=C=O ⟷ C̅H$_2$-C≡O̅

23.3   2.7 194.6

(Chemical shifts in ppm)

(C$_6$H$_5$)$_2$C=N̅=N̅   (C$_6$H$_5$)$_2$C=C=O

62.3   48.6 201.4

Isocyanates.   The ^{13}C chemical shifts of isocyanate carbons appear at 118 to 132 ppm (Table 5.27). Conjugation results in deshielding of isocyanate resonances. The chemical shifts of the product of reaction with an alcohol (carbamate) are in the region 150 to 160 ppm (see section on ureas, pp. 152-153). Cyanate ion appears at slightly lower field.

TABLE 5.27.   CHEMICAL SHIFTS (IN PARTS PER MILLION) OF ISOCYANATE CARBONS[34a,36a,47a]

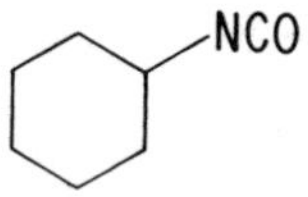

Cyclohexyl—NCO	123.6
Et NCO	122.5
CH₃ NCO	121.4
Ph NCO	129.5

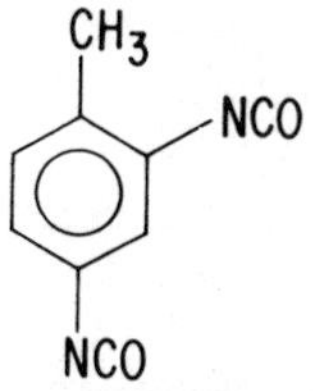

(2-methylphenyl, 4-NCO) —NCO	125.0
$CCl_3—\overset{O}{\overset{\|}{C}}—NCO$	130.2
$(K)\overset{\oplus}{N}CO^{\ominus}$	129.7

Thio- and Selenocarbonyls.   Equivalent thiocarbonyl and some selenocarbonyl compounds exist for many of the foregoing classes of carbonyl compounds. Thioamides and thioureas as well as other sulfur-containing functions such as thiocyanates and thioisocyanates are common. In general, the replacement of a carbonyl by a thiocarbonyl results in a deshielding of the nucleus. Carbon dioxide appears at 135 ppm and carbon disulfide, at 192.8 ppm. Table 5.28 lists the thio and seleno functional-group chemical shifts for several compounds. Comparative studies of carbonyl and thiocarbonyl compounds have been published,[61] and a correlation between the two has been suggested.[62]

TABLE 5.28.   [13]C CHEMICAL SHIFTS (IN PARTS PER MILLION) OF THIO- AND SELENO FUNCTIONAL-GROUP CARBONS[35a,36a,47a,54,60,61]

COMPOUND	X = S	X = Se	X = 0
CH_3CNH_2 (X=)	207.2		172.7
$CH_3CH(CH_3)_2$ (X=)	199.6	202.0	170.7
EtNHCNHEt (X=)	182.8		158.2
CH_3CNHPh (X=)	200.4		168.3
PhCNHCH$_3$ (X=)	199.7	204.6	168.8
H_2NCNH_2 (X=)	176.7		161.1
$CH_3NHCNHCH_3$ (X=)	182.7		159.5
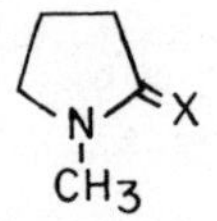 pyrrolidinone N-CH$_3$, =X	200.7	202.1	175.2
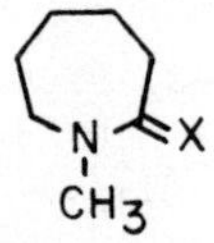 azepanone N-CH$_3$, =X	205.7	208.3	176.0
BuXCN	112.1		
CH_3XCN	113.5		
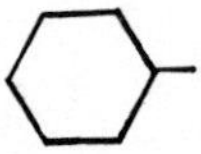 cyclohexyl—NCX	132.3		123.6

165

TABLE 5.28 (Continued).

	X = S	X = Se	X = O
EtNCX	130.7		122.5
CH_3NCX	128.6		121.4
PhNCX	135.7		125.2
K^+NCX^-	134.3		129.7

	X = S	X = Se
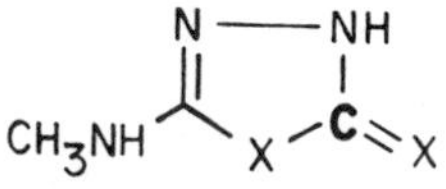	181.1	171.7

REFERENCES

1. P. C. Lauterbur, *J. Chem. Phys.*, **26**, 217 (1957).
2. G. E. Maciel, *ibid.*, **42**, 2746 (1965).
3. L. M. Jackman and D. P. Kelley, *J. Chem. Soc. (B)*, **1970**, 102.
4. G. E. Maciel, P. D. Ellis, and D. C. Hofer, *J. Phys. Chem.*, **71**, 2160 (1967).
5. F. J. Weigert and J. D. Roberts, *J. Am. Chem. Soc.*, **92**, 1338, 1347 (1970).
6. (a) J. B. Grutzner, M. Jautelat, J. B. Dence, R. A. Smith, and J. D. Roberts, *ibid.*, 7107; (b) J. B. Stothers and C. T. Tan, *Can. J. Chem.*, **53**, 581 (1975); (c) R. Bicker, H. Kessler, A. Steigel, and G. Zimmermann, *Chem. Ber.*, **111**, 3215 (1978).
7. D. H. Marr and J. B. Stothers, *Can. J. Chem.*, **45**, 225 (1967).
8. J. B. Stothers and P. C. Lauterbur, *ibid.*, **42**, 1563 (1964).
9. E. Lippmaa, T. Pehk, J. Paasivirta, N. Belikova, and A. Platé, *Org. Magn. Resonance*, **2**, 581 (1970).
10. (a) G. L. Nelson and G. C. Levy, unpublished results; (b) D. H. Marr and J. B. Stothers, *Can. J. Chem.*, **43**, 596 (1965).
11. E. V. Dehmlow, R. Zeisberg, and S. S. Dehmlow, *Org. Magn. Resonance*, **7**, 418 (1975).
12. (a) K. S. Dhami and J. B. Stothers, *Can. J. Chem.*, **43**, 479 (1965); (b) D. Leibfritz, *Chem. Ber.*, **108**, 3014 (1975); (c) K. S. Dhami and J. B. Stothers, *Can. J. Chem.*, **43**, 498 (1965).

13. (a) G. E. Maciel and J. J. Natterstad, *J. Chem. Phys.*, **42**, 2752 (1965); (b) G. E. Maciel and G. C. Ruben, *J. Am. Chem. Soc.*, **85**, 3903 (1963).

14. S. Veji and M. Nakamura, *Tetrahedron Lett.*, 2549 (1976).

15. W. H. de Jeu, *J. Phys. Chem.*, **74**, 822 (1970).

16. R. A. McClelland and W. F. Reynolds, *Can. J. Chem.*, **54**, 718 (1976).

17. (a) B. Tiffon and J. E. DuBois, *Org. Magn. Resonance*, **11**, 295 (1978); (b) T. T. Nakashima, D. D. Traficante, and G. E. Maciel, *J. Phys. Chem.*, **78**, 124 (1976); (c) A. Fratiello, R. Kubo, and S. Chow, *J. Chem. Soc., Perkin Transact. II*, **1976**, 1205; (d) R. Radeglia, *Z. Chem.*, **10**, 389 (1978).

18. (a) S. Berger and A. Rieker, *Tetrahedron*, **28**, 3123 (1972); (b) *Chem. Ber.*, **109**, 3252 (1976); (c) G. Höfle, *Tetrahedron*, **32**, 1431 (1976).

19. (a) I. A. McDonald, T. J. Simpson, and A. F. Sierakowski, *Aust. J. Chem.*, **30**, 1727 (1977); (b) A. Arnone, G. Fronza, R. Mondelli, and J. St. Pyrek, *J. Magn. Resonance*, **28**, 69 (1977).

20. W. Städeli, R. Hollenstein, and W. von Philipsborn, *Helv. Chim. Acta*, **50**, 948 (1977).

21. A. Mathias, *Tetrahedron*, **22**, 217 (1966).

22. P. A. Couperus, A. D. H. Clague, and J. P. C. M. van Dongen, *Org. Magn. Resonance*, **11**, 590 (1978).

23. R. Hagen and J. D. Roberts, *J. Am. Chem. Soc.*, **91**, 4504 (1969).

24. (a) G. E. Maciel and D. D. Traficante, *ibid.*, **88**, 220 (1966); (b) G. E. Maciel and D. D. Traficante, *J. Phys. Chem.*, **69**, 1030 (1965).

25. (a) M. Christl, H. J. Reich, and J. D. Roberts, *J. Am. Chem. Soc.*, **93**, 3453 (1971); (b) S. W. Pelletier, Z. Djarmati, and C. Pape, *Tetrahedron*, **32**, 995 (1976).

26. (a) G. L. Nelson, unpublished results; (b) F. J. Koer, A. J. de Hoog, and C. Altona, *Rec. Trav. Chim. Pays-Bas.*, **94**, 75 (1975); (c) F. J. Koer and C. Altona *ibid.*, **75**, 127 (1975).

27. (a) F. I. Carroll, G. N. Mitchell, J. T. Blackwell, A. Sobti, and R. Meck, *J. Org. Chem.*, **39**, 3890 (1974); (b) M. J. Gasic, Z. Djarmati, and S. W. Pelletier, *ibid.*, **41**, 1219 (1976).

28. G. Gray, P. D. Ellis, D. D. Traficante, and G. E. Maciel, *J. Magn. Resonance*, **1**, 41 (1969).

168   Organic Functional Groups

29. (a) G. C. Levy and G. L. Nelson, unpublished results;
(b) G. C. Levy and G. L. Nelson, *J. Am. Chem. Soc.*, **94**,
4897 (1972); (c) R. Buchi and E. Pretsch, *Helv. Chim.
Acta*, **58**, 1573 (1975); (d) G. Severini-Ricca, P. Manitto,
D. Monti, and E. W. Randall, *Gazz. Chim. Ital.*, **105**, 1273
(1975); (e) A. K. Bose and P. R. Srinivasan, *Org. Magn.
Resonance*, **12**, 34 (1979).
30. P. S. Pregosin and E. W. Randall, *Chem. Commun.*, **1971**, 399.
31. H. Fritz, P. Hug, H. Sauter, T. Winkler, and E. Logemann,
Org. Magn. Resonance, **9**, 108 (1977).
32. (a) R. G. Jones and J. M. Wilkins, *Org. Magn. Resonance*,
11, 20 (1978); (b) J. A. Lepoivre, R. A. Dommisse, and
F. C. Alderweireldt, *ibid.*, **7**, 422 (1975).
33. (a) J. Feeney, P. Partington, and G. C. K. Roberts, *J. Magn.
Resonance*, **13**, 268 (1976); (b) R. A. Newmark and J. R. Hill,
ibid., **21**, 1 (1976).
34. (a) G. L. Nelson, unpublished results; (b) S. Combrisson,
J. -P. Lautié, and M. Olomuczki, *Bull. Soc. Chim. Fr.*,
1975, 2769.
35. (a) H. O. Kalinowski and H. Kessler, *Org. Magn. Resonance*,
6, 305 (1974); (b) M. P. Sibi and R. L. Lichter, *J. Org.
Chem.*, **44**, 3017 (1979) and references cited therein.
36. (a) L. F. Johnson and W. C. Jankowski, *Carbon-13 Spectra*,
Wiley-Interscience, New York, 1972; (b) W. Voelter, G. Jung,
E. Breitmaier, and E. Bayer, *Z. Naturforsch.*, **26b**, 213 (1971);
(c) Y. Nomura, N. Masai, and Y. Takeuchi, *J. Chem. Soc.
Chem. Commun.*, **1975**, 307.
37. (a) A. R. Tarpley, Jr., and J. H. Goldstein, *J. Am. Chem.
Soc.*, **93**, 3573 (1971); (b) J. W. Triplett, G. A. Digenis,
W. J. Layton, and S. L. Smith, *Spectrosc. Lett.*, **10**, 141
(1977).
38. (a) A. Fratiello, M. Mardirossian, and E. Chavez, *J. Magn.
Resonance*, **12**, 221 (1973); (b) R. C. Long, Jr. and J. H.
Goldstein, *ibid.*, **16**, 228 (1974); (c) F. I. Carroll and
C. G. Moreland, *J. Chem. Soc., Perkin Transact. II*, **1974**,
374; (d) J. Okada and T. Esaki, *Yakugaku Zasshi*, **93**, 1014
(1973); *ibid.*, **95**, 56 (1975); *Chem. Pharm. Bull.*, **22**,
1580 (1974); (e) P. R. Srinivasan and R. L. Lichter,
unpublished results; (f) S. Icli and L. D. Colebrook, *Metu
J. Pure Appl. Sci.*, **9**, 39 (1976).
39. D. B. Bigley, C. Brown, and R. H. Weatherhead, *J. Chem.
Soc., Perkin Transact. II*, **1976**, 701.
40. (a) G. E. Maciel, J. W. McIver, Jr., N. S. Ostlund, and
J. A. Pople, *ibid.*, **92**, 1 (1970); (b) D. W. Ewing, *Org.
Magn. Resonance*, **3**, 279 (1971).
41. S. L. Rock and R. M. Hammaker, *Spectrochim. Acta*, **27A**, 1899 (1971).

42. A. R. Tarpley, Jr., and J. H. Goldstein, *Molec. Phys.*, **21**, 549 (1971).
43. (a) W. G. Espersen and R. B. Martin, *J. Phys. Chem.*, **80**, 741 (1976); (b) U. Vögeli, W. von Philipsborn, K. Nagarajan, and H. D. Nair, *Helv. Chim. Acta*, **61**, 607, (1978).
44. R. E. Wasylishen, *Annu. Rep. NMR Spectrosc.*, **7**, 245 (1977).
45. (a) J. L. Marshall and D. E. Miiller, *J. Am. Chem. Soc.*, **95**, 8305 (1973); (b) J. L. Marshall, L. G. Faehl, A. M. Ihrig, and M. Barfield, *ibid.*, **98**, 3406 (1976).
46. (a) R. E. London, T. E. Walker, V. H. Kollman, and N. A. Matwiyoff, *J. Am. Chem. Soc.*, **100**, 3723 (1978); (b) W. Haar, S. Fermandjiau, J. Vicar, K. Blaha, and P. Fromageot, *Proc. Nat. Acad. Sci. (USA)*, **72**, 4948 (1975).
47. (a) G. E. Maciel and D. A. Beatty, *J. Phys. Chem.*, **59**, 3920 (1965); (b) G. A. Gray, G. E. Maciel, and P. D. Ellis, *J. Magn. Resonance*, **1**, 407 (1969); (c) W. McFarlane, *Molec. Phys.*, **10**, 603 (1966).
48. (a) F. W. Wehrli, J. W. de Haan, A. I. M. Keulemans, O. Exner, and W. Simon, *Helv. Chim. Acta*, **52**, 103 (1969); (b) G. L. Lebel, J. D. Laposa, B. G. Sayer, and R. A. Bell, *Anal. Chem.*, **43**, 1500 (1971); compare with A. Colligiani, R. Ambrosetti, and L. Guibe, *J. Chem. Phys.*, **54**, 2105 (1971).
49. J. Bromilow, R. T. C. Brownlee, and D. J. Craik, *Aust. J. Chem.*, **30**, 351 (1977).
50. I. Morishima, A. Mizuno, T. Yonezawa, and K. Goto, *Chem. Commun.*, **1970**, 1321.
51. (a) G. L. Nelson, unpublished results; (b) N. Naulet, M. L. Filleux, G. J. Martin, and J. Pornet, *Org. Magn. Resonance*, **7**, 326 (1975).
52. (a) C. A. Bunnell and P. L. Fuchs, *J. Org. Chem.*, **42**, 2614 (1977); (b) R. R. Fraser, K. L. Dhawan, and K. Taymaz, *Org. Magn. Resonance*, **11**, 269 (1978); (c) J. E. Arrowsmith, M. J. Cook, and D. J. Hardstone, *ibid.*, **11**, 160 (1978).
53. F. A. L. Anet and I. Yavari, *Org. Magn. Resonance*, **3**, 327 (1976).
54. J. Firl, W. Runge, W. Hartman, and H. Utikal, *Chem. Lett.*, **41** (1975).
55. J. Firl and W. Runge, *Angew. Chem. Internat. Ed.*, **12**, 668 (1973).
56. J. P. C. M. van Dougen, M. J. A. de Bie, and R. Steur, *Tetrahedron Lett.*, **1973**, 1371.
57. E. A. Williams, J. D. Cargioli, and A. Evo, *J. Chem. Soc., Chem. Commun.*, **1975**, 366.
58. G. A. Olah and P. W. Westerman, *J. Am. Chem. Soc.*, **95**, 3706 (1973).
59. T. A. Albright and W. J. Freeman, *Org. Magn. Resonance*, **9**, 75 (1977).

60. M. L. Filleux-Blanchard, *Org. Magn. Resonance*, **9**, 125 (1977).
61. (a) I. W. J. Still, N. Plavac, D. M. McKinnon, and M. S.
     Chauhan, *Can. J. Chem.*, **54**, 280 (1976); (b) C. Piccinni-
     Leopardi, O. Fabre, D. Zimmerman, J. Reisse, F. Cornea, and
     C. Fulea, *ibid.*, **55**, 2649 (1977); *Org. Magn. Resonance*, **8**,
     536 (1976); (d) F. I. Carroll, A. Philip, and C. G. Moreland,
     J. Med. Chem., **19**, 521 (1976); (e) I. D. Rae, *Aust. J.
     Chem.*, **32**, 567 (1979); (f) J. R. Bartels-Keith, M. T.
     Burgess, and J. M. Stevenson, *J. Org. Chem.*, **42**, 3725 (1977).
62. (a) H. -O. Kalinowski and H. Kessler, *Angew. Chem. Internat.
     Ed. Engl.*, **13**, 90 (1974); (b) B. S. Pedersen, S. Scheibye,
     N. H. Nilsson, and S. -O. Lawesson, *Bull. Soc. Chim. Belg.*,
     87, 223 (1978).

Chapter 6
Ions, Radicals, and Complexes

The ^{13}C nmr properties of carbocations, carbanions, radicals, and organometallic complexes are examined in this chapter. Especially for carbocations, probably the species of this type most intensively studied by ^{13}C nmr, the discussion is necessarily selective. More detailed reviews are available.[1]

CARBOCATIONS

Olah and coworkers have applied ^{13}C nmr techniques to the study of carbocations. Indeed, much of the seminal work in the area has come from this group, whose publications provide many leading references.[2-4] A comprehensive survey of static C-3 to C-8 alkyl cations has been published,[2a] and other examples include haloacylium[2b] and diacetoacylium[3] ions and acyclic dications.[4]

Carbon-13 resonances for aliphatic carbocations are the lowest field absorptions known. The ^{13}C chemical shift of the cationic center in isopropyl cation is 320.6 ppm in SO_2ClF-SbF_5 solution.[2] Replacement of a hydrogen with a methyl group (*tert*-butyl cation) results in a chemical shift of 335.2 ppm,[2a] or a deshielding of 14.6 ppm. This α effect is somewhat larger than that observed in neutral alkanes (9.4 ppm). By contrast, the chemical shift of the 2-methyl-2-butyl cation is 335.4 ppm; that is, the β effect is < 1 ppm (for a series of such cations, the β effect is 0.8 ± 0.6 ppm). Despite the difference in magnitude from neutral alkanes, the methyl substituent effects are additive and have been used to predict shifts of equilibrating cations.[2a] In isopropyl cation, $^1J_{CH}$ = 171.3 Hz for the central carbon, indicative of its sp^2-hybridization.[2a,5-10]

Table 6.1 gives ^{13}C chemical shifts for cationic centers in a variety of static ions,[2a] which exhibit a wide range of values. If σ-delocalized structures are included, the chemical-shift range approaches 300 ppm. Chemical shifts and coupling constants of

171

adjacent carbons may also be expected to reflect the charge at the cationic center. However, these differences do not necessarily reflect the *magnitudes* of adjacent charges.[11] Indeed, despite attempts to use ^{13}C shifts of carbocations as probes of charge density, delocalization, thermodynamic stability, and so on, enough ambiguities still exist to cast doubt on the validity of these procedures.[12-13] A simple example is seen in comparison of the heats of formation of protonated ketones with ^{13}C shifts of the cationic centers. The ΔH_f values indicate that protonated

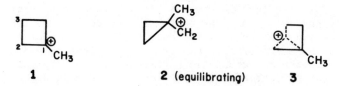

	$\left[CH_3-\underset{\substack{\| \\ OH}}{C}-CH_3\right]^+$	$\left[CH_3-\underset{\substack{\| \\ OH}}{C}-C_6H_5\right]^+$	$\left[C_6H_5-\underset{\substack{\| \\ OH}}{C}-C_6H_5\right]^+$	
δ_{C^+}	250.3	220.4	210.0	ppm
ΔH_f	-19.1	-18.9	-16.9	kcal/mole

acetone is the most stable ion and protonated benzophenone, the least. With the assumption that increased shielding of the indicated carbons implies greater charge delocalization (hence greater stability), the ^{13}C shifts of the ions would seem to imply the reverse order of stability. The same inference results if the difference in chemical shift between each neutral and protonated ketone is taken.[13] ^{13}C Shifts and electronic charge in π-delocalized cations, however, do correlate because charge and bond order terms are parallel and approximately equal.[14]

A very careful determination of the ^{13}C spectrum of the 1-methylcyclobutyl cation[15] (1) also underscores the need for caution in drawing inferences from ^{13}C data. Based on the observation of a single *averaged* peak at approximately 100°C for the methylene carbons of this cation, a rapidly equilibrating cyclopropylcarbinyl structure (2) or σ-delocalized species (3) had been

proposed.[16] However, at -145°C this peak separates into two peaks at 71.1 ppm (C-2) and at -3 ppm (C-3). Over the same temperature range the methyl and cationic carbon resonance positions remain unchanged ·at 25.1 and, strikingly, 162.0 ppm, respectively. From the temperature-dependent spectra and from estimates of shifts expected for 2, an equilibrium between 1 and 2 may exist, but it is dominated (>99.5%) by 1 and is characterized by an activation barrier ΔG = 5.8 kcal/mole. To rationalize the unusually high field

Structure	Shift	Conditions		Structure	Shift	Conditions
$(CH_3)_2\overset{+}{C}-Et$	335.4	-80°C(a)		$Ph_2\overset{+}{C}-Ph$	212.7	-60°C(c)
$(CH_3)_2\overset{+}{C}-CH_3$	335.2	-80°C(a)		$Ph_2\overset{+}{C}-H$	200.2	-60°C(c)
$(CH_3)_2\overset{+}{C}-H$	320.6	-80°C(a)		$CH_3-\overset{+}{C}(OH)(OH)$	196.2	-30°C(b)
cyclopropyl–$\overset{+}{C}(CH_3)_2$	281.5	-60°C(c)		$H-\overset{+}{C}(OH)(OH)$	177.0	-30°C(b)
$Ph-\overset{+}{C}(CH_3)_2$	255.7	-60°C(c)		$HO-\overset{+}{C}(OH)$	166.6	-50°C(b)
$(CH_3)_2\overset{+}{C}-OH$	250.3	-50°C(b)		$Ph-\overset{+}{C}=O$	154.6	
$(CH_3)(H)\overset{+}{C}-OH$	237.2	-50°C(b)		$CH_3-\overset{+}{C}=O$	152.0	
$(H)(H)\overset{+}{C}-OH$	223.8	-50°C(b)		$FCH_2\overset{+}{C}=O$	145.3(e)
cyclohexyl$\overset{+}{}$–OH	251.1	-115°C(d)		$ClCH_2\overset{+}{C}=O$	146.4(e)
$CH_3\overset{+}{C}NH$	109.6			$BrCH_2\overset{+}{C}=O$	147.1(e)
				$ICH_2\overset{+}{C}=O$	150.6(e)

[a]Solvent: $SO_2ClF-SbF_5$.
[b]Solvent: $SO_2-FSO_3H-SbF_5$.
[c]Solvent: SO_2-SbF_5.
[d]Ref. 15.
[e]Ref. 17.

Ions, Radicals, and Complexes

cationic carbon chemical shift, the possibility of sp^3-hybridization for this cation was tentatively suggested.[15]

π-Delocalized Carbocations.   The charge in the triphenylmethyl cation in Table 6.1 would be expected to be extensively delocalized throughout the aromatic system. The cmr behavior of *para*-substituted triphenylmethyl cations has been reported.[4,17] Generated from the parent carbinols, the cations show a linear relationship between the ^{13}C chemical shift at the central carbon (176.0 to 210.9 ppm over seven cations) and σ^+. The ^{13}C chemical shifts were interpreted in terms of charge densities. The charge densities derived from ^{13}C chemical shifts for triphenylmethyl cations correlate well with those calculated by the CNDO method, although the agreement is not quantitative.[17] Of particular interest is the fact that the ^{13}C shifts suggest that whereas the positive charge is considerably delocalized into the aromatic ring, the aromatic point of attachment shows a small buildup of negative charge. This is a feature predicted by CNDO and a general feature of substituted aromatics, as discussed in Chapter 4. However, the uncertainties outlined previously in drawing inferences about charge distribution from ^{13}C shifts should be kept in mind.

Protonation of benzene, naphthalene, and anthracene with SbF_5-FSO_3H at -80 to -120°C produces the corresponding cations **4-6** ("Wheland intermediates").[18] The ^{13}C chemical shifts of the static cations are given with the structures. The shifts are consistent with largely allylic charge distributions. Moreover, the spectra show no evidence for the presence of the valence isomeric bicyclo [3.1.0] hexenyl cations.[19]

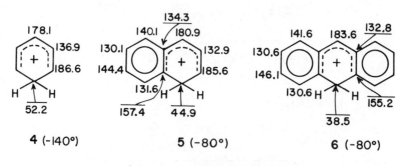

(chemical shifts in ppm)

Other reported aromatic cations include substituted cyclobutadiene (**7**),[20] cyclooctatetraene (**8**),[21] and benzocyclobutadiene (**9**)[22] dications. Representative shifts are given with the structures.

CH₃ CH₃
(+2)
18.8
CH₃ 209.7 CH₃

33.5
CH₃
182.7
CH₃ (+2) 170.0
CH₃
CH₃

27.9
178.8 CH₃
(+2) 186.9
169.7
CH₃
136.2

7 (-60°)   **8 (-70°)**   **9 (-80°)**

(chemical shifts in ppm)

Equilibrating or Bridged Ions.   Questions concerning σ-delocal-
ization and equilibrium processes have been of considerable interest in organic chemistry; cmr techniques have been applied to these
questions.

For the 2,3-dimethyl-2-butyl cation,[2a] the ^{13}C chemical shift
observed for the central carbons is 197.8 ppm ($^{1}J_{CH} = 67.5$ Hz). If
the spectrum is that of a rapidly equilibrating pair of ions, the
observed shift should be the average of the static shifts of the
two sites. A good model for estimation of these shifts is the
methyl ^{13}C shift and the central carbon shift of *tert*-butyl cation;

the average of these values is 191.1 ppm.[2a]   From studies on static
cations, additional methyl substituent effects may be applied,
leading to a corrected value of 201.7 ppm. Thus the ^{13}C chemical
shift observed is consistent with a rapidly equilibrating pair of
ions. The coupling constant can also be estimated by using *tert*-
butyl cation as a model; from direct and long-range coupling constants the estimate is 64 Hz.

For dimethyl-*tert*-butylcarbinyl cation[2a] a similar situation is
observed. The ^{13}C shift for the central carbon is 197.3 ppm, consistent with an equilibrating structure. Cyclopentyl cation[5] is a
particular example of a multiply degenerate rearrangement. The ^{13}C

INDOR spectrum shows a 10-line **multiplet** at 99.2 ppm ($^{1}J_{CH} =$
28.5 Hz). By using cyclopentane (25.6 ppm) and isopropyl cation
(320.6 ppm) as models, the average of isopropyl cation and four
methylene shifts is found to be 86.6 ppm; if allowance is made for
two of the methylenes being adjacent to a positive charge, 99 ppm
is consistent. Similarly, with isopropyl cation ($^{1}J_{CH} = 171$ Hz)
and cyclopentane ($^{1}J_{CH} = 131$ Hz) as models, the coupling constant

resulting from the complete scrambling of nine protons on five carbons can be calculated: $(171 + 8 \times 131)/(9 \times 5) = 27$ Hz.[11]

Because of the degenerate nature of the preceding rearrangements, the ^{13}C spectra are independent of temperature, providing
equilibration is rapid. In contrast, equilibration between non-
degenerate species yields temperature-dependent spectra. Each
rearrangement results in distinct cations, which generally are of
unequal thermodynamic stability. Hence the populations, and therefore the spectra, change with temperature. An example is 2,3-
dimethyl-2(3)-pentyl cation. From -50 to -130°C the C-2 resonance
changes from 198.6 to 175.6 ppm, whereas that for C-3 varies from
197.9 to 221.2 ppm. Thus the cationic character at C-3 increases

$$
\begin{array}{ccc}
\underset{\diagdown}{CH_3} \quad \underset{\diagup}{CH_2CH_3} & & \underset{\diagdown}{CH_3} \quad \underset{\diagup}{CH_2CH_3} \\
\overset{2}{CH} - \overset{\oplus}{C}{}^{3} & \rightleftharpoons & \overset{\oplus}{C} - CH \\
\underset{\diagup}{CH_3} \quad \underset{\diagdown}{CH_3} & & \underset{\diagup}{CH_3} \quad \underset{\diagdown}{CH_3}
\end{array}
$$

with decreasing temperature, suggesting qualitatively that the
3-pentyl cation is the more stable.[2a]

For the ethylenephenonium (benzenium) ion (**10**) at -80 to -90°C,
the ^{13}C chemical shifts of C-1 and C-α are 68.8 and 60.7 ppm, respectively; for C-α, $^1J_{CH.}$ = 178.4 Hz[23] (cf. ethylene oxide,

10

$^1J_{CH}$ = 175.7 Hz). These values are not consistent with equilibrating open cations or with π-bridged species and support the
σ-bridged structure, (**10**). The same conclusion holds for the
ethylenebromonium ion.[5]

The effect of π-bridging is displayed by the 7-norbornenyl
cation, (**11**).[24] At -75°C in FSO_3H-SO_2ClF, **11** gives rise to four

11

resonances at 26.7, 34.0, 58.0, and 125.9 ppm. The large $^1J_{CH}$ of 218.9 Hz and a lower intensity allows assignment of the resonance at 34.0 ppm to C-7. The presence of double doublets at 26.7 ppm in the proton-coupled spectrum allows identification of C-5 and C-6. The remaining resonance at higher shielding is attributable to C-1 and C-4, and the signal at 125.9 ppm, to C-2 and C-3. The data appear to be consistent only with a π-bridged species in which the charge is substantially delocalized by the 2,3 π-bond.[24]

How ^{13}C nmr may be used to address the question of nonclassical σ-bridged ions, for example, the 2-norbornyl cation, has been extensively debated in the literature. Brown,[25] Olah,[26] and Farnum[1b] have made important contributions. Olah's assignments at -150°C for the *five* resonances of the 2-norbornyl cation (**12**) are consistent with a nonclassical σ-bridged species. On the other hand, the 2-methyl-2-norbornyl cation (**13**), whose shifts at -80°C are also shown,[11] is considered to be static. The reader is referred to the original literature[25,26] for details of the debate!

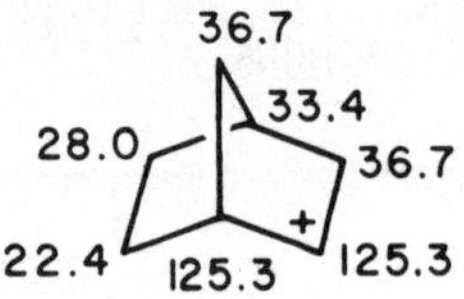

12 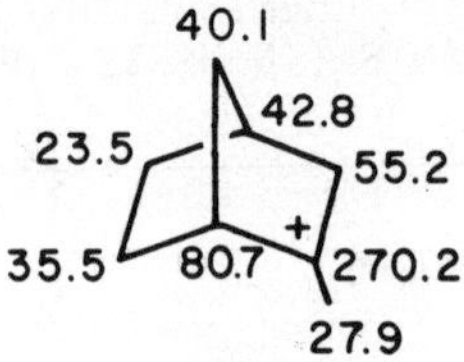   **13**

(Chemical shifts in ppm)

Other reported polycyclic carbocations include dehydroadamantyl,[26a,d] manxyl,[26b] cyclopropylcarbinyl,[16c] and brexyl.[27] Carbon-13 spectra of apical cations and dications of the types $(CH)_5^+$ and $(CCH_3)_6^{2+}$ have also been reported. Ion **14** displays shifts consistent with the bridged structure shown.[28a] In particular, the large $^1J_{CH}$ for the apical carbon, 220 Hz, should be noted. This value, which perhaps coincidentally is very close to that for C-7 of **11**, is consistent with the *sp*-hybridization that has been predicted theoretically for this center.[28a]

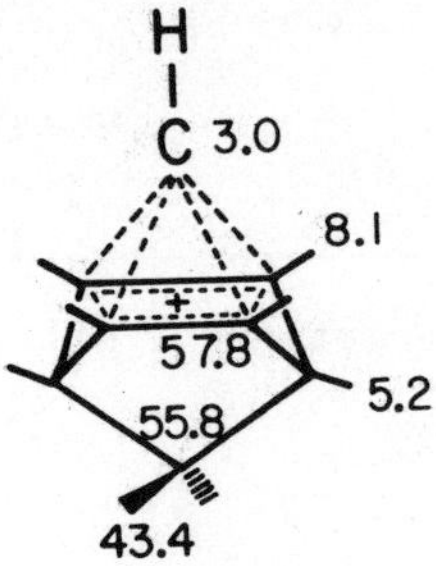

14

Finally, the first reports of carbocation ^{13}C spectra in the *solid state* have appeared.[29] The possibility of such studies allows determination of the effects of solvent on the shifts and, conceivably, on the structures of the ions themselves.

CARBANIONS

Carbon-13 chemical shifts of carbanions reflect gross changes in charge distribution, and in conjugated systems ·an increase in shielding is often displayed by carbons at which a charge buildup is expected. However, chemical shifts of carbanions depend measureably on solvent, concentration, temperature, and nature of the cation.[30-35] Table 6.2 presents data for some alkyl- and alkenyllithiums, and Table 6.3 gives values for some phenylmethyl car-

TABLE 6.2. ^{13}C CHEMICAL SHIFTS (IN PARTS PER MILLION) AND COUPLING CONSTANTS (IN HERTZ) OF ANIONIC CARBONS[a,30,31,33]

COMPOUND	C-1	$^1J_{CH}$
CH_3Li	-15.3	98
$CH_3CH_2CH_2CH_2Li$[b]	11.8	--
$CH_3CH_2CH(Li)CH_3$[b]	17	--
$(CH_3)_2CHLi$[b]	10.2	--
$(CH_3)_3CLi$[b]	10.7[c]	
▷CHLi	- 9.4	105
$CH_2=CHLi$	183.4	88
$\overset{2}{C}H_2\text{---}\overset{1}{C}H\text{---}CH_2Li$	51.2(C-1) 147.2(C-2)	146 133
$\overset{2}{C}H_2\text{---}\overset{1}{C}H\text{---}CH_2K$	52.8(C-1) 144.0(C-2)	

[a] Solvent tetrahydrofuran (THF).
[b] Solvent benzene.
[c] $^1J_{CLi}$ = 10.1 Hz.

banions. The alkyl anion α-carbons are shielded compared with their corresponding hydrocarbons. By contrast, the α-carbons of the phenylmethyl anions are all deshielded with respect to their hydrocarbon precursors, indicating that hybridization and charge delocalization dominate the shielding. A change to sp^2-hybridization is also indicated by the larger values of $^1J_{CH}$. The ring carbon shifts indicate that most of the delocalized charge is concentrated at the *para* carbons. It is interesting

TABLE 6.3. ^{13}C CHEMICAL SHIFTS (IN PARTS PER MILLION) AND COUPLING CONSTANTS (IN HERTZ) OF PHENYLMETHYL CARBANIONS[a,30,33]

COMPOUND	C-α	$^1J_{CH}$	C (*para*)	$^1J_{CH}$
PhCH$_2$Li	36.7	134	104.2	148
PhCH$_2$K	52.7		95.7	
Ph$_2$CHLi	76.8	141	107.3	149
Ph$_2$CHNa	74.8		108.4	
Ph$_2$CHK	78.5		108.4	
Ph$_3$CLi	90.2		113.0	150
Ph$_3$CK	88.3		114.2	

[a]Solvent THF.

that only *tert*-butyllithium displays carbon-lithium spin-spin coupling,[31] with a pattern indicating coupling to four equivalent lithiums. Since this compound is known to be tetrameric in benzene, rapid intramolecular exchange of lithium is indicated. Absence of coupling in the other alkyllithiums may be ascribed to a combination of carbon-lithium bond exchange and ^7Li quadrupole relaxation.[32]

Although direct correlations between the carbanion chemical shifts and local electron densities are tenuous, a correlation between the *total average* chemical shifts of each member of a series of compounds and the *total average* charge density has been suggested.[33a] The possible generality of this concept is indicated by the inclusion of carbo*cations* in the correlation.

Even if no direct correlation with local electron density exists, changes in chemical shift as a function of solvent, cation, and so on can serve as a probe for details of structure. For example, the carbons of diphenylmethyllithium and diphenyl-methylsodium are shielded with increasing temperature, whereas essentially no temperature variation is found with the potassium, rubidium, or cesium salts.[33b]   At the higher temperature the

Ph_2CHNa	Ph_2CHLi
74.8 ppm (48°C)	76.5 ppm (54°C)
79.1 ppm (-36°C)	80.7 ppm (-54°C)

equilibria consist mainly of contact ion pairs, and hence the charge is polarized extensively to C-α. The magnitude of the po-larization (and hence the increase in shielding at C-α) is related to the size of the cation: the smaller the ionic radius, the tighter the ion pairing, and the more shielded C-α becomes (the values for the K^+, Rb^+, and Cs^+ salts in THF are 79.1, 79.9, and 81.5 ppm, respectively, and the temperature independence indicates that there is little change in the extent of solvation). In addi-tion, shielding of C-α decreases as the solvating power of the solvent increases (2-methyltetrahydrofuran < THF < glyme). Because the cation is better solvated in glyme than in THF, for example, less charge is polarized to C-α. This solvent dependence neces-sarily decreases as the cation becomes larger. At the lower tem-peratures the α-carbon shifts approach values appropriate for a solvent-separated ion pair. Chemical shifts of these species are less solvent dependent because the solvated cation plays a lesser role in influencing shifts.[33b]

A similar effect of solvent is seen in the indenyl anion (**15**).[34] Whereas the chemical shift of cyclopentadienyllithium is relatively constant at 103.1 to 103.7 ppm in a variety of solvents,

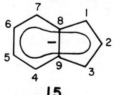

15

	Ether	Diglyme
C-1,3	91.8	93.2
C-2	114.5	117.8
C-4,7	120.6	118.4
C-5,6	116.4	111.1
C-8,9	127.5	131.0
Average	114.1	113.9

the carbons of indenyllithium show a marked variation. Increasing
solvent polarity deshields carbons 1, 2, 3, 8, and 9 and shields
C-4 to C-7. Assuming that the cation lies preferentially over the
five-membered ring in the contact ion pair in the less polar solvent, enhanced solvation of the cation reduces the ion pairing and
allows charge to move into the six-membered ring.

Peoples and Grutzner have found substantial differences in the
[13]C chemical shifts for C-7 and C-*para* of 7-phenylnorbornyllithium
in THF compared with the potassium and cesium salts.[35] They suggest that the lithium compound is pyramidal, with an inversion
barrier of 11 ± 1 kcal/mole at 25°, while the potassium and cesium
salts are planar. Their work also presents useful cautionary
notes for the interpretation of anion ^{13}C spectra.

Carbon-13 shifts have been used as a probe for homoconjugation
in carbanions. The chemical shifts of several cyclohexadienyl
anions are given below.[36] These ions might have been expected to
undergo homoconjugative interaction to give a $4n + 2$ Hückeloid
arrangement. However, the chemical shifts are quite characteristic of normal pentadienyl anions with alternating charge distributions. It is striking that the shifts of the carbons at which
little charge buildup is expected in fact do not differ much from
those of the neutral analogues or from the corresponding carbons
in the cations.

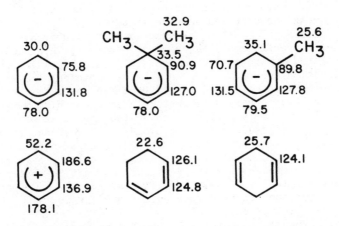

Phenyllithium and vinyllithium (Table 6.2) define a different
class of carbanion from those discussed previously. In these compounds the excess electron density is in a nonbonding type orbital
orthogonal to the π system. In both cases the anionic carbons are
deshielded relative to the parent hydrocarbon. This change is
analogous to that exhibited by the ^{15}N shift of the isoelectronic
pyridinium ion when it is converted to pyridine.[37] This effect
has been explained in terms of an additional contribution to σ_p

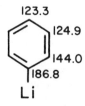

arising from the lone-pair n-π^* excitation energy in pyridine and phenyl anion, respectively.

Other systems that have been reported include sulfur-stabilized carbanions,[38] enolates,[39] substituted alkyl-[40] and pentadienyllithiums,[41] and trimethylenecyclopropane dianions.[42] Detection of dianionic organic reaction intermediates by ^{13}C nmr has also been reported, although no spectral data have been given.[43]

ORGANOMETALLICS

Many organometallic compounds and metal complexes have been studied by ^{13}C nmr (early papers are summarized in three reviews[44-46]). Carbon nmr has been especially useful in elucidation of carbon-metal bonding and in understanding fluxional (rapid migration) processes. Most of these organometallic studies utilize standard ^{13}C spectroscopic techniques. However, in a number of cases isolated nonprotonated carbons (e.g., carbonyls) require spin-relaxation agents to shorten T_1s for efficient ^{13}C FT spectroscopy. Both Cr(acac)$_3$ and Cr(dpm)$_3$ have been used for this purpose (see Chapter 8). A very different relaxation problem may occur in organometallics. Metals with *abundant* isotopes that have spin quantum numbers exceeding 1/2 can broaden ^{13}C resonances for directly attached (and sometimes remote) carbons in organometallic compounds. In some cases where this *scalar* spin-spin relaxation occurs (see Chapter 8) carbon-13 signals may not be observed at all.

Recently, studies of organometallic systems by ^{13}C spectroscopy have been supplemented by direct metal nmr,[47] especially where spin 1/2 nuclides are present.

Metal carbonyl ^{13}C shieldings, including many organic ligands, have been summarized by Gansow and Vernon.[44] Several metal carbonyl shifts are given in Table 6.4. In general, organic ligand substitution deshields metal carbonyl resonances. Low-temperature studies of metal cluster carbonyl compounds can determine mechanisms of ligand scrambling or exchange.[48-50] The dinuclear η^5-dienyliron or ruthenium carbonyls (**16** to **19**) are representative:

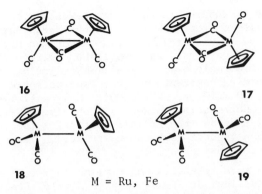

16

17

18   M = Ru, Fe   **19**

Early ^{13}C work showed that $cis \rightleftharpoons trans$ (**16** $\rightleftharpoons$ **17**) and bridged $\rightleftharpoons$ non-bridged (**16** $\rightleftharpoons$ **18**) and **17** $\rightleftharpoons$ **19**) equilibria proceeded simultaneously, at varying rates. More recent ^{13}C work[49] has probed the mechanism for these intramolecular rearrangements. The authors postulate

TABLE 6.4.   METAL CARBONYL SHIELDING (IN PARTS PER MILLION)[44]

COMPOUND	SOLVENT	(C=O)
$Cr(CO)_6$	CH_2Cl_2	212.5
	$CHCl_3$	211.2
$(Mesitylene)Cr(CO)_3$	CH_2Cl_2	235.5
$Mo(CO)_6$	CH_2Cl_2	202.0
$(Mesitylene)Mo(CO)_3$	CH_2Cl_2	223.7
$W(CO)_6$	CH_2Cl_2	192.1
$(Mesitylene)W(CO)_3$	CH_2Cl_2	212.6
$Mn_2(CO)_{10}$	CH_2Cl_2	cis 212.9, trans 231.1
$Re_2(CO)_{10}$	THF	cis 192.7, trans 183.7
$Fe(CO)_5$	neat	209.0
$Fe(CO)_{12}$	CH_2Cl_2	212.5
$Ru_3(CO)_{12}$	$CHCl_3$	cis 188.7, trans 198.8
$Rh_4(CO)_{12}$	CD_2Cl_2	bridging 228.8
		terminal 183.4, 181.8
	$(-60°)$	175.5
$Ni(CO)_4$	neat	193.6

that metal-metal bonding in these complexes may be directional and that the barrier to rotation in the nonbridged isomers may be due in part to hindered internal rotation about the metal-metal bond.

Brown and co-workers[52] have published self-consistent charge and configuration molecular-orbital (MO) calculations of ^{13}C para-magnetic shielding constants for the isoelectronic series $C_6H_6Cr(CO)_3$, $C_5H_5Mn(CO)_3$, $C_4H_4Fe(CO)_3$, $C_3H_5(\equiv allyl)$ $Co(CO)_3$, and $C_2H_4Ni(CO)_3$. Correlation between the observed downfield shifts for the CO ligands and the upfield shifts for the carbon atoms of the hydrocarbon ligands with the calculated σ_p values is noted without need to consider σ_d.

Carbon-13 spin-relaxation mechanisms in methyl-transition metal compounds include largely dipole-dipole interactions with spin-rotation-relaxation contributions.[53] In $CH_3Re(CO)_5$ scalar relaxation with the quadrupolar rhenium dominates T_1 (for dis-cussions of these relaxation mechanisms, see Chapter 8). Ob-served methyl T_1s at room temperature range from 3 to 14 sec.

Some representative ^{13}C organometallic studies are summarized in Table 6.5.[54-70] Although these examples are by no means com-prehensive, they do indicate some of the types of potential studies.

TABLE 6.5.   SELECTED ^{13}C NMR STUDIES OF ORGANOMETALLIC COMPOUNDS

CLASS OF COMPOUNDS	REFERENCES
Aryltrimethyltins	54
Organomercurials	55,56
Amino acid complexes (Pt)	57
Porphyrins [Fe(III)]	58
Ferrocenes, ferrocenium ions	59,60
Aryl complexes (Pt)	61
(Sb,Bi)	62a
(Ga,In)	62b
Acetylenic and olefinic complexes (Rh)	63
(Pt)	64,65
(Fe)	66
Metal (Cr)-stabilized carbocations	67
Organosilver complexes	68
Organolead compounds	69,70

In studies of organometallic compounds that have spin-1/2
metals of high isotopic abundance, spin-spin coupling constants
may be more useful than shielding parameters. For example,
^{207}Pb - ^{13}C one-bond couplings reflect hybridization at the lead
atom, ranging from less than 200 Hz to over 2000 Hz in some 20
compounds studied. The largest couplings were interpreted in
terms of hexacoordinate lead.[69]
   In 1972 Leibfritz, Wagner, and Roberts[71] demonstrated the
anionic character of Grignard reagents, using benzene substituent
effects to monitor charge movements. Figure 6.1 summarizes their
results (compare with data in Table 6.3).

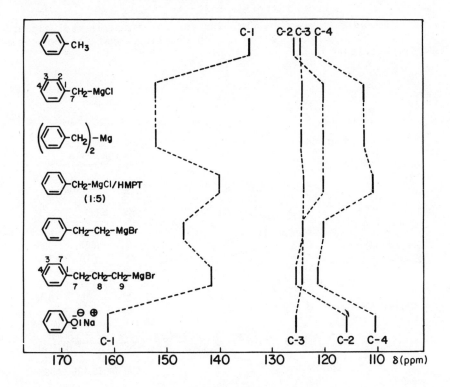

Figure 6.1.[71] ^{13}C Shifts in Grignards

RADICALS

   As described previously, organometallic compounds in which the
metal is diamagnetic display ^{13}C chemical shifts over the same
range as other organic molecules. If the metal is paramagnetic,

however, chemical shifts occur over a much wider range. The con-
tributors to such shifts are the *contact shift* resulting from the
transmission of unpaired σ- and π-type electron spin density to
the ligand and the *pseudocontact shift* resulting from electron-
nucleus dipolar interactions. The theory and the general applications have been reviewed.[72]

Specific cmr applications include studies relating to pyridine-
type bases and triphenylphosphine complexed to nickel(II) and
cobalt(II) acetylacetonates,[73-74] nickel(II)-*N,N*-di(*p*-tolyl)-
aminotroponiminate,[75] metallocenes,[76] and amino acid and ethylene-
diamine complexes of nickel(II).[77] Effects on solvent molecules
of ferric trisacetonylacetonate and diphenylpicrylhydrazyl
radicals[78] as well as nitroxide radicals[79] have been reported
(see also Chapter 8). The cmr behavior of nitroxide radicals and
the sign and magnitude of the carbon coupling constants have been
reported (Figure 6.2).[80]

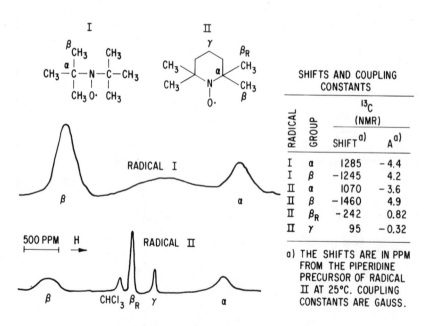

RADICAL	GROUP	^{13}C (NMR)	
		SHIFT[a]	A[a]
I	α	1285	-4.4
I	β	-1245	4.2
II	α	1070	-3.6
II	β	-1460	4.9
II	β_R	-242	0.82
II	γ	95	-0.32

a) THE SHIFTS ARE IN PPM
FROM THE PIPERIDINE
PRECURSOR OF RADICAL
II AT 25°C. COUPLING
CONSTANTS ARE GAUSS.

Figure 6.2.[80]   *Spectra of radicals I and II. The intensities*
of the center lines in the spectrum of radical
II are reduced because of partial saturation.

REFERENCES

1. (a) G. J. Martin, M. L. Martin, and S. Odiot, *Org. Magn. Resonance*, **7**, 2 (1975); (b) D. G. Farnum, *Adv. Phys. Org. Chem.*, **11**, 123 (1975); (c) R. N. Young, *Progr. NMR Spectrosc.*, **12**, 261 (1979) (d) G. L. Nelson and E. A. Williams, *Progr. Phys. Org. Chem.*, **12**, 229 (1976).

2. (a) G. A. Olah and D. J. Donovan, *J. Am. Chem. Soc.*, **99**, 5026 (1977); (b) G. A. Olah, A. Germain, and H. C. Lin, *J. Am. Chem. Soc.*, **97**, 5481 (1975).

3. G. A. Olah, A. Germain, H. C. Lin, and K. Dunne, *J. Am. Chem. Soc.*, **97**, 5477 (1975).

4. G. A. Olah, J. L. Grant, R. J. Spear, J. M. Bollinger, A. Serianz, and G. Sipos, *J. Am. Chem. Soc.*, **98**, 2501 (1976).

5. G. A. Olah and A. M. White, *J. Am. Chem. Soc.*, **91**, 5801 (1969).

6. G. E. Maciel, *ibid.*, **93**, 4375 (1971).

7. (a) G. A. Olah and A. M. White, *ibid.*, **90**, 1884 (1968); (b) G. A. Olah and M. B. Comisarow, *ibid.*, **88**, 1818 (1966).

8. G. A. Olah and R. D. Porter, *ibid.*, **93**, 6876 (1971).

9. R. Ditchfield and D. P. Miller, *ibid.*, 5287.

10. H. Hogeveen, E. L. Mackor, P. Ros, and J. H. Schachtschneider, *Rec. Trav. Chim. Pays-Bas*, **87**, 1057 (1968).

11. D. P. Kelly, G. R. Underwood, and P. F. Barron, *J. Am. Chem. Soc.*, **98**, 3106 (1976).

12. (a) H. C. Brown and E. N. Peters, *J. Am. Chem. Soc.*, **99**, 1712 (1977); (b) H. C. Brown, M. Ravindranathan, and E. N. Peters, *J. Org. Chem.*, **42**, 1073 (1977).

13. J. W. Larsen, *J. Am. Chem. Soc.*, **100**, 330 (1978).

14. Y. Tokuhiro and G. Fraenkel, *J. Am. Chem. Soc.*, **91**, 5005 (1969).

15. R. P. Kirschen and T. S. Sorensen, *J. Am. Chem. Soc.*, **99**, 6687 (1977).

16. (a) M. Saunders and J. Rosenfeld, *J. Am. Chem. Soc.*, **92**, 2548 (1970); (b) G. A. Olah, C. L. Jeuell, D. P. Kelly, and R. D. Porter, *ibid.*, **94**, 146 (1972); (c) G. A. Olah, R. J. Spear, P. C. Hiberty, and W. J. Hehre, *ibid.*, **98**, 7470 (1976).

17. (a) G. A. Olah, P. W. Westerman, and J. Nishimura, *J. Am. Chem. Soc.*, **96**, 3548 (1974) and references cited therein; (b) G. A. Olah, G. K. Surya Prakash, and G. Liang, *J. Org. Chem.*, **42**, 2666 (1977).

18. G. A. Olah, J. S. Staral, G. Asencio, G. Liang, D. A. Forsyth, and G. D. Mateescu, *J. Am. Chem. Soc.*, **100**, 6299 (1978).

19. G. A. Olah, G. Liang, and S. P. Jindal, *J. Org. Chem.*, **40**, 3259 (1975).

20. G. A. Olah and J. S. Staral, *J. Am. Chem. Soc.*, **98**, 6290 (1976).

21. G. A. Olah, J. S. Staral, G. Liang, L. A. Paquette, W. P. Melega, and M. J. Carmody, *J. Am. Chem. Soc.*, **99**, 3349 (1977).

188   Ions, Radicals, and Complexes

22. G. A. Olah and G. Liang, *J. Am. Chem. Soc.*, **99** 6045, (1977).
23. G. A. Olah, R. J. Spear, and D. A. Forsyth, *J. Am. Chem. Soc.*, **98**, 6284 (1976).
24. G. A. Olah and G. Liang, *J. Am. Chem. Soc.*, **97**, 6803 (1975).
25. (a) D. P. Kelly and H. C. Brown, *J. Am. Chem. Soc.*, **97**, 3897 (1975); (b) H. C. Brown, M. Ravindranathan, and F. N. Peters, *J. Org. Chem.*, **42**, 1073 (1977).
26. (a) G. A. Olah *et al.*, *J. Am. Chem. Soc.*, **98** 576 (1978); (b) G. A. Olah, G. Liang, P. v. R. Schleyer, W. Parker, and C. I. F. Watt, *ibid.*, **99**, 966 (1977); (c) G. A. Olah, G. K. S. Prakash, and G. Linag, *ibid.*, 5683 (1977); (d) G. A. Olah, G. Liang, D. B. Ledlie, and M. G. Costopoulos, *ibid.*, 4196 (1977); (e) G. A. Olah *et al.*, *ibid.*, **100**, 1494 (1978); (f) G. A. Olah, G. K. Surya Prakash, T. N. Rawdah, D. Whittaker, and J. C. Rees, *ibid.*, **101**, 3935 (1979).
27. R. S. Bly, R. K. Bly, J. B. Hamilton, J. N. C. Hsu, and P. K. Lillis, *J. Am. Chem. Soc.*, **99**, 216 (1977).
28. (a) H. Hart and R. Willer, *Tetrahedron Lett.* **1978**, 4189; (b) C. Giordano, R. F. Heldeweg, and H. Hogeveen, *J. Am. Chem. Soc.*, **99**, 5181 (1977).
29. J. R. Lyerla, C. S. Yannoni, D. Bruck, and C. A. Fyfe, *J. Am. Chem. Soc.*, **101**, 4770 (1979).
30. J. P. C. M. van Dongen, H. W. D. van Dijkman, and M. J. A. deBie, *Rec. Trav. Chim. Pays-Bas*, **93**, 29 (1974).
31. S. Bywater, P. Lachance, and D. J. Worsfold, *J. Phys. Chem.*, **79**, 2148 (1975).
32. G. Fraenkel, A. M. Fraenkel, M. J. Geckle, and F. Schloss, *J. Am. Chem. Soc.*, **101**, 4745 (1979).
33. (a) D. H. O'Brien, A. J. Hart, and C. R. Russell, *J. Am. Chem. Soc.*, **97**, 4410 (1975); (b) D. H. O'Brien, C. R. Russell, and A. J. Hart, *J. Am. Chem. Soc.*, **98**, 7427 (1978).
34. U. Edlund, *Org. Magn. Resonance*, **29**, 593 (1977).
35. P. R. Peoples and J. B. Grutzner, private communication, 1980.
36. G. A. Olah, G. Asensio, H. Mayr, and P. v. R. Schleyer, *J. Am. Chem. Soc.*, **100**, 4347 (1978).
37. G. C. Levy and R. L. Lichter, *Nitrogen-15 Nuclear Magnetic Resonance Spectroscopy*, Wiley-Interscience, New York, (1979), p. 77.
38. R. Lett, G. Chassaing, and A. Marguet, *J. Organometal Chem.*, **111**, C17 (1976).
39. (a) H. O. House, A. V. Prabhu, and W. V. Phillips, *J. Org. Chem.*, **41**, 1209 (1976); (b) M. Raban and D. P. Haritos, *J. Am. Chem. Soc.*, **101**, 5178 (1979).
40. H. Bauer, M. Angrick, and D. Rewicki, *Org. Magn. Resonance*, **12**, 624 (1979).
41. D. H. Hunter, R. E. Klinck, R. P. Steiner, and J. B. Stothers, *Can. J. Chem.*, **54**, 1464 (1976).
42. T. Fukunaga, *J. Am. Chem. Soc.*, **98**, 610 (1976).
43. J. B. Lambert and S. M. Wharry, *J. Chem. Soc. Chem. Commun.*, **1978** 172.

44.   O. A. Gansow and W. D. Vernon, in *Topics in Carbon-13 NMR Spectroscopy*, Vol. 2, G. C. Levy, Ed., Wiley-Interscience, New York, 1976.

45.   B. E. Mann, *Adv. Organometal. Chem.*, **12**, 135 (1974).

46.   L. J. Todd and J. R. Wilkinson, *J. Organometal. Chem.*, **77**, 1 (1974).

47.   R. K. Harris and B. E. Mann, *NMR and the Periodic Table*, Academic, New York, 1978.

48.   F. A. Cotton, R. J. Haines, B. E. Hanson, and J. C. Sekutow-ski, *Inorg. Chem.*, **17**, 2010 (1978); F. A. Cotton, B. E. Hansen, J. R. Kolb, P. Lahuerta, G. G. Stanley, B. R. Stults, and A. J. White, *J. Am. Chem. Soc.*, **99**, 3673 (1977); F. A. Cotton, B. E. Hanson, J. D. Jamerson, and B. R. Stults, *ibid.*, **99**, 3293 (1977).

49.   O. A. Gansow, A. R. Burke, and W. D. Vernon, *J. Am. Chem. Soc.*, **98**, 5817 (1976).

50.   L. Kruczynski and J. Takats, *Inorg. Chem.*, **15**, 3140 (1976).

51.   D. Seyferth, G. H. Williams, C. S. Eschbach, M. O. Nestle, J. S. Merola, and J. E. Hallgren, *J. Am. Chem. Soc.*, **101**, 4867 (1979).

52.   D. A. Brown, J. P. Chester, N. J. Fitzpatrick, and I. J. King, *Inorg. Chem.*, **16**, 2497 (1977).

53.   R. F. Jordan and J. R. Norton, *J. Am. Chem. Soc.*, **101**, 4853 (1979).

54.   C. D. Schaeffer, Jr. and J. J. Zuckerman, *J. Organometal. Chem.*, **99**, 407 (1975).

55.   N. K. Wilson, R. D. Zehr, and P. D. Ellis, *J. Magn. Resonance*, **21**, 437 (1976); G. Singh, *J. Organometal. Chem.*, **99**, 251 (1975).

56.   A. J. Brown, O. W. Howarth, and P. Moore, *J. Chem. Soc. Dalton*, **1976**, 1589.

57.   L. E. Erickson, J. E. Sarneski, and C. N. Reilley, *Inorg. Chem.*, **17**, 1711 (1978).

58.   H. Goff, *Biochim Biophys. Acta*, **542**, 348 (1978).

59.   A. A. Koridze, P. V. Petrovskii, S. P. Gulsin, V. I. Sokolov, and A. I. Mokhov, *J. Organometal. Chem.*, **136**, 65 (1977).

60.   A. A. Koridze, P. V. Petrovskii, S. P. Gubin, and E. I. Fedin, *J. Organometal. Chem.*, **93**, C26 (1975).

61.   D. R. Coulson, *J. Am. Chem. Soc.*, **98**, 3111 (1976).

62.   (a) A. Ouchi, T. Uehiro, and Y. Yoshino, *J. Inorg. Nucl. Chem.*, **37**, 2347 (1975); (b) W. J. Freeman, S. B. Miller, and T. B. Brill, *J. Magn. Resonance*, **20**, 378 (1975).

63.   L. J. Todd, J. R. Wilkinson, M. D. Rausch, S. A. Gardner, and R. S. Dickson, *J. Organometal. Chem.*, **101**, 133 (1975).

64.   L. E. Manzer, *Inorg. Chem.*, **15**, 2354 (1976).

65.   M. A. M. Meester, D. J. Stufkens, and K. Vrieze, *Inorg. Chim. Acta*, **15**, 137 (1975).

66.   J. P. Hickey, J. R. Wilkinson, and L. J. Todd, *J. Organometal. Chem.*, **99**, 281 (1975).

67. D. Seyferth, J. S. Merola, and C. S. Eschbach, *J. Am. Chem. Soc.*, **100**, 4124 (1978).
68. P. M. Henrichs, S. Sheard, J. J. H. Ackerman, and G. E. Maciel, *J. Am. Chem. Soc.*, **101**, 3222 (1979).
69. R. H. Cox, *J. Magn. Resonance*, **33** 61 (1979).
70. C. S. Baxter, H. E. Wey, and A. D. Cardin, *Toxic. Appl. Pharmacol.*, **47**, 477 (1979).
71. D. Leibfritz, B. O. Wagner, and J. D. Roberts, *Liebigs Ann. Chem.*, **763**, 173 (1972).
72. D. A. Eaton and W. D. Phillips, *Adv. Magn. Resonance*, **1**, 103 (1965).
73. D. Doddrell and J. D. Roberts, *J. Am. Chem. Soc.*, **92**, 6839 (1970).
74. I. Morishima, T. Yonezawa, and K. Goto, *ibid.*, **92**, 6651 (1970).
75. D. Doddrell and J. D. Roberts, *ibid.*, **92**, 4484 (1970).
76. P. K. Burkert, H. P. Fritz, P. H. Kohler, and H. Rupp, *J. Organometal. Chem.*, **24**, C59 (1970).
77. G. E. Strouse and N. A. Matwiyoff, *Chem. Commun.*, **1970**, 439
78. O. W. Howarth, *Molec. Phys.*, **21**, 949 (1971).
79. I. Morishima, K. Endo, and T. Yonezawa, *J. Am. Chem. Soc.*, **93**, 2048 (1971); *Chem. Phys. Lett.*, **9**, 371 (1971); I. Morishima, K. Okuda, and T. Yonezawa, *J. Am. Chem. Soc.*, **94**, 1425 (1972); I. Morishima, T. Inubishi, K. Endo, T. Yonezawa, and K. Goto, **94**, *ibid.*, 4812 (1972).
80. G. F. Hatch and R. Kreilick, *Chem. Phys. Lett.*, **10**, 490 (1971).

Chapter 7

Polymers

Proton nmr has contributed significant information concerning configurational sequences and dynamics of synthetic polymers.[1] A large amount of literature has developed through the use of techniques that are now standard and routine. Configurational sequences of up to five repeating units (pentads) can be resolved for many polymers. Pulsed ^{1}H nmr studies have examined dynamic processes that occur along polymer chains, such as fast rotations around C-C bonds. The greater chemical-shift range of ^{13}C nmr has been applied to the problems of structure and conformation in polymer systems for both solution and solid analyses. This chapter describes some applications of cmr techniques to synthetic polymer systems.

SPECTRAL CHARACTERISTICS

Carbon-13 techniques can be applied easily to problems dealing with oligomers. Figure 7.1[2] presents the ^{13}C nmr spectra of poly-(2,6-dimethyl-1,4-phenylene oxide) and the acetate of a tetramer model compound. The spectrum of the polymer (molecular weight ≈ 20,000) in deuteriochloroform solvent displays five sharp lines. It is interesting to point out that one of the polymer lines (3') shows broadening. This has been ascribed to restricted rotation around the diphenyl ether bond (see p.208). For the tetramer, the carbons of the end monomer units are plainly visible, as are differences for carbons 1', 3', and 4' for the two interior residues [an expanded spectrum is shown in Figure 7.1c; similar behavior has been observed for oligomers of poly(m-diethynylene-phenylene)[3]].

Resolution similar to that in Figure 7.1a can be obtained for many polymer systems. For other systems, however, dipolar broadening contributes to deterioration of spectral quality. In proton nmr spectra of many high polymers dipolar broadening obscures most [1]H-^{1}H spin-spin coupling information and lowers the effective resolution of the spectra. Linewidths of >10 Hz are commonly observed. Dipolar broadening arises from incomplete averaging of the localized magnetic fields surrounding magnetic nuclei (e.g., ^{1}H nuclei

191

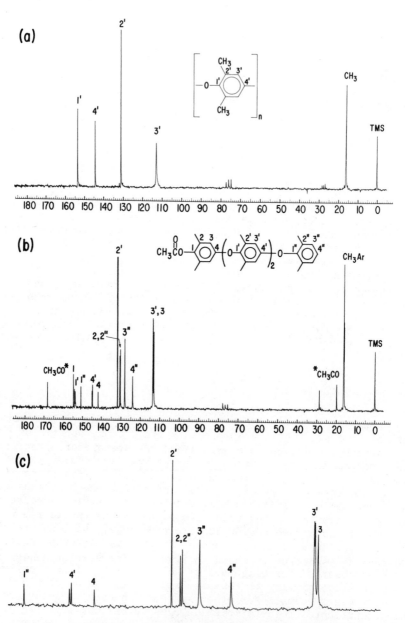

FIGURE 7.1.[2]   Carbon-13 spectra of (a) poly(2,6-dimethyl-1,
4-phenylene oxide) and (b) the acetate of a
tetramer. The expanded low-field region of
the tetramer is shown in (c).

and ^{13}C nuclei) in solution. Averaging is incomplete because the size of a polymer molecule restricts microscopic molecular motion even when the polymer is dissolved in a solvent (for a discussion of molecular motion in solution, see Chapter 8).

In ^{13}C nmr dipolar broadening is less restrictive for several reasons:

1. Dipolar broadening is proportional to the square of the magnetogyric ratios of the nuclei involved $[\gamma(^{13}C) \cong (1/4)\gamma(^1H)]$. Therefore, carbon lines are not subject to as much dipolar broadening as in the case of proton lines.

2. Since carbons are in the polymer backbone, they tend to be shielded from significant intermolecular and non-nearest-neighbor intramolecular C-H dipolar interactions. In general, each carbon nucleus sees fewer magnetic nuclei, leading to an overall decrease in dipole-dipole interaction and less line broadening.[4]

3. The larger frequency range for ^{13}C nuclei (e.g., 10 kHz vs. 2 kHz for ^{1}H nuclei, both at 4.7 T) means that resolution of ^{13}C spectra would be superior to that of ^{1}H spectra even if ^{13}C lines had comparable dipolar broadening.

4. Carbon-13 spectra are usually obtained without coupling information present, thus reducing spectral line complexity and overlap.

For polymers of high stereoregularity, spectra can be interpreted in terms of a single repeating unit whose chemical shift reflects the average molecular environment of the unit over all conformations of the chain. Provided that the repeating unit is of relatively simple structure, the problem of spectral interpretation is not unlike that encountered for low-molecular-weight molecules. Prediction of ^{13}C nmr chemical shifts by the methods described in Chapter 3 has been successfully applied to problems of monomer sequences, particularly for ethylene-propylene copolymers, which form complex systems at various levels of propylene incorporation.

The effect of substituent variation in polymer systems is similar to that described for small molecules in Chapters 3 to 5. A correlation diagram for the methylene and methine carbons of substituted vinyl polymers of the form $-(CH_2-CHX)_n-$ is given in Figure 7.2. The effect of relative local stereochemistry is also shown and is dealt with in detail in the following paragraphs. In copolymers it is usually possible to characterize block lengths (if present) as well as other features of chain stereoregularity.[8]

PROBLEM 7.1.   *Using the methods described in Chapter 3, confirm the ^{13}C chemical shifts for the methylene and methine carbons for the polymers given in Figure 7.2.*

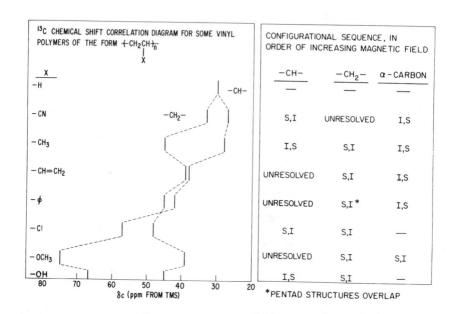

FIGURE 7.2.   *Correlation diagram for the methylene and methine*
carbons of some vinyl polymers (left). The effect
of microtacticity is shown (right) where S and I
represent syndiotactic and isotactic placements,
respectively.

CONFORMATIONAL SEQUENCES

For the substituted vinyl polymers presented in Figure 7.2, each repeating unit possesses an asymmetric center; thus the polymer may have stereochemical differences depending on the relative stereochemistry of adjacent repeating units throughout the polymer chain. Carbon-13 spectra can distinguish such differences. The spectrum of a mixture of meso- and racemic-2,4-dichloropentane is given in Figure 7.3a.[9] Assignments of meso (m) or racemic (r) sites are made for each carbon resonance. Figures 7.3b and 7.3c present the spectrum of poly(vinyl chloride), showing three multiplets for the methine carbon and two multiplets for the methylene carbon.[9] A more recent example of differentiated polymer stereochemistry is shown in Figure 7.4[10]. This spectrum of poly(methyl methacrylate) shows the effect of polymer stereochemistry on both the main-chain CH_2 and CR_4 carbons and the chain C-**CH_3** and ester carbonyl groups.

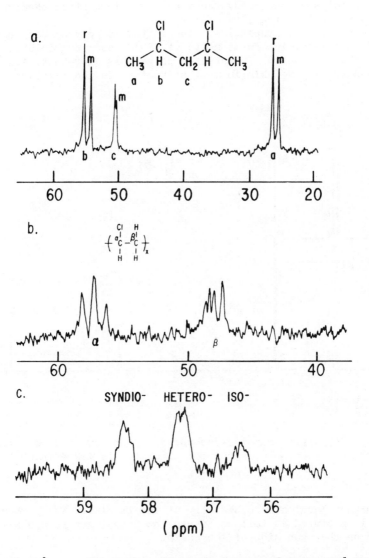

Figure 7.3.[9]   (a) Carbon-13 nmr spectrum of meso- and racemic-
2,4-dichloropentane, assignments of meso (m) and
racemic (r) are made for each carbon resonance;
(b) cmr spectrum of poly(vinyl chloride) in
chlorobenzene showing three multiplets for the
methine carbon and two multiplets for the methylene carbon; (c) the methine carbon resonances
for poly(vinyl chloride) showing fine structure.

195

Some polymers can be synthesized in various stereochemical forms. For polystyrene, two forms, isotactic and atactic, have different physical properties. The isotactic polymer can be obtained in crystalline form, melts at 250°C, and is insoluble in most common solvents. The atactic polymer is amorphous, low melting, and readily soluble in most solvents. In the first case the

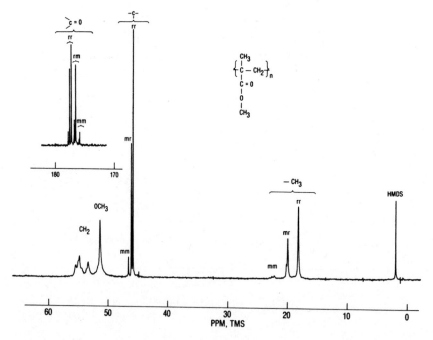

Figure 7.4.[10]   Proton-noise-decoupled ^{13}C nmr spectrum at 25.2 MHz of poly(methyl methacrylate) in 1,2,4-trichlorobenzene at 120°C.

predominant local stereochemistry of the imaginary fully stretched chain is pictured as follows, where the phenyl groups are all arranged on the same side of the chain:

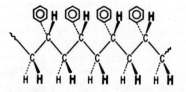

Isotactic polystyrene

For atactic (or heterotactic) polystyrene, the phenyl groups are completely at random, both below and above the plane of the backbone. A third possibility, syndiotactic, designates polymers possessing alternate orientations for the adjacent asymmetric centers.

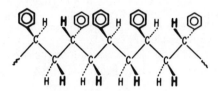

Atactic polystyrene

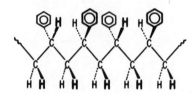

Syndiotactic polystyrene

Although a polymer may have a predominant overall tacticity (isotactic, syndiotactic, atactic, etc.), the ^{13}C nmr spectrum will reflect the sum of relative local stereochemistry at each carbon along the chain, the *microtacticity* (isotactic, syndiotactic, or heterotactic). For a particular asymmetric carbon, the next adjacent asymmetric carbons determine gross ^{13}C nmr behavior, and the second adjacent asymmetric carbons and beyond determine fine structure often observable in ^{13}C nmr spectra.

By focusing on a particular asymmetric carbon, the relative configuration of adjacent asymmetric centers can be described using the meso (m) and racemic (r) nomenclature. For three adjacent asymmetric carbons (triads), the relative configurations can be mm, mr, and rr. For five adjacent asymmetric carbons (pentads), the 10 possibilities are mmmm, mmmr, mmrm, mmrr, mrrm, mrmr, rmmr, rrrm, rrmr, and rrrr. In the following heterotactic vinyl polymer segment the triad and pentad structures for the starred asymmetric carbon are rr and rrrm, respectively:

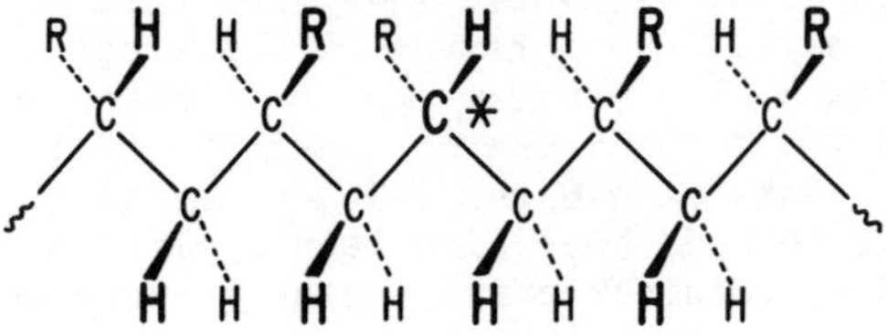

The local tacticity of the three methine multiplets (triads) in poly(vinyl chloride) is assigned in Figure 7.3c. Further line structure due to the pentads is visible. The cmr behavior of stereochemical sequences of eight vinyl polymers is given in Figure 7.2; S refers to syndiotactic (r), and I refers to isotactic (m) placements (for the CH carbon, S and I represent rr and mm triads), respectively. Figure 7.4 shows easy resolution of three different triads for the α-methyl group and also for the quartemary carbon of poly(methyl methacrylate). For the carbonyl carbon, there is further splitting into pentad structure. Figures 7.5 and 7.6 (where HMDS = hexamethyldisiloxane) show ^{13}C nmr spectra of

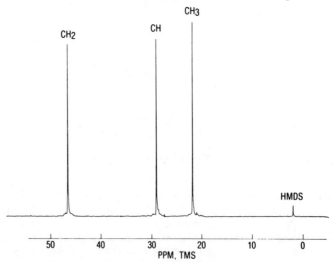

Figure 7.5.[10]   *Proton-noise-decoupled ^{13}C spectrum at 25.2 MHz of a crystalline polypropylene in 1,2,4-trichlorobenzene at 120°C.*

isotactic and atactic polypropylenes. The stereoregular isotactic polymer shows only three resonances, whereas the amorphous polymer resonances show fine structure due to tacticity. The methyl region is especially well resolved for polypropylene. The 10 possible *pentad* sequences are largely resolved in the expanded plot of amorphous polypropylene shown in Figure 7.7.[10]

Small amounts of head-to-head (H-H) and tail-to-tail (T-T) monomer addition may also be present in vinyl polymers such as polypropylene. Such placements may also be observed in ^{13}C spectra. Figure 7.8[10] shows the cmr spectrum of a polypropylene that has large amounts of H-H and T-T monomer addition.

Carbon-13 nmr methods have also been applied to polyelectrolytes, for example, to atactic and isotactic poly(methacrylic

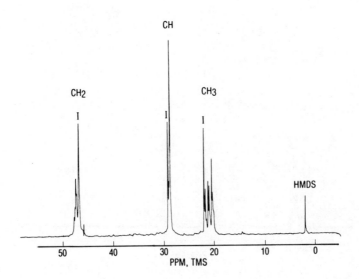

Figure 7.6.[10]   *Proton-noise-decoupled* [13]*C nmr spectrum at 25.2*
MHz of an amorphous polypropylene in 1,2,4-
trichlorobenzene at 120°C.

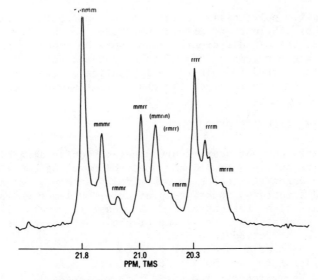

Figure 7.7.[10]   *Expanded methyl region of the amorphous polypropyl-*
ene shown in Figure 7.6.

199

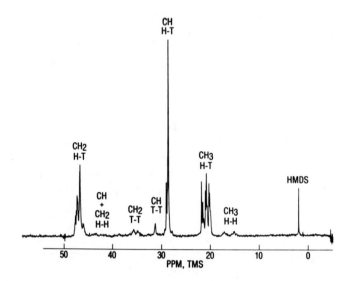

Figure 7.8.[10]   *Proton-noise-decoupled ^{13}C nmr spectrum at 25.2 MHz of a polypropylene containing inverted monomer additions in 1,2,4-trichlorobenzene at 120°C.*

acid), atactic poly(acrylic acid), and a completely alternating copolymer of ethylene and maleic anhydride.[11]   The stereochemical differences between different segments of the atactic chains were distinguishable by cmr and found to be pH dependent. This pH dependence is related to changes in ionization of the substituent; the local stereochemistry of the chain remains unchanged.

QUANTITATIVE ANALYSIS OF POLYMERS

Synthetic polymer systems are generally quite amenable to quantitative analysis by ^{13}C nmr spectroscopy. Carbon-13 nmr can be used to quantify local stereochemical placements, with pentad and heptad structures sometimes available. Polymer end group analyses are also straightforward, at least to levels $\gtrsim 1\%$; with modern instrumentation, end-group or chain-defect analyses can *sometimes* be performed to 0.01%.

It is, of course, necessary to account for ^{13}C spin-relaxation characteristics (T_1 and NOE) to perform meaningful quantitative analyses. Early reports indicated that stereochemical placements did not result in variation of ^{13}C T_1s but recent studies have shown that large T_1 differentials may be observed for mm, mr, and rr placements. For example, in poly(butyl methacrylate) T_1s for mr and rr placements can vary by approximately 50% in an individual polymer[12] and by larger amounts in polymers having differing overall tacticities.[13] These differences in T_1s reflect variation in chain dynamics and often conformational flexibility (see Chapter 8). In synthetic polymers, NOEs for individual carbons may vary. In some polymer samples all carbon NOEs may be equivalent (either full, NOEF = 2.0, or reduced to similar values, NOEF $\lesssim$ 1.2),[14] but in other molecules polymer main-chain NOEs may be reduced whereas side-chain carbon NOEs have values ranging from NOEF = 0.5 to 2.0.[15]

Variation in T_1s and NOEs place requirements on quantitative analysis of polymer systems, as in other molecules. However, dipolar relaxation is generally effective for polymer carbons, reducing requirements for gated decoupling and/or pulse delays used for other quantitative ^{13}C nmr measurements. For some quantitative ^{13}C nmr polymer analyses it is not necessary to use any pulse delays. For example, in analysis of eight polypentenamer samples, *cis* and *trans* placements were quantified under conditions of rapid data acquisition.[16]

Polyethylene is one of the most studied polymer systems by ^{13}C nmr.[17-22] Low-density polyethylenes (LDPE) have been examined with the goal of quantitative determination of the various types of short side chain present. Carbon-13 nmr has also been used to evaluate long-chain branching and end-group concentrations in these polymers.

In one respect LDPEs are quite favorable for quantitative studies, since main- and side-chain carbons all show full Overhauser enhancements. On the other hand, most ^{13}C analyses of these polymers did not anticipate large gradations in ^{13}C T_1s. Thus rapid-pulse conditions resulted in underestimation of long T_1 side-chain carbon groupings.[17a,18a,19,21] *For ^{13}C quantitative analyses, it is always necessary to characterize relaxation behavior.* In LDPE T_1 values at 120°C vary from about 1 sec to more than 7 sec[20a]:

(^{13}C T_1s in sec)

Thus in these analyses pulse intervals of the order of 30 sec are required to avoid some selective saturation of side-chain methyl carbons (if these peaks are to be used in the analyses).

Carbon-13 nmr has enabled characterization of unusual branching patterns in LDPE at levels below one branch per 1000 main-chain carbons.[20b] These studies provide insight into the polymerization process.

Side-chain branches can be evaluated in other vinyl polymers, but generally only after chemical modification of the system to eliminate tacticity complications. For example, poly(vinyl chloride) can be reduced with $LiAlH_4$ or with $(n\text{-}Bu)_3 SnH$ to produce a polyethylene-like system that shows chain branching.[22] Further, it is possible to incorporate deuterium into the polymers by using labeled reducing agents.[22]

Randall has recently reported[23] high-sensitivity 50-MHz ^{13}C spectra of high-density (linear) polyethylenes. The use of modern high-sensitivity, high-dynamic-range FT nmr spectrometers allows determination of chain branches and end groups to the level of 0.01%. Figure 7.9 shows the ^{13}C nmr spectrum of a Phillips high-density polyethylene sample, easily demonstrating end group analysis at 0.013% (number average molecular weight, 22,000). The ^{13}C spectrum shows two distinct types of end group: saturated and allylic (peak a). Also note that peaks α and β arise solely from polymer chain branching while peaks 1s and 2s represent both saturated end group and long chain *branch* ends. Thus accurate peak area measurements make full characterization possible for these polymers.

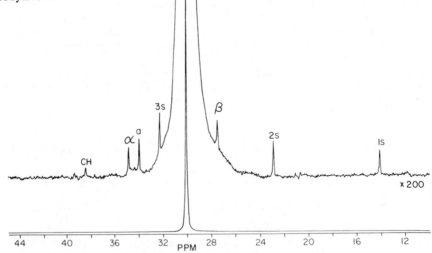

Figure 7.9.[23]   *50-MHz ^{13}C nmr spectrum of a polyethylene in a melt at 150°C (courtesy of Nicolet Technology Corporation and J. C. Randall).*

SPIN RELAXATION AND NOE

Detailed discussion of ^{13}C spin relaxation is deferred to Chapter 8. However, it is useful to indicate qualitative relaxation trends for polymer samples at this time. Polymer chains undergo complex molecular motions in solution with relatively rapid group segmental motions superimposed on slow overall reorientation of the chains themselves. Furthermore, the nature of the segmental motion is complex in some systems, combining rotational and librational (torsional) motions often with inclusion of conformational restrictions.

In some cases polymer motional dynamics is easily interpreted, whereas in other cases only very approximate pictures are as yet available. When polymer dynamics is simply described, the T_1s and NOEs are predicted from standard formulas (Chapter 8 and ref. 24). This behavior is observed in polyethylene and polyethylene oxide. When polymer dynamics is not described by the standard equations, new formalisms are required. Such is the case for many vinyl polymers: polystyrene, poly(methylmethacrylate), polybutadienes, and so on. In these systems ^{13}C T_1s become *relatively* insensitive to changes in reorientational mobility, and NOEFs are not directly calculable from observed T_1 values.

For CH carbons in vinyl polymers, T_1s typically range from 0.05 to 0.15s and nonprotonated carbon T_1s from 0.5 to several seconds (both of these ranges assume measurements at ^{13}C frequencies below 30 MHz; at higher frequencies, measured T_1s will be longer).

Typically, Overhauser enhancements will be similar (but not usually at the maximum value NOEF = 2.0) for all carbons in a polymer sample. Exceptions include systems that have (1) rapidly reorienting side-chain groups and (2) nonprotonated sp- or sp^2-carbons measured at higher ^{13}C frequencies. In the first case observed NOEFs will be a complex function of the various motions present, and in the second case NOEFs will be reduced as a result of competitive shift anisotropy relaxation (Chapter 8).

Polymer carbon T_1s have been shown to be remarkably insensitive to chain molecular weight,[25,26] showing the dominance of local segmental motions. Spin-spin relaxation times, as determined from ^{13}C linewidths (in favorable cases) sometimes depend on molecular weight, reflecting their higher sensitivity to slower motions of the polymer chains.

Polymer ^{13}C spin relaxation is discussed in a recent review dealing largely with low-molecular-weight molecules.[27]

BULK POLYMER ANALYSES

It was realized quite early that high-resolution ^{13}C nmr spectra could be obtained from some *solid* polymer samples with ordinary techniques. Using the CW technique available in the late 1960s, Duch and Grant[28a] obtained ^{13}C spectra of natural *cis*- and *trans*-1,4-polyisoprene (Figure 7.10). With the exception of some line broadening, the bulk and solution spectra for the *cis* material are essentially the same. For the *trans* polymer, the bulk spectrum required a very large number of scans and results were inferior.

For the *cis* material, it was noted that the temperature of the probe (45°C) was above the region of the glass transition temperature (T_g) at which the onset of motion of chain segments occurs in the amorphous regions of the polymer. Below the glass-transition temperature the segments of the amorphous polymer undergo mainly vibrational motions about fixed positions in a disorganized quasilattice. Above this temperature the segments exhibit translational and diffusional motions and the properties change from glassy to rubbery. Thus at 45°C the molecular motion of the bulk *cis* material is sufficient to effectively average the magnetic environment. Motional narrowing as it is present in the rubbery state is adequate for obtaining significantly improved resolution in ^{13}C spectra of bulk polymers.

The resonance linewidths of the bulk *trans* polymer are approximately two to three times that of the *cis* case. Whereas the T_g of the *cis* isomer is below the temperature at which the spectra were obtained, the T_g of the two polymorphs of the *trans* isomer occur above this temperature (45°C). Below the true thermodynamic melting point the bulk *trans* polymer may be expected to contain small regions of crystallinity. Presence of highly immobile segments of the *trans* polymer chain would result in a loss of intensity in the high-resolution component of the spectrum and could also account indirectly for the increased linewidths of the amorphous polymer. Determination of crystallinity by loss of signal in this way for *trans* polyisoprene agreed with X-ray determination of the degree of crystallinity in the solid polymer.

The average molecular weight of *trans*-1,4-polyisoprenes is between 42,000 and 100,000; for naturally occurring *cis*-1,4-polyisoprenes, molecular weights range from 200,000 to 400,000. Nevertheless, the ^{13}C nmr spectra are readily interpreted in terms of a single repeating unit in which the contributions of monomer residues to observed shifts are additive.

Problem 7.2.   *Using the techniques described in Chapter 3, estimate the ^{13}C nmr chemical shifts for all carbons in* cis- *and* trans-1,4-polyisoprene. *(Compare your answers with the observed spectra.)*

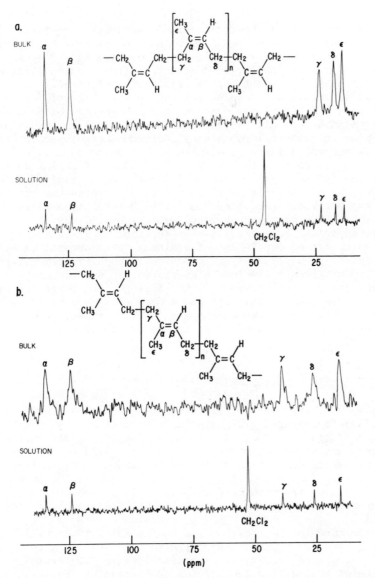

Figure 7.10[28a]   (a) The proton-decoupled ^{13}C nmr spectra of bulk
and solution samples of cis-1,4-polyisoprene.
The bulk spectrum represents the time-averaged
accumulation of 24 scans; (b) the proton-decoupled
^{13}C nmr spectra of bulk and solution samples of
trans-1,4-polyisoprene.   The bulk spectrum represents the time-averaged accumulation of 426 scans.

More recently it has been determined that most largely non-crystalline polymers give high-resolution ^{13}C spectra at temperatures sufficiently above T_g (typically 75°C above T_g).[28b]

In cases where synthetic polymer solubility is quite limited it is often possible to obtain high-resolution ^{13}C spectra on swollen gels, obtained by adding solvent to the bulk polymer in an nmr tube. Schaefer[29] has observed the ^{13}C spectra of gels formed from high-molecular-weight poly(vinyl chloride) in a moderately poor solvent (chlorobenzene) and from a copolymer of ethylene and maleic anhydride cross-linked with 3 mole% of a difunctional monomer. Despite the restricted mobility of chains in these systems, the ^{13}C nmr spectra are sufficiently well resolved to reveal details of the polymer microstructure. Proton nmr spectra of the same materials reveal no usable, high-resolution signals. Other authors have examined hydrogel systems by ^{13}C nmr.[30]

It is important to emphasize that the studies discussed previously were carried out by use of ordinary spectroscopic conditions. Recently, newly designed experiments have greatly extended ^{13}C nmr capabilities to include high-resolution spectroscopy of glassy and crystalline polymers in the bulk phase. These latter experiments require special high-power decoupling and rapid sample spinning capabilities not found on most ^{13}C instrumentation (but instruments with these added capabilities are commercially available today).

GLASSY AND CRYSTALLINE POLYMERS

In glassy or crystalline polymers local motions are markedly decreased and unaveraged ^{13}C-H dipolar interactions result in ^{13}C linewidths of order 10 kHz. The slowing of motion additionally gives broadening from anisotropies in carbon shielding tensors. Two techniques circumvent these two peak-broadening mechanisms: (1) high-power *dipolar* decoupling, and (2) "magic angle" spinning (these techniques are described in Chapter 10). Implementation of these techniques is generally straightforward, although not routine.[31-36]

In some cases polymer carbon T_1s in the solid state will be quite long, and for these systems a third technique is also required: cross polarization (also described in Chapter 10).

In one of the earlier CPMAS (cross-polarization, magic-angle spinning) studies Schaefer and Stejskal showed a largely resolved solid-state spectrum of polysulfone (Figure 7.11).[31,36] Other studies have appeared also,[32,33] including variable-temperature CPMAS studies.[34]

In some cases information not available from the liquid is present in the solid-state spectrum. For example, the broadened resonance of poly(2,6-dimethy-1,4-phenylene oxide) (C-3', Figure 7.1a) is observed as a *doublet* in the CPMAS spectrum of the

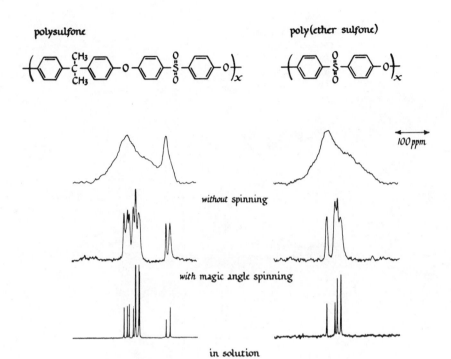

Figure 7.11.[31,36]   Carbon-13 nmr spectra of solid polysulfone
and poly(ether sulfone) samples obtained
under varying conditions.

solid polymer (Figure 7.12)[32a,36] due to restricted rotation
around the nonlinear diphenyl ether bond. Presumably this rotation is rapid in solution, but not rapid enough to eliminate some
peak broadening.

A solid sample of low crystallinity, high-molecular-weight
polyethylene showed two resolved peaks, arising from crystalline
and noncrystalline regions. The noncrystalline signal was observed at 31.7 ppm, and the crystalline component gave a resonance
at 34.1 ppm.[35] Under high power-decoupling the 34.1-ppm signal
showed a linewidth of 1 to 2.5 Hz. The peak representing non-
crystalline regions was approximately 14 Hz wide, presumably
determined by molecular mobility, with contributions from isotropic chemical-shift dispersion. An excellent review of polymer
solid-state ^{13}C nmr appeared in 1979.[36]

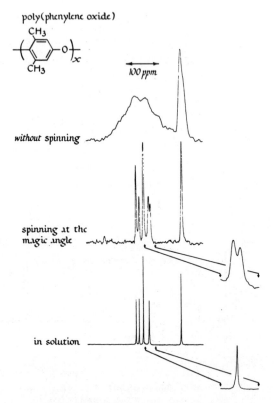

poly(phenylene oxide)

100 ppm

without spinning

spinning at the
magic angle

in solution

Figure 7.12.[32a,36]   Cross-polarization ^{13}C nmr spectra of
poly(phenylene oxide), with and without
magic-angle spinning. The cross-polarization spectra are compared to a FT spectrum
of the polymer in solution (with solvent
lines omitted for clarity of presentation).

HIGH-FIELD ^{13}C NMR OF SYNTHETIC POLYMERS

The higher sensitivity of modern high-field ^{13}C spectrometers
is well suited for polymer studies. Proportionate increases in
spectral dispersion, however, are not always realized. Furthermore, solid-state CPMAS studies are more difficult to perform at
high magnetic fields.

REFERENCES

1. F. A. Bovey, *High Resolution NMR of Macromolecules*, Academic, New York, 1972.
2. D. M. White, *Polym. Prepr.*, **13** (1), (1972).
3. D. M. White and G. C. Levy, *Macromolecules*, **5**, 526 (1972).
4. J. Schaefer, Eastern Analytical Symposium, New York, November 1971, unpublished results.
5. J. M. Saunders and R. A. Komoroski, *Macromolecules*, **10**, 1214 (1977); G. J. Ray, P. E. Johnson, and J. R. Knox, *ibid.*, **10**, 773 (1977); A. E. Tonelli, *ibid.*, **11**, 634 (1978).
6. F. A. Bovey, M. C. Sacchi, and A. Zambelli, *Makromol. Chem.*, **7**, 752 (1974); K. F. Elgert and W. Ritter, *Makromol. Chem.*, **177**, 281 (1976).
7. (a) W. O. Crain, Jr., A. Zambelli, and J. D. Roberts, *Macromolecules*, **4**, 330 (1971); (b) A. Zambelli, G. Gatti, C. Sacchi, W. O. Crain, Jr., and J. D. Roberts, *ibid.*, **4**, 475 (1971); (c) C. J. Carman and C. E. Wilkes, *Rubber Chem. Tech.*, **44**, 781 (1971).
8. E. A. Williams, J. D. Cargioli, and S. Y. Hobbs, *Macromolecules*, **10**, 682 (1977).
9. (a) C. J. Carman, A. R. Tarpley, Jr., and J. H. Goldstein, *J. Am. Chem. Soc.*, **93**, 2864 (1971); (b) *ibid.*, *Macromolecules*, **4**, 445 (1971).
10. J. C. Randall, *Polymer Sequence Determination, Carbon-13 NMR Method*, Academic, New York, 1977.
11. J. Schaefer, *Macromolecules*, **4**, 98 (1971).
12. G. C. Levy, *J. Am. Chem. Soc.*, **95**, 6117 (1973).
13. (a) J. R. Lyerla, Jr., T. T. Honkawa, and D. E. Johnson, *J. Am. Chem. Soc.*, **99**, 2463 (1973); (b) Y. Inoue, T. Konno, R. Chujo, and A. Nishoka, *Makromol. Chem.*, **178**, 2131 (1977).
14. J. Schaefer, *Macromolecules*, **4**, 107 (1971).
15. G. C. Levy, D. E. Axelson, R. Schwartz, and J. Hochmann, *J. Am. Chem. Soc.*, **100**, 410 (1978).
16. C. J. Carman, *Macromolecules*, **7**, 40 (1974).
17. (a) D. E. Dorman, E. P. Otocka, and F. A. Bovey, *Macromolecules*, **5**, 574 (1978); (b) F. A. Bovey, F. C. Schilling, F. L. McCracken, and W. L. Wagner, *ibid.*, **9**, 76 (1976); (c) H. N. Cheng, F. C. Schilling, and F. A. Bovey, *ibid.*, **9**, 363 (1976)..
18. (a) J. C. Randall, *J. Polym. Sci., Polym. Phys. Ed.*, **11**, 275 (1973); (b) J. C. Randall, *J. Appl. Polym. Sci.*, **22**, 585 (1978).
19. M. E. A. Cudby and A. Bunn, *Polymer*, **17**, 345 (1976).
20. (a) D. E. Axelson, L. Mandelkern, and G. C. Levy, *Macromolecules*, **10**, 557 (1977); (b) D. E. Axelson, G. C. Levy, and L. Mandelkern, *ibid.*, **12**, 41 (1979).
21. T. N. Bowmer and J. H. O'Donnell, *Polymer*, **18**, 1032 (1977).
22. F. A. Bovey, F. C. Schilling, and W. H. Starnes, Jr., *Polym. Prepr.*, **20**, 160 (1979).
23. J. C. Randall, *Polym. Prepr.*, **20**, 235 (1979).

24. J. Schaefer, *Topics in Carbon-13 NMR Spectroscopy*, Vol. 1, G. C. Levy, Ed., Wiley-Interscience, New York, 1974, p. 150.
25. A. Allerhand and R. K. Hailstone, *J. Chem. Phys.*, **56**, 3718 (1972).
26. Y. Inoue, et al., *J. Polym. Sci.*, *Polym. Phys.* **11**, 2240 (1973).
27. D. A. Wright, D. E. Axelson, and G. C. Levy, *Topics in Carbon-13 NMR Spectroscopy*, Vol. 3, G. C. Levy, Ed., Wiley-Interscience, New York, 1979, p. 3.
28. (a) M. W. Duch and D. M. Grant, *Macromolecules*, **3**, 165 (1970); (b) D. E. Axelson and L. Mandelkern, *J. Polym. Sci.*, *Polym. Phys. Ed.*, **16**, 1135 (1978).
29. J. Schaefer, *Macromolecules*, **4**, 110 (1971).
30. K. Yokota, A. Abe, S. Hosaka, I. Sakai, and H. Saito, *Macromolecules*, **11**, 95 (1978).
31. J. Schaefer and E. O. Stejskal, *J. Am. Chem. Soc.*, **98**, 1031 (1976).
32. (a) J. Schaefer, E. O. Stejskal, and R. Buchdahl, *Macromolecules*, **10**, 384 (1977); (b) M. D. Sefcik, E. O. Stejskal, R. A. McKay, and J. Schaefer, *ibid.*, **12**, 423 (1979).
33. M. M. Maricq, J. S. Waugh, A. G. MacDiarmid, H. Shirakawa, and A. J. Heeger, *J. Am. Chem. Soc.*, **100**, 7729 (1978).
34. J. R. Lyerla, H. Vanni, C. S. Yannoni, and C. A. Fyfe, *Polym. Prepr.*, **20**, 255 (1979): C. A. Fyfe, J. R. Lyerla, W. Volksen, and C. S. Yannoni, *Macromolecules*, **12**, 757 (1979).
35. W. L. Earl and D. L. Vanderhart, *Macromolecules*, **12**, 762 (1979).
36. J. Schaefer and E. O. Stejskal, *Topics in Carbon-13 NMR Spectroscopy*, Vol. 3, G. C. Levy, Ed., Wiley-Interscience, New York, (1979), Chapter 4.

Chapter 8

Relaxation Studies

The development of FT nmr has encouraged a strong interest in ^{13}C spin-relaxation measurements and their application to organic chemistry problems. Minor variations of standard pulsed FT nmr techniques allow rapid, quantitative determination of spin-lattice (longitudinal) relaxation times (T_1) whereas somewhat more elaborate experiments are utilized to measure spin-spin (transverse) relaxation times (T_2). Usually, T_1 measurements yield the desired information; therefore, the more difficult T_2 measurements are used less often, especially in studies of organic compounds. Most of this chapter is devoted to results and applications of T_1 measurements. However, a short section on T_2 experiments is included.

THE NMR EXPERIMENT IN THE ROTATING FRAME OF REFERENCE

The pulse nmr experiment can be considered in a way that simplifies explanations of 90° and 180° pulses, and spin-lattice and spin-spin relaxation: the so-called rotating frame of reference.[1] In the rotating frame the entire coordinate system rotates at the Larmor frequency ω_0, corresponding to the laboratory magnetic field B_0. The rf field, B_1, oscillating at ω_0, is then fixed in the rotating frame.

Figure 8.1 depicts nmr excitation and relaxation in the rotating frame. The characteristic of the nuclear spins (in this case assumed to be ^{13}C nuclei with no magnetic nuclei present) that is considered in Figure 8.1 is M, the net magnetization of the sample. Here M corresponds to the *sum of all the individual nuclear magnetic moments*. When a sample is introduced into the magnetic field, B_0, no polarization of nuclear spins is initially present (i.e., the populations of the upper and lower energy levels are equal; $M = 0$, Figure 8.1a). Interactions between the ^{13}C nuclei and the lattice (spin-lattice relaxation) eventually result in establishment of an equilibrium excess of ^{13}C nuclei in the lower energy level, according to the Boltzmann distribution law. The equilibrium results in a small *net* magnetization aligned with the

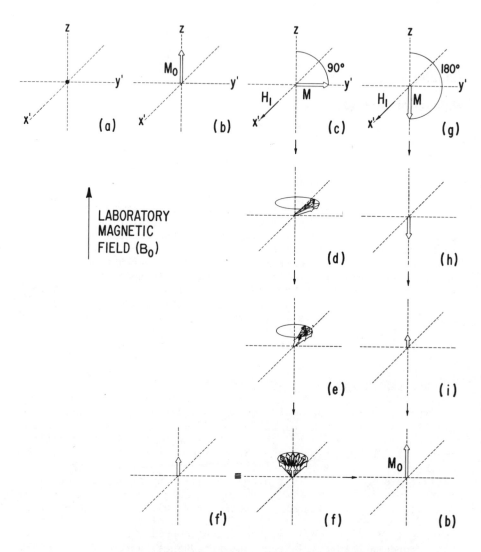

Figure 8.1.   The pulse nmr experiment in the rotating frame.
Sequences c,d,e,f,b and g,h,i,b demonstrate
relaxation processes.

direction of B_0 (the z axis). This equilibrium net magnetization
is referred to as M_0 (Figure 8.1b). The net magnetization remains
equal to M_0 only until rf excitation of the sample is initiated.
When the sample is irradiated, the rf field is applied along
the x axis in the rotating frame, as shown in Figures 8.1c and
8.1g. The magnetic component of the rf field turns M out of

alignment with B_0 (the z axis) and toward the y axis. In pulse nmr
this process is very rapid, normally precluding significant relaxation during the pulse excitation. The pulse can be applied for
an experimentally determined time (usually 1 to 100 μsec) to
result in tipping M by 90° (Figure 8.1c), or the *width* of the pulse
in *time* may be twice as long, causing M to completely invert (Figure 8.1g). A 180° pulse applied to a fully relaxed spin system
$(M = M_0)$ leads to a so-called negative spin temperature since immediately following the pulse a Boltzmann excess of nuclear spins
is in the *higher energy level*.

Immediately following every excitation pulse the process of
spin-lattice relaxation begins. In the context of the rotating
frame, spin-lattice relaxation is relaxation along the z axis
whereas spin-spin relaxation corresponds to relaxation in the $x'y'$
plane. The former can be considered as an energy process and the
latter an entropy process. Figures 8.1h and 8.1i show the return
of M to M_0 following a single 180° pulse. In this case only spin-
lattice relaxation is possible since there is no $x'y'$ magnetization, and M returns to M_0 according to first-order kinetics with a
rate constant $1/T_1$. No free induction decay is observed following
a 180° pulse such as in Figure 8.1g. *The signal detected in nmr*
spectrometers is the magnetization in the $x'y'$ plane, which is
zero in this case. By contrast, a single 90° pulse causes M to
coincide with the y' *axis, in the $x'y'$ plane* (Figure 8.1c). The
decay of the $x'y'$ magnetization as a function of time is the
free induction decay (FID) discussed in Chapter 1. Fourier transformation of the FID results in an nmr spectrum as a function of
frequency. Following the 90° pulse both spin-lattice and spin-
spin relaxation processes begin (Figures 8.1d to 8.1f). As before,
M tends to return vertically toward the equilibrium value M_0 (spin-
lattice relaxation). The $x'y'$ magnetization, on the other hand,
"dephases" as a function of the spin-spin relaxation time T_2. No
signal is observed after the $x'y'$ magnetization is completely de-
phased, even if z-axis relaxation is incomplete (Figure 8.1f; this
corresponds to T_2 shorter than T_1).

Free induction decay signals normally decay as a function of T_2^* and
not T_2; T_2^* contains contributions to dephasing from magnetic-field
inhomogeneities, incomplete decoupling, and so on. Spectral resolution in most FT spectra is determined by T_2^*. The full linewidth
at half-height $(\upsilon_{1/2})$ of a single line is given by

$$\upsilon_{1/2} = \frac{1}{\pi T_2^*}$$

SPIN-LATTICE RELAXATION MECHANISMS

As mentioned in Chapter 1, spin-lattice relaxation is the
process of energy exchange between individual nuclear spins and
the surrounding liquid or solid "lattice". This relaxation allows

the lattice to act as a *heat sink* to establish and restore thermal equilibrium for the nuclear spins, following initial placement of the sample in the magnet and following rf excitation, respectively.

A mechanism for *coupling* the nuclear spins and the lattice is required for efficient energy transfer. All the potential ^{13}C spin-lattice relaxation mechanisms share one characteristic: dependence on the presence of fluctuating localized magnetic fields at or near the nucleus being relaxed. The relaxation mechanisms that can be specifically important for ^{13}C spin-lattice relaxation are:

1. Dipole-dipole relaxation (DD):
   a. With ^{1}H nuclei.
   b. With other nuclei (e. g., ^{19}F, other ^{13}C, etc.).
   c. With unpaired spins (e. g., dissolved O_2, relaxation agents, etc.).
2. Spin-rotation relaxation (SR).
3. Chemical-shift anisotropy (CSA).
4. Scalar relaxation (SC).

Dipolar relaxation with protons generally dominates for ^{13}C nuclei in organic molecules. However, in specific cases any of these relaxation interactions can give rise to efficient ^{13}C relaxation. Detailed treatments of ^{13}C spin-relaxation mechanisms have been published elsewhere.[2-4] However, the ^{13}C relaxation mechanisms are considered briefly here, with particular note to structural features that favor their operation.

Spin-Rotation Relaxation.   A molecule or a group possessing angular velocity is a rotating charge system, and this can give rise to a magnetic field at the resonant nucleus. Fluctuations in this field result from modulation of both the magnitude and the direction of the angular momentum vector associated with the rotating molecule or group. In liquids, interruptions of the angular velocity generally occur too often for this process to be efficient as a ^{13}C relaxation interaction (i.e., modulation of the interaction is at frequencies much higher than the ^{13}C resonance frequency ω_0). However, with very small molecules, or for freely spinning groups (e.g., the CH_3 group of toluene; see later) the time scale of the spin-rotation interaction becomes long enough to approach ω_0; in these cases spin-rotation relaxation becomes relatively efficient ($T_1^{SR} \approx$ 10 sec). For spherical molecules, the spin-rotation relaxation rate is given by Equation 8.1:

$$R_1^{SR} = \frac{1}{T_1^{SR}} = \frac{2\pi kT}{\hbar^2} I\, C^2 \tau_{SR} \qquad (8.1)$$

where I_m is the moment of inertia, C is the isotropic spin-rotation interaction constant, and τ_{SR} is the appropriate correlation time; τ_{SR} and the molecular reorientational correlation time τ_c appropriate for dipole-dipole relaxation, have a reciprocal relationship. Thus T_1^{DD} and T_1^{SR} processes each operate at the expense of the other. An increase in temperature will increase the mobility of a dissolved small organic molecule, thus leading to a shortening of T_1^{SR} and a lengthening of T_1^{DD}. When the two mechanisms together dominate spin-lattice relaxation for a carbon, changes in sample temperature may not substantially change the observed T_1 (although the NOE will change dramatically).

Chemical-Shift Anisotropy Relaxation.   Anisotropy (directionality) in the screening of a magnetic nucleus by surrounding electrons results in a shielding tensor σ having directional components that undergo modulations as a result of molecular reorientation (rotation). The functional form of this relaxation mechanism for a molecule of cylindrical symmetry (e.g., acetylene) is given by Equation 8.2:

$$\frac{1}{T_1^{CSA}} = \frac{2}{15} \gamma_C^2 B_0^2 (\sigma\|- \sigma\bot)^2 \tau_c \qquad (8.2)$$

where γ_C is the magnetogyric ratio for ^{13}C, B_0 is the static (spectrometer) magnetic field, $\sigma\|$ and $\sigma\bot$ are the respective values of σ parallel and perpendicular to the symmetry axis, and τ_c is the molecular reorientation time.

Chemical-shift anisotropy relaxation may be identified by its quadratic dependence on B_0. At one time[2] it was thought that CSA relaxation would be rare for ^{13}C, but recent use of higher magnetic fields has markedly changed the situation. At ^{13}C observation frequencies near 68 MHz T_1^{CSA} commonly contributes $\gtrsim 50\%$ to relaxation of nonprotonated sp^2 or sp carbons;[5] at very high fields ($^{13}C \gtrsim 100$ MHz) T_1^{CSA} should become a significant factor even for protonated sp^2 and sp carbons. Rarely, T_1^{CSA} dominates ^{13}C relaxation even at low magnetic fields ($^{13}C \sim 25$ MHz).[6]

Scalar Relaxation.   Rapid modulation of the spin-spin (scalar) coupling between ^{13}C and another spin (nucleus or electron) may in principle effect ^{13}C spin-lattice relaxation. Two sources of modulation can be considered: (1) chemical exchange and (2) rapid relaxation of the spin-coupled nucleus or electron. For a scalar-coupled nucleus X, the first case could occur by way of diffusion-controlled or rapid fluxional exchange processes and so on; the second case would result if X were undergoing rapid quadrupolar relaxation.

In fact, scalar spin-lattice relaxation is rarely significant for ^{13}C, as a result of the need for ^{13}C and X nuclei to resonate at closely spaced frequencies. The equation for scalar spin-lattice relaxation of a carbon shows this:

$$\frac{1}{T_1^{SC}} = \frac{8\pi^2 J^2}{3} S(S+1) \frac{\tau_{SC}}{1+(\omega_X - \omega_C)^2 \tau_{SC}^2} \tag{8.3}$$

where J is the scalar coupling constant, S is the spin quantum number of X, ω_X and ω_C represent the Larmor frequencies of ^{13}C and X (in rad/sec), and τ_{SC} is the scalar relaxation correlation time.[2,3] The denominator in Equation 8.3 grows formidable when the resonance frequencies for ^{13}C and X nuclei differ substantially. (However, with very large spin-spin coupling constants $(\omega_X - \omega_C)$ of order 10^6 to 10^7 can still allow SC relaxation.[7]) The only common case for scalar spin-lattice relaxation of ^{13}C is for carbons attached to one or more bromines.[8]   (The Larmor frequency for ^{79}Br is almost coincident with that of ^{13}C; the ^{81}Br resonance frequency differs by a few MHz at moderate magnetic fields.)
    Scalar ^{13}C *spin-spin* relaxation, by contrast, is relatively common. For the T_2 process there is an additional term that does not have the $(\omega_X - \omega_C)^2 \tau_{SC}^2$ denominator. Thus ^{13}C nuclei can show *line broadening* if they are spin-spin coupled to ^{14}N or to other nuclei undergoing rapid quadrupolar relaxation.

Dipole-Dipole Relaxation.   Dipole-dipole relaxation arises from local magnetic fields associated with magnetic nuclei (nuclei with spin $I \neq 0$). If two neighboring magnetic nuclei are placed in an external magnetic field B_0, each nucleus sees a total magnetic field comprised of B_0 and a contribution from the local magnetic field of the other nucleus. The strength and the direction of this localized interaction depends on the magnetic moments and the internuclear separation of the two nuclei and their internal orientation *relative* to B_0. When molecular motions are rapid, as they are in a liquid, the relative orientation of the two nuclei with respect to B_0 is constantly changing. The rapid internal reorientation of the two nuclei gives rise to rapid fluctuations of the localized (and thus of the total) magnetic field as seen by each nucleus; these fluctuations effect relaxation of the nuclei.
    Relaxation of ^{13}C nuclei in organic molecules usually results from dipole-dipole interactions with protons. This relaxation follows the general Equation 8.4:

$$\frac{1}{T_1^{DD}} = \sum_{i=1}^{n} \frac{\gamma_C^2 \gamma_H^2 \hbar^2}{r_{CH_i}^6} \tau_c^{eff} \tag{8.4}$$

where γ_C and γ_H are the magnetogyric ratios for ^{13}C and ^1H nuclei, r_{CHi} is the through-space distance between the ^{13}C nucleus and proton i, and τ_c^{eff} is the effective reorientational correlation time (rigorously τ_c for a symmetric tumbler such as adamantane). Dipole-dipole interactions are summed for all relevant protons i.

The r_{CHi}^{-6} distance dependence for the ^{13}C dipolar relaxation rate means that *intermolecular* contributions may usually be ignored. In fact, for protonated carbons, only *bonded* hydrogens make major dipolar contributions to T_1^{DD}. In the case of protonated carbons the dipolar relaxation rate is then simplified, following Equation 8.5:

$$\frac{1}{T_1^{DD}} = N_H \gamma_C^2 \gamma_H^2 \hbar^2 r_{CH}^{-6} \tau_{eff} \qquad (8.5)$$

Where N_H is the number of directly attached hydrogens (each one has a magnetic field), and r_{CH} is now represented by the C-H bond distance, assumed to be constant at 1.09×10^{-8} cm.*

The average frequency (or, more accurately, the frequency distribution) of a localized fluctuating magnetic field determines its effectiveness in causing spin relaxation. Frequencies of motions for molecules in solution range from very slow to very fast. In the case of very small molecules, most of the individual molecules will be "rotating" at rates of $>10^{12}$ revolutions per second while a few will be "rotating" much more slowly. Actually, the movements of a molecule in solution are very complex; characteristic rotation rates cannot be dissected easily from overall tumbling motions. Fortunately, it is generally not necessary to consider separation of the various molecular motions. It is usually sufficient to describe an effective correlation time τ_{eff}, which represents the *average* time for a molecule to rotate through 1 radian (2π rad/sec = 1 revolution/sec or 1 Hz). Dipole-dipole relaxation is best effected by rotational motions with frequencies comparable with the resonance frequency, which for ^{13}C nuclei is 10^7 to 10^8 Hz or 10^8 to 10^9 rad/sec. A correlation time close to the reciprocal of the resonance frequency in rad/sec leads to most efficient dipolar spin-lattice relaxation of a ^{13}C nucleus. A graphical representation of the effect of τ_c on T_1 values is shown in Figure 8.2. Figure 8.2 also shows what happens to the spin-spin relaxation time T_2 as τ_c changes.

The correlation time for most molecules falls on the left-hand side of Figure 8.2. Typical τ_c values for small molecules would be 10^{-12} to 10^{-13} sec; for reasonably large organic molecules

---

*This assumption does not always prove valid, but its use has been shown to be satisfactory for many studies. In certain cases C-H bond length may be calculated accurately.[9]

τ_c may be as long as 10^{-10} sec. Near room temperature only very large molecules are likely to have values approaching the elbow in Figure 8.2 ($\sim 10^{-9}$). Several molecules are shown in Figure 8.2 at positions indicating τ_c for these compounds.

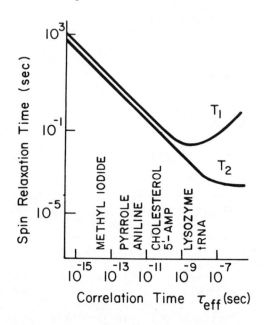

Figure 8.2.   *Relationship between T_1 (and T_2) and molecular tumbling as represented by τ_{eff} (assumes dipolar relaxation).*

It is clear from Figure 8.2 that for typical organic molecules *any change that shortens* τ_c *results in a lengthening of* T_1. For example, lowering the solution viscosity and raising the temperature of the sample both shorten τ_c; lower viscosity loosens constraints on molecular tumbling, and higher temperature raises the average thermal (kinetic) energy of the molecules.

Large organic molecules (e.g., mw > 300) tumble relatively slowly in solution because of their size (inertial effects) and the extensive reordering of solvent molecules that are swept aside during molecular rotation (this second effect also depends on molecular symmetry). Figure 8.2 indicates that these large molecules will relax much more efficiently than will small, rapidly tumbling molecules. Of course, *slow* and *rapid* are meaningful only in the context of Figure 8.2.

If τ_c exceeds 10^{-8} to 10^{-9} sec, the dependence of T_1 on viscosity and temperature reverses; this is not commonly observed except with some synthetic or biopolymer molecules, or in solids.

Spin Relaxation in Very Large Molecules: The Extreme Narrowing Approximation.

Equations 8.1 to 8.5 are valid only for molecules undergoing rotational reorientation that is within the region of extreme spectral narrowing, that is, rapid rotation relative to the inverse of the ^{13}C resonance frequency. Strictly speaking, the condition of spectral narrowing is fulfilled when $(\omega_C + \omega_H)$ times $(\tau_c)^2 \ll 1$. This condition is violated with very large molecules that reorient as a unit, such as large natural products (e.g., vitamin B-12) and with biopolymers. Synthetic polymers and highly associated low-molecular-weight molecules may also reorient too slowly to meet the extreme narrowing definition. Also, recent studies[10] have indicated that some small molecules show T_1 behavior deviating from the extreme narrowing limit despite overall motion that is still sufficiently rapid.

In general, when a molecule reorients slowly enough to violate the spectral narrowing condition, ^{13}C relaxation behavior is defined by more complex equations.[11] Furthermore, ^{13}C relaxation parameters become dependent on the spectrometer magnetic field. Figure 8.3 expands the region of Figure 8.2 corresponding to longer correlation times and further shows the effect of magnetic field on ^{13}C T_1s. Modern use of very high field spectrometers places additional compounds outside of the region of extreme spectral narrowing.

For molecules deviating from the extreme narrowing condition, the NOE associated with ^{13}C-^1H dipole-dipole relaxation is reduced and is also field dependent. In the limit of very slow solution reorientation $^{13}C\{^1H\}$ NOEFs approach 0.1 (vs. 1.99 for full $^{13}C\{^1H\}$ dipolar relaxation within extreme narrowing).

Carbon-13 Electron-Nuclear Spin Relaxation.

Unpaired spins are quite effective in relaxing ^{13}C nuclei. The magnetic moment of the electron is over two orders of magnitude larger than nuclear magnetic moments, and thus dipole-dipole interactions between ^{13}C nuclei and unpaired spins are more efficient than nuclear-nuclear dipolar interactions by a factor exceeding 10^5, assuming identical motion and distance factors. At levels of $10^{-3}M$ to $10^{-4}M$ of unpaired spins, electron-nuclear dipole-dipole relaxation efficiently competes with normal diamagnetic dipolar relaxation. The degree of replacement of normal diamagnetic relaxation depends on (1) the specific relaxation agent, (2) the magnitude of T_1 in the absence of the paramagnetic spins (T_1^{dia}), and (3) the presence of specific chemical or physical (e.g., electrostatic) interactions between the diamagnetic molecules and the paramagnetic ion or molecule.[12] Normal diamagnetic relaxation is not completely replaced with addition of relaxation agents unless the achievable T_1^e is *much*

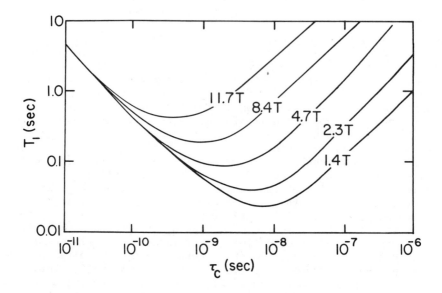

FIGURE 8.3.   Dipolar T_1 for a CH carbon at field strengths of
1.4 to 11.7 Tesla (Note spectral narrowing region
at left).

shorter than $T_1{}^{dia}$ as from Equation 8.6:

$$\frac{1}{T_1{}^{e}} = \frac{1}{T_1{}^{para}} - \frac{1}{T_1{}^{dia}} \qquad (8.6)$$

where $T_1{}^{para}$ and $T_1{}^{dia}$ are the *observed* T_1s in the paramagnetic
sample and diamagnetic sample, respectively, and $T_1{}^{e}$ represents the
electron-nuclear contribution from the relaxation agent.

Typical paramagnetic relaxation agents (PARRs) include complexed Gd^{3+} or simple Mn^{2+} salts used in aqueous solutions and
$Cr(acac)_3$,[13] $Cr(dpm)_3$,[14] and $Gd(dpm)_3$[15] for use in organic
solvents. These reagents at required concentrations do not produce significant changes in substrate chemical shifts, although
small shift changes can be observed for carbons close to PARR
binding sites.[16] Nevertheless, relaxation reagents can be mis-
used. For one thing, all these agents except perhaps $Cr(dpm)_3$
may selectively bind to one site, leading to variable substrate

T_1s.[12,14] Furthermore, it is not always practical to add sufficient PARR to fulfill the requirement for *complete* suppression of
$^{13}C\{^1H\}$ NOEs (relaxation fully controlled by T_1^e, see Equation 8.6).

Separation and Identification of Relaxation Contributions.   For
most applications, to obtain meaningful information from ^{13}C spin-
relaxation data, it is *absolutely necessary* to know the mechanism(s)
responsible for the observed T_1s. When more than one mechanism is
contributing, a semiquantitative, or, better, a quantitative assessment of at least one mechanistic contribution is necessary.   In
fact, this is usually a straightforward procedure.   The observed
^{13}C relaxation *rate* is dissected according to Equation 8.7:

$$R_1^{obs} = \frac{1}{T_1^{obs}} = \frac{1}{T_1^{DD}} + \frac{1}{T_1^{SR}} + \frac{1}{T_1^{CSA}} + \frac{1}{T_1^{SC}} + \frac{1}{T_1^e} \qquad (8.7)$$

In fact, several other separations such as *intramolecular* and *intermolecular* ^{13}C-^1H dipolar terms, may be effected, depending on the
degree of identification possible.
  The identification of mechanistic contributions to T_1 requires
a number of experiments. Assuming that extreme spectral narrowing
condition applies, $T_1^{DD}(^{13}C$-$^1H)$ is identified from the observed
NOEF in $^{13}C\{^1H\}$ experiments:

$$T_1^{DD} = T_1^{obs} \cdot \frac{1.98}{(NOEF)_{obs}} \qquad (8.8)$$

  Spin-rotation relaxation is identified by its temperature
dependence, T_1^{SR} shortening with increasing temperature, whereas
T_1^{DD} and T_1^{CSA} increase with applied heat (T_1^{CSA} is easily quantified from measurements at two or more magnetic fields). Scalar
and/or electron-nuclear spin-lattice relaxation processes are not
identified from any one characteristic, but from a combination of
structural effects and temperature and frequency dependencies.

Problem 8.1.   (a) *using the following data, calculate the contribution from atmospheric oxygen to the observed T_1
for benzene in a sample that is not degassed.*

BENZENE (38°C, 25.2 MHz)	T_1^{obs} (SEC)	NOEF
Not degassed	23.0	1.3
Degassed	29.3	1.6

(b) *What is the origin of the remaining nondipolar
relaxation contribution? How could you test this?*

Problem 8.2. *Using the information provided, estimate the individual relaxation mechanism contributions at 30°C and at 25 MHz for the carbons indicated.*

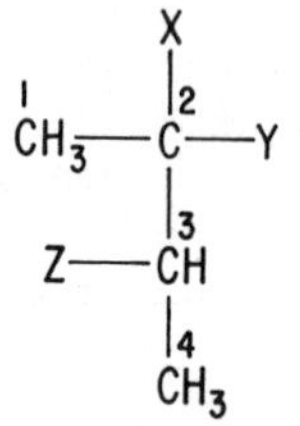

$$
\begin{array}{c}
\quad\ \ \overset{X}{|} \\[2pt]
\overset{1}{CH_3} \!-\! \overset{2}{\underset{\underset{\ \ |_4}{\underset{CH_3}{}}}{\overset{|}{\underset{|_3}{C}}}} \!-\! Y \\
Z \!-\! CH
\end{array}
$$

(X, Y, and Z are unspecified groups or atoms.)

		T_1 (sec)	NOE (η, 2.0 max)
30°C Sample not degassed (25 MHz)	C-1	9.0	1.1
	C-2	53	1.4
	C-3	4.2	0.6
	C-4	6.5	2.0
60°C Sample not degassed (25 MHz)	C-1	8.8	
	C-2	78	
	C-3	1.3	0.2
	C-4	10.4	2.0
30°C Sample degassed (25 MHz)	C-2	62	1.8
30°C Sample not degassed (75 MHz)	C-2	26	

Cross-Correlation Effects.   It has recently been shown that the simple picture of proton-decoupled ^{13}C spin relaxation is not entirely accurate. Various theoretical and experimental treatments have shown that significant deviations from predicted behavior may result from cross-correlation interaction terms.[17] This at once presents difficulty and opportunity for ^{13}C T_1 applications. Difficulty arises from the lack of an easily defined *exact* correspondence between observation and theory; opportunity arises because with very simple molecular systems (e.g., CH_2X_2) selective multipulse experiments give insight into details of spin-relaxation processes.

These considerations will not usually apply to organic chemistry applications of ^{13}C spin relaxation studies, where large spin manifolds tend to minimize deviations from simple behavior.

EXPERIMENTAL MEASUREMENT OF ^{13}C SPIN-RELAXATION PARAMETERS

Experimental determinations of ^{13}C NOEs and spin-lattice and spin-spin relaxation times are all practical. With modern FT instrumentation T_1 and NOE measurements on moderately dilute solutions are straightforward. Pulse T_2 measurements in natural-abundance ^{13}C nmr, however, must still be classified as nonroutine.

For successful ^{13}C relaxation measurements, it is critical that experimental conditions be well controlled. This section discusses available methods for efficient measurement of ^{13}C relaxation parameters.

Measurement of T_1.   As noted earlier, spin-lattice relaxation is the process that establishes or restores the longitudinal or *z* magnetization of a spin system to its equilibrium value, M_0. Measurements of T_1 are designed to monitor M_z as a function of a time parameter, to describe its approach to M_0.

Pulse FT methods for T_1 measurements include inversion-recovery,[18] fast-inversion-recovery,[19] progressive saturation,[20] saturation-recovery,[21] and dynamic NOE/gated-decoupled[22] pulse sequences. Table 8.1 summarizes the characteristics of these pulse sequences for T_1 measurements. Generally, the slower sequences are less useful for natural-abundance ^{13}C T_1 measurements. The fast-inversion-recovery Fourier transform (FIRFT) and saturation-recovery Fourier transform (SRFT) pulse sequences can be recommended for routine use, but the progressive saturation Fourier transform (PSFT) sequence gives large errors with misset 90° pulses and should be avoided except under exceptional conditions. (A recent paper[23] concludes that the FIRFT sequence generally has the highest efficiency.) Other pulse sequences have also been described.[24]

Several papers discuss experimental methods for T_1 measurements, treating sources of errors and outlining requirements for accurate determinations.[25-27] Other papers discuss statistical treatment of the experimental data to give accurate T_1s in shorter times.[28-30]

One common method for measurement of T_1 is based on a two-pulse sequence and is called the *inversion-recovery* method, which utilizes a (180°-τ-90°) pulse sequence. The 180° pulse inverts the two ^{13}C energy-level populations, instantaneously producing a Boltzmann excess of nuclei in the higher energy level (refer to Figure 8.1). Following the 180° pulse, the nuclei immediately begin to relax to reestablish the normal Boltzmann distribution (excess nuclei in the lower energy state). The 90° pulse is applied after a waiting period τ, which is varied in successive experiments. A free induction decay (FID) results from the 90° pulse and is digitized and stored. If τ is very long relative to the longest T_1 for any nucleus being observed, the FID yields a fully relaxed spectrum, indistinguishable from the spectrum that results from a single 90° pulse experiment. If on the other hand, τ is not much longer than T_1 for all nuclei, the FID that results from the 90° pulse will yield data (after FT) as shown in Figure 8.4. Figure 8.4 shows three cases in addition to the fully relaxed spectrum.

1. $τ \ll T_1$. The signal for this nucleus (signal A) appears inverted, indicating that little relaxation occurred following the 180° inverting pulse.
2. $τ \sim T_1 \cdot \ln 2 \approx 0.69 \, T_1$.   Signal B is nearly nulled.
3. $τ \sim T_1$. Signal C is small and positive.
4. $τ \gg T_1$. Signal D is positive and is "full" intensity, as discussed previously.

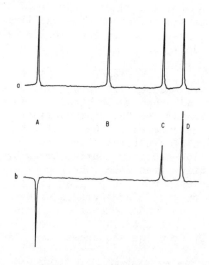

Figure 8.4.   *Simulated (a) normal and (b) IRFT ^{13}C nmr spectrum.*

TABLE 8.1   CARBON-13 FT NMR T_1 PULSE SEQUENCES

METHOD	PULSE SEQUENCE REPETITION RATE (SEC)	STRENGTHS AND LIMITATIONS
Inversion-recovery (*IRFT*) $(T-180°-\tau-90°$ FID$)_n$	$T \gtrsim 4\,T_1$	Gains factor of 2 in "dynamic range,"[a] pre-evaluation of T_1 necessary, slow sequence
Fast inversion-recovery (*FIRFT*) $(T-180°-\tau-90°$ FID$)_n$	$T \gtrsim 1$ sec	"Dynamic range"[a] self-optimizing, fast repetition
Progressive saturation (*PSFT*) $(90°$ FID$-\tau)_n$	$T \gtrsim 0.3$ sec for high resolution	Pulse angles must be accurate, resolution $\propto$ (pulse interval)$^{-1}$, fastest sequence
Saturation-recovery (*SRFT*) $(z^*-90°-z^*-\tau-90°$ FID$)_n$	$T \gtrsim 1$ sec	Fast repetition, z-homospoil (z^*) or pulse burst required
Dynamic NOE (*DNOEFT*) $[T(\tau\text{-decouple})-90°$ FID$]_n$	$T \gtrsim 10\,T_1$	Slowest sequence, determines T_1 and NOE simultaneously

[a] "Dynamic range" corresponds to the total change in M_z that is being monitored. A factor of 2 in dynamic range may be considered as a factor of 2 in S:N.

A spectrum such as that in Figure 8.4b may be called a *partially relaxed FT* (PRFT) spectrum. A set of these spectra with variation of τ allows calculation of the T_1 values for all the different nuclei in a molecule.

In practice, it is necessary to accumulate many FIDs to achieve usable S:N ratios. In most relaxation studies a pulse sequence such as $(T\text{-}180°\text{-}\tau\text{-}90°)_n$ is used. Here, the new waiting time T allows the sample to recover completely from the previous pulse sequence, thereby simulating an isolated $180°\text{-}\tau\text{-}90°$ pulse sequence. In practice it is necessary to wait only three to four times longer than the longest T_1 value to be determined in the IRFT experiment.

Provided that certain conditions are met, it is *not*, in fact, necessary to use a long T waiting period for T_1 measurements. For example, if the first FID in the FIRFT sequence is deleted, the pulse-sequence delay T can be much shorter than T_1, provided it exceeds T_2* (typically $\gtrsim 1$ sec).[19]

Current FT nmr computer programs implement nonlinear least-squares analysis to determine T_1s without requiring equilibrium spectra (to obtain the value of M_0) in FIRFT or SRFT experiments. These exponential fitting procedures save time by a factor of 2 to 3 for accurate measurement of long T_1s, where fully relaxed spectra require extended time averaging. Two-parameter fits to Equation 8.9, however, have been shown to be unreliable as a result of pulse inaccuracies, frequency offsets, and so on. In Equation 8.9 $S(\tau)$ is the signal following the FIRFT 90° pulse and

$$S(\tau) = M_0 \left[1 - 2 \left(1 - \exp\left(\frac{-T}{T_1}\right) \exp\left(\frac{-\tau}{T_1}\right) \right) \right] \qquad (8.9)$$

T is the waiting time between repetitions of the FIRFT pulse sequence. Instead of Equation 8.9, a generalized three-parameter fit is useful with pulse imperfections and the like accounted for in parameter B of Equation 8.10:

$$S(\tau) = A + B \exp\frac{-\tau}{T_1} \qquad (8.10)$$

The saturation recovery (SRFT) pulse sequence also obviates the need for long delays between repetitions of the sequence, using pulsed elimination of spin magnetization.[21] In both the FIRFT and SRFT pulse sequences speed is significantly increased, even though some of the gain is lost to the smaller range of magnetization measured ["dynamic range," arising from the maximum value of $(\delta M_z/\delta t)$]. The FIRFT pulse sequence is superior in this respect since its dynamic range is *automatically optimized* for multiline spectra that have dispersed T_1s.

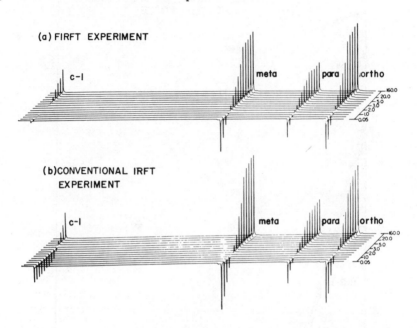

(a) FIRFT EXPERIMENT

c-1   meta   para   ortho

(b) CONVENTIONAL IRFT
EXPERIMENT

c-1   meta   para   ortho

FIGURE 8.5.[19]   *Conventional IRFT and FIRFT spectral sets for*
                       80% w/v phenol: D_2O solution. Spectral window,
                       5 kHz; 8K transform, exponentially weighted to
                       produce 0.6-Hz line broadening. Five scans were
                       accumulated; total data acquisition time 35 min
                       for (a) and 4.2 hr for (b).

Figure 8.5 compares conventional IRFT and FIRFT ^{13}C T_1 experiments on phenol. Note that the inversion of lines in the FIRFT sequence varies from close to null (C-1) to > 75% of the inversion obtained in the IRFT experiment. Figure 8.6 shows a conventional semilogarithmic graphical treatment of these data. Note that the slopes of lines are identical for IRFT and FIRFT experiments, giving identical T_1s; only the intercepts differ.

Use of three-parameter exponential fitting procedures[28-30] further improves the time advantage of FIRFT and SRFT pulse sequences, where τ_∞ spectra require disproportionate data acquisition times. In fact, τ values need range only to $\approx 1 \cdot T_1$ to $2 \cdot T_1$ to give useful data (typically ±10 to 20% for the shorter τ limit and ±5 to 10% with τ values ranging to $2 \cdot T_1$).[29]

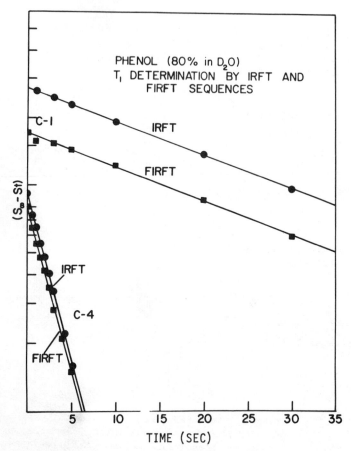

Figure 8.6.   Comparison of FIRFT and IRFT pulse sequences

Measurement of Nuclear Overhauser Enhancements.   Carbon-13 NOEs
may be evaluated by using the dynamic NOE pulse sequence (Table
8.1) or from simple gated decoupling (decoupler *on* during data
acquisition and *off* during delays exceeding $10 \cdot T_1$.[31] The DNOE
experiment has a higher degree of accuracy, since the growth of
magnetization gives multiple data points; it is also amenable to
nonlinear regression analysis.[32]

Measurement of T_2.   Pulse ^{13}C T_2 measurements may be performed
if suitable instrumentation is available and a number of experimental requirements are met.[33] These measurements are by no means
routine. Related parameters such as $T_1\rho$ (T_1 in the rotating frame,
essentially equivalent to T_2 for liquids)[34] or off-resonance $T_1\rho$
may also be measured.[35]

Carbon-13 T_2s may be easily obtained in cases where observed ^{13}C spectral linewidths may be unambiguously assigned to the T_2 process. However, this is generally not the case as chemical-shift dispersion and other effects may contribute to the observed lines.

APPLICATIONS OF RELAXATION STUDIES

Relaxation studies have several applications. Measurements of T_1 in small molecules often yield information about very rapid internal group rotations ($\gtrsim 10^{13}$ sec^{-1}) and the nature of molecular symmetry and steric and bonding interactions; whereas T_1 measurements in large molecules can be used to facilitate spectral assignments in very complex ^{13}C nmr spectra and also to gain insight into molecular configuration. As in small molecules, these T_1 measurements may be used to detect fast internal motions, such as CH_3 group rotations. It is also possible to learn a great deal about the geometry of short-lived "charge-transfer" and hydrogen-bonded molecular complexes.

One other application area, not directly of interest to organic chemists, is the use of ^{13}C spin relaxation measurements to evaluate basic theories of liquid dynamics, including concepts of solution structure.

Molecular Dynamics.   As discussed earlier, the efficiency of ^{13}C-^{1}H dipolar relaxation depends on the rate of molecular reorientation. For very small and fast moving molecules, dipolar relaxation is so inefficient as to be augmented or largely replaced by spin-rotation relaxation. However, in larger molecules $T_1{}^{DD}$ is often virtually the exclusive T_1 process, particularly for protonated carbons. As a result of the intimate relationship between $T_1{}^{DD}$ and motion, $T_1{}^{DD}$ is a powerful probe for evaluation of molecular reorientational mobility. Furthermore, group internal rotations and/or librational motions effect $T_1{}^{DD}$. Thus it is also possible to evaluate rapid ($\approx 10^{-9}$- to 10^{-13}-sec) conformational changes in diverse structures.

Table 8.2 gives ^{13}C dipolar T_1s for a number of molecules having molecular dynamics affected by (1) molecular symmetry, (2) electrostatic interactions, (3) intermolecular hydrogen-bonding interactions, (4) ion-pair and ion-solvent interactions, and (5) group internal mobility.

Several trends may be noted from Table 8.2. In general ^{13}C T_1s are longer in smaller, nonassociating, symmetrical molecules (e.g., benzene and cyclobutane). Within molecules, protonated carbon T_1s are typically shorter than T_1s for nonprotonated carbons by an order of magnitude.

To directly compare mobilities of CH, CH_2, and CH_3 it is necessary to multiply T_1 by the number of directly attached hydrogens, N (giving so-called NT_1 values). In several of the larger

TABLE 8.2. C-13 DIPOLAR RELAXATION TIMES IN SMALL AND INTERMEDIATE-SIZED MOLECULES

COMPOUND	CARBON	T_1^{obs} (SEC)	NOEF	T_1^{DD}	OTHER MECHANISMS, COMMENTS	REFERENCE
$\overset{1}{C}H_3-N-\overset{2}{C}H_3$, O=CH (1CH3, 2CH3, N, O=CH3)	1	18.1	1.3	29	T_1^{SR}; electrostatic ordering, internal CH$_3$ motion	36
	2	11.1	1.7	13		
	3	20.2	1.4	29		
 (benzene)		29	1.6	35	T_1^{SR}: high degree of symmetry	37
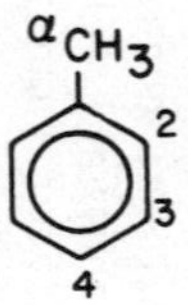 (toluene, αCH$_3$)	1	89	0.56		T_1^{SR} for C-1 and C-α; high degree of symmetry; rapid internal rotation of CH$_3$ gives short T_1^{SR}	37
	2	24	1.6			
	3	24	1.7			
	4	17	1.6			
	α	16	0.61			
(phenol, OH)	1	18.4	$\gtrsim$1.8		Fully dipolar; H bonding; preferred rotation with shorter T_1 for C-4; dilution gives longer T_1	37
	2	2.8	2.0			
	3	2.8	2.0			
	4	1.9	2.0			
(in CCl$_4$)						

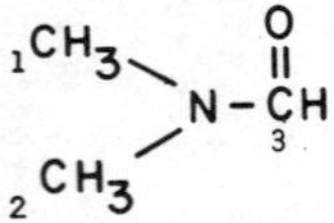

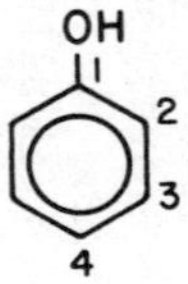

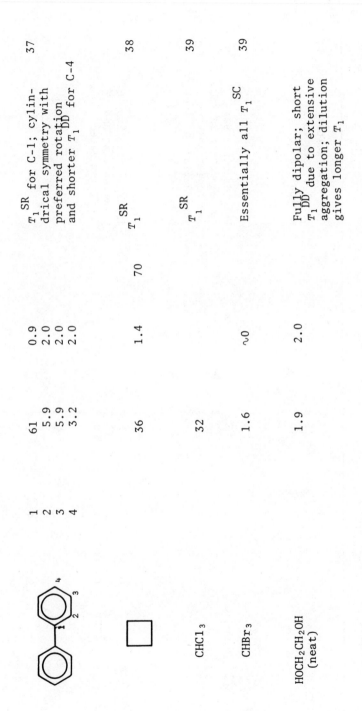

						Ref.
	1	61		0.9	T_1 SR for C-1; cylindrical symmetry with preferred rotation and shorter T_1 DD for C-4	37
	2	5.9		2.0		
	3	5.9		2.0		
	4	3.2		2.0		
		36	70	1.4	T_1 SR	38
$CHCl_3$		32			T_1 SR	39
$CHBr_3$		1.6		~0	Essentially all T_1 SC	39
$HOCH_2CH_2OH$ (neat)		1.9		2.0	Fully dipolar; short T_1 DD due to extensive aggregation; dilution gives longer T_1	39

TABLE 8.2, Continued

COMPOUND	CARBON	T_1^{obs} (SEC)	NOEF	T_1^{DD}	OTHER MECHANISMS, COMMENTS	REFERENCE
4 3 2 1 $CH_3CH_2CH_2CH_2NH_2$	1 2 3 4	13.4 13.4 15 12.1			Essentially fully dipolar; segmental motion present	40
4 3 2 1 + $CH_3CH_2CH_2CH_2NH_3$	1 2 3 4	1.5 2.3 3.1 4.0			Full dipolar; seg-mental motion marked due to ion/solvent interactions to $^+NH_3$; micellar structure	40
Nicotine	2 3 4 5 6 2' 3' 4' 5' CH₃	5.0 39.5 4 4.5 3.0 4.5 2.0 2.0 2.0 1.5			Internal rotations of pyridine ring and N-CH₃ indicated	41

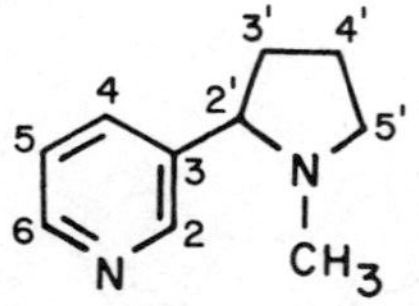

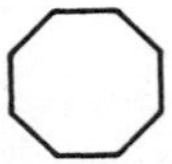		10.3	2.0		38

$CH_3(CH_2)_8CH_2OH$ (numbered 10 ... 1)

Position				Fully dipolar; marked segmental motion; C-1 anchored by H bonding	42
1	0.65				
2	0.77				
3	0.77				
4-6	~0.8				
7	1.1				
8	1.6				
9	2.2				
10	3.1				

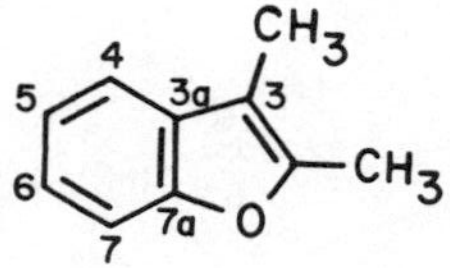 Position				Note primarily T_1^{DD} for nonprotonated carbons despite $T_1 \gtrsim 100$ sec	43
2	83	1.4			
3	92	1.6			
4-7	4 to 6	~2			
3a	114	1.5			
7a	117	1.3			
2-CH₃	6.4	1.9			
3-CH₃	7.7	1.6			

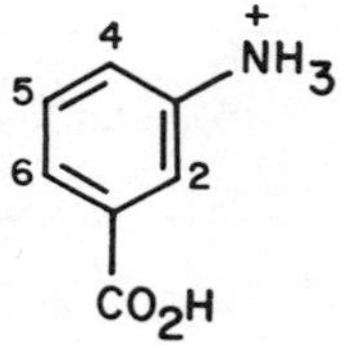 Position				Fully dipolar; dimers (CO_2H) strong ion-solvent and ion-ion interactions (NH_3) restrict mobility	44
2	0.42	2.0			
4	0.26	2.0			
5	0.37	2.0			
6	0.32	2.0			

TABLE 8.2, Continued

COMPOUND	CARBON	T_1^{obs} (SEC)	NOEF	T_1^{DD}	OTHER MECHANISMS, COMMENTS	REFERENCE
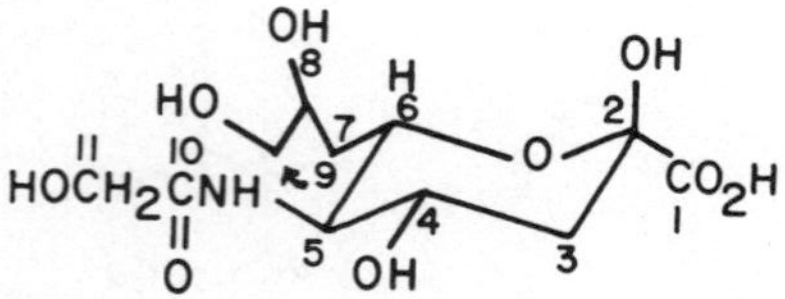	1	15	1.3		Essentially fully di-polar; carboxyl rota-tion, motion of C-11 side chain	45
	2	5.0	1.7			
	3	0.13	2.0			
	4-8	0.3	2.0			
	9	0.22	2.0			
	10	6.6	1.7			
	11	1.4	1.6			

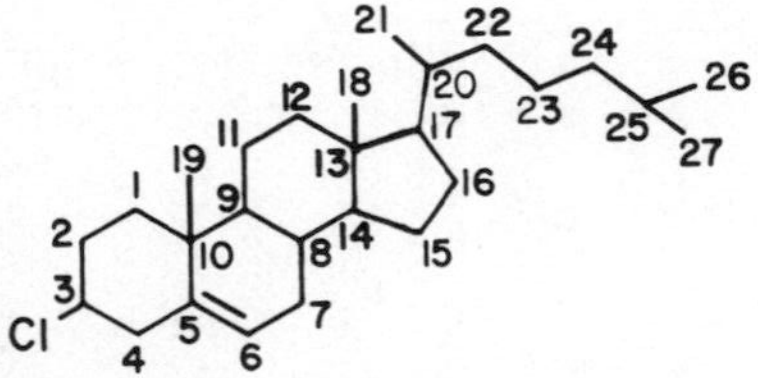 CARBON	T_1^{obs} (SEC)	NOEF	REFERENCE
1,2,4,7,15,16,11,12	0.39 ± 0.06	1.9 ± 0.2	5
3,6,8,9,17	0.70 ± 0.16	1.9 ± 0.2	
10,13	4.5 ± 0.20	1.9 ± 0.2	
5	5.6	1.6	

molecules in Table 8.2 (and also in Table 2.6) NT_1 values for CH
and CH_2 carbons are quite comparable, consistent with relatively
rigid structures undergoing motion that is not highly anisotropic.
In the rigid, roughly globular molecule cholesteryl chloride, for
example, T_1 values for the ring CH and CH_2 carbons are easily
differentiated ($\sim$0.7 and 0.35 sec, respectively). Angular methyl
groups in cholesteryl chloride rotate freely relative to the generally isotropic overall tumbling of the molecule as evidenced by
T_1 values ($\approx$2 sec) found for C_{18} and C_{19} (numbering system; see
Table 8.2). In the study of cholesteryl chloride longer relax-
ation times found for the side-chain carbons are due to segmental
motions.

For a given geometry, degree of dispersion in NT_1^{DD} values for
different protonated carbons gives a measure of the degree of motional anisotropy of the system. The geometry determines how
sensitive NT_1 values will be to motional anisotropy. Quantitative
treatments are available,[2-4] but this aspect is not covered here.
Carbon-13 T_1 data for biphenyl indicate highly anisotropic reorientation, with preferred reorientation around the long biphenyl axis.
This has been observed in other systems:[37]

Anisotropic tumbling results from unequal molecular dimensions for
the butadiyne; for the anilinium ion it results from ion-pair and
ion-solvent interactions and subsequent restrictions on reorientation of the phenyl-NH_3^+ axis. In both of these molecules the *para*
C-H vector (bond) is *exactly* coincident with the axis for fastest
rotation. Thus this C-H pair does not "see" the fast motion, and
hence τ_c is effectively longer, giving a shorter T_1.

Anisotropic tumbling of the nonplanar cyclohexanol molecule
results in NT_1 values that vary by a factor of only ~2.[46] But
this geometry is less favorable than in the planar phenyl ring, and
the cyclohexanol motional anisotropy is in fact comparable with that
of the anilinium ion.

In acyclic systems group segmental motions are common. These rotational or librational motions lead to shorter effective correlation times and hence longer T_1 values. Spin-lattice relaxation times in 1-decanol show evidence of segmental motion. The hydrogen-bonded hydroxyl group effectively anchors the CH_2OH end of the molecule in the liquid matrix, whereas the hydrocarbon end of the chain is relatively mobile.[42]

It is not necessary to have hydrogen bonding or other chemical interactions to see segmental motion in organic molecules. Often the mass of a molecular fragment is sufficient to effectively restrict the motion of attached groups. The molecular mass requirement may be of comparable importance with chemical interactions.

It is often quite difficult to distinguish among anisotropic overall tumbling and internal rotational and librational motions. In favorable cases semiquantitative separations are possible, as in the *p*-aminobiphenyl salt system, where the molecular symmetry is useful:[47]

3.2s 3.3s   2.8s 2.8s

1.4s (4') (3') (2')   (2) (3) (4) —NH$_2$ (in CCl_4)

2.3s 2.3s   2.0s 2.0s

0.60s —NH$_3^+$ $CF_3CO_2^-$ (in CH_3OH)

0.75s 0.75s 0.52s 0.53s

0.22s —NH$_3^+$ $CH_3CO_2^-$ (in CH_3CO_2H)

The nonsubstituted phenyl ring shows internal motion (T_1 for C-2' and C-3' longer than T_1 for C-2 and C-3); in this case rapid librational motion ($\pm30°$ to $60°$ amplitude). The biphenylammonium acetate in particular is heavily restricted in overall reorientation in acetic acid solution; here the libration of the phenyl ring is particularly evident.

Variable-temperature studies of ^{13}C T_1^{DD} values allow calculations of apparent activation energies for molecular reorientation. These E_a terms typically range from 2 to 8 or more kcal/mole, with the higher values observed in cases where intermolecular association is interrupted during molecular reorientation. It has been noted that T_1 measurements evaluate solution microviscosity, which may deviate significantly from viscosity as evaluated by macroscopic measurements.[2]

Restrictions on overall molecular tumbling in solution may come from subtle structural differences. For example,

OH
3.2s
Cl   Cl
2.9s

OH
Cl   Cl
6.ls
6.7s

longer ^{13}C T_1s indicate that the 2,6-dichlorophenol tumbles more rapidly than does the 3,5-substituted compound. Intermolecular association is restricted in the 2,6-dichlorophenol as a result of *intramolecular* H···Cl hydrogen-bonding, and also steric hindrance of the hydroxyl group by the bulky *ortho* substituents.[48]

Another example is found in 2-aminoethanol and 3-aminopropanol. Intramolecular association for 2-aminoethanol is not as favorable as in the larger molecule where six-membered ring formation is possible; thus T_1s are shorter, reflecting greater *inter*molecular association for the ethanolamine.[49]

O   H   N—H
        H   H
CH$_2$—CH$_2$
0.6ls   0.7ls

O   H   N—H
        H   H
1.ls CH$_2$ CH$_2$ 1.3s
        CH$_2$
        1.2s

During the last few years ^{13}C T_1s and other methodologies have been used to examine basic theories of molecular reorientation in liquids.[50] Also, deviations from the single-correlation time model for molecular tumbling have been observed for synthetic high polymers (see Chapter 7), in biomolecules (Chapter 9), and also in associated small molecules.[10,51] Carbon-13 and ^2H relaxation times have been used to determine molecular dynamics and to calculate ^2H quadrupole coupling constants.[52,53]

<u>Steric Effects on Group Rotations</u>.   It was pointed out earlier that the angular methyl groups in cholesteryl chloride rotate freely. Methyl group rotational rates have been studied in various systems, including steroids,[54] methyl alkanes,[55] polymethylbenzenes[56] and other methylated aromatic compounds,[57] and methylated bicyclic compounds.[58] Calculations have been developed to relate T_1SR terms to barriers for rotation of methyl groups.[59]

It is important to note that a steric interaction does not *intrinsically* affect group rotational mobility. Rather, this is controlled by *all* stereoelectronic factors and their combined effect on the energies of available rotameric conformations. Thus in 9-methylanthracene,[57] 1-methylnaphthalene,[57] and hemimellitene,[56] only those methyl groups that have opposing steric interactions (giving sixfold symmetry) spin freely.

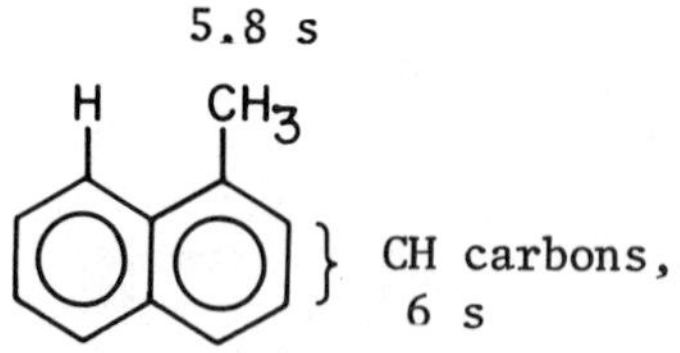

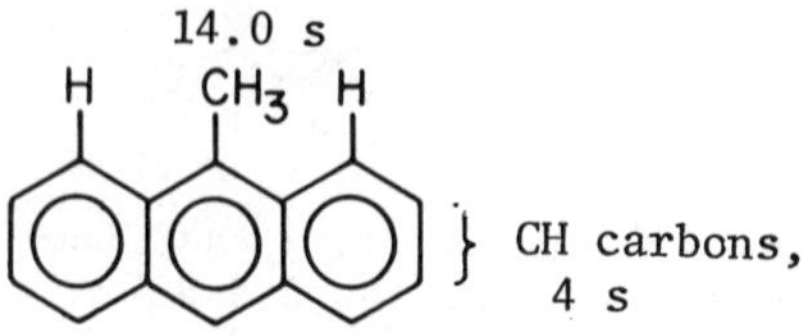

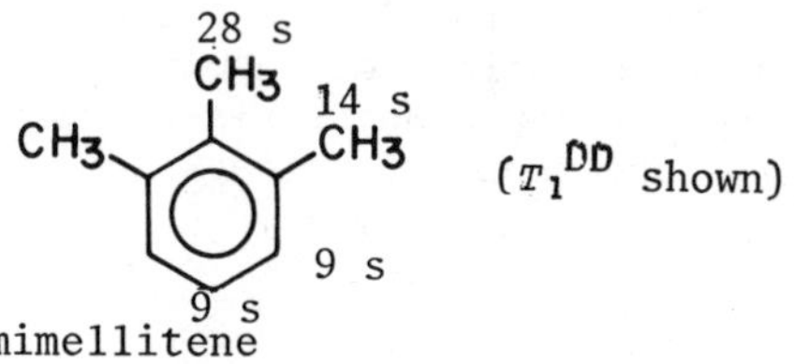

Hemimellitene

For idealized methyl geometry (tetrahedral carbon), free internal spinning results in $T_1{}^{DD}$ three times longer than $T_1{}^{DD}$ for a rigid CH carbon (assuming isotropic overall motion). A rigid methyl group will have $T_1{}^{DD} = 1/3 \, T_1{}^{DD}$ for a C-H carbon (from Equation 8.5), and methyl groups that undergo rapid but not free-spinning (compared to overall reorientational rates, in these cases $\sim 10^{-11}$ to 10^{-12} sec) show intermediate $T_1{}^{DD}$ behavior. It has been recently pointed out that exact methyl geometries play a role in calculations of methyl group rotational barriers.[60]

Applications to Organic Structure Analysis.   Carbon-13 relaxation times and NOEs may be used in a number of ways to support other ^{13}C spectral data in determinations of organic structures. Some of these methods are quite facile, whereas other techniques are subtle and require careful experimental design. Possibilities include:

1. Partially relaxed FT spectra to distinguish nonprotonated, CH, and CH_2 carbons;

2. Accurate $T_1{}^{DD}$ determination to identify individual nonprotonated carbons;

3. Selective $^{13}C\{^1H\}$ or $^{13}C\{X\}$ experiments, especially on nonprotonated carbons;

4. Selective quenching of diamagnetic relaxation (T_1 and NOEF) by use of lanthanide shift reagents or nonshifting PARRs.

Partially Relaxed Spectra: To distinguish nonprotonated carbons from CH$_x$-carbons in complex molecules, it is not necessary to
actually measure T_1s.   A single inversion-recovery FT nmr spectrum
obtained with a relatively short τ ($\gtrsim$ 1 sec for moderately large
molecules; ~0.5 sec for very large molecules) allows rapid identification of nonprotonated carbons, as in Figure 8.7.   With large,
rigid molecules, PRFT spectra can also distinguish between CH and
CH$_2$ carbons, but in this application two or three PRFT spectra
should be obtained ($\tau \approx$ 0.1 to 0.5 sec).

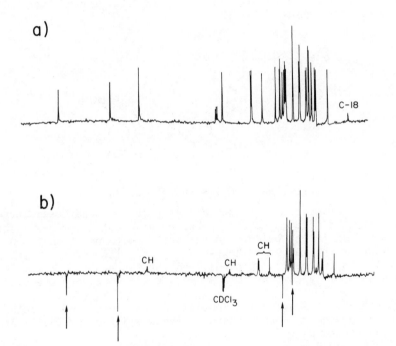

a)

C-18

b)

CH

CH

CH

CDCl$_3$

Figure 8.7.   (a) Normal and (b) PRFT ^{13}C nmr spectra of
cholesteryl acetate in CDCl$_3$.   Spectrum b was obtained with a τ value of 0.2 sec.   Nonprotonated
carbons are inverted in the PRFT spectrum (the CDCl$_3$
carbon is also inverted as a result of its long T_1).
Note that additional qualitative information is
present.   Methine carbons give small positive resonances in (b), as compared with the lines observed
for methylene carbons.   The motionally free 18-CH$_3$
group resonance is nulled in the PRFT spectrum.
(Spectra courtesy of Bo Norden.)

Spectral Assignments: These are difficult for nonprotonated carbons in complex molecules where coupled ^{13}C spectra may not be useful. One way of identifying individual nonprotonated carbons is from comparisons of observed T_1s. In these large molecules T_1 is often entirely dipolar, even for nonprotonated carbons (at low magnetic fields; at high fields NOEs will be reduced, and T_1^{CSA} will contribute for sp^2 or sp nonprotonated carbons). Accurate T_1 (or better, T_1^{DD}) measurements differentiate nonprotonated carbons from their dependence on nearby nonbonded protons for dipolar relaxation.[57,61]

For example, in reserpine the observed ^{13}C T_1s confirmed assignments made from chemical shifts:

Reserpine (^{13}C T_1s in seconds[61])

Note, for example, that the four aromatic carbons that bear methoxy groups have T_1s related to the number of adjacent CH groups. The longest T_1 (12.8 sec) belongs to the carbon with no vicinal protons, whereas the indole methoxy-substituted ring carbon T_1 (at 3 sec) reflects two geminal protons, and the two equivalent remaining methoxy-bound ring carbons with T_1 = 4.8 sec each have one vicinal proton.

In some studies selective deuteration has been used to retard relaxation of nearby nonprotonated carbons and thereby facilitate ambiguous assignments.[62]

Selective NOE Measurements: With careful low-power proton irradiation it is possible to produce selective Overhauser enhancements for individual carbons. These may supplement information from selective collapse of long-range scalar couplings.

Paramagnetic Relaxation and Selective Augmentation of Diamagnetic Relaxation: So-called inert paramagnetic relaxation reagents are seldom neutral to organic substrates.[12] In general, weak ordering effects are present, thus resulting in preferential relaxation of carbons close to the site of interaction between the paramagnetic agent and the substrate. Both lanthanide shift reagents and shiftless relaxation agents operate in this manner. Table 8.3 summarizes the applicability of these reagents for selective relaxation (or nmr spin-labeling) experiments. The inner sphere reagent,

TABLE 8.3.   CHARACTERISTICS OF PARAMAGNETIC REAGENTS IN ^{13}C SPECTROSCOPY

COMPOUND TYPE	COMMERCIAL	TYPICAL CONCENTRATION (M)	INTERACTING FUNCTIONAL GROUPS
Lanthanide shift reagents			
(Lu, Pr, Yb chelates)	Yes	10^{-1}	C=O, ether, amine, alcohol, etc.
Tris[acetylacetonato-chromium(III)], Cr(acac)$_3$	Yes	10^{-2} to 10^{-1}	Outer-sphere coordination; ligands hydrogen-bond to all acidic hydrogens, OH, NH$_2$, etc.; electrostatic interactions with polar groups
Tris[dipivaloylmethanato-chromium(III)], Cr(dpm)$_3$	No[a]	10^{-2} to 10^{-1}	Inert as tested[b]
Tris[dipivaloylmethanato-gadolinium(III)], Gd(dpm)$_3$	Yes	10^{-4} to 10^{-3}	Inner-sphere coordination to basic nitrogens, etc.

[a]Synthesis: W. C. Fernelius and J. E. Blanch, Inorg. Syn., **5**, 130 (1957).
[b]See ref. 12.

Gd(dpm)$_3$, will result in strong polarized ^{13}C T_1s across a molecule that is binding to the reagent. Outer-sphere agents such as Cr(acac)$_3$ show smaller T_1 gradations across substrates. Carbon-13 and ^{15}N nmr spin-labeling techniques have been discussed.[63]

Copper(II) acetylacetonate has been used as a T_2 (line-broadening) reagent for ^{13}C spectral assignments in amines;[64] Cr(dpm)$_3$ has been used to evaluate the translational diffusion process in CCl$_4$.[65]

Finally, it should be pointed out that samples may *unintentionally* contain paramagnetic materials that can ruin otherwise carefully planned experiments. In some cases 10^{-7}-M concentrations of paramagnetic metal ions dominate ^{13}C relaxation (e.g., imidazole[66]). It is *always* advisable to measure NOEs in conjunction with T_1 studies to confirm dipolar relaxation contributions.

Applications for $T_{1\rho}$ and T_2 Measurements.   The experimental difficulties of ^{13}C $T_{1\rho}$ or T_2 measurements have discouraged many researchers.[33,34] Nevertheless, the ability to extend dynamic nmr measurements through the intermediate frequency range ($\approx 10^{-4}$ to

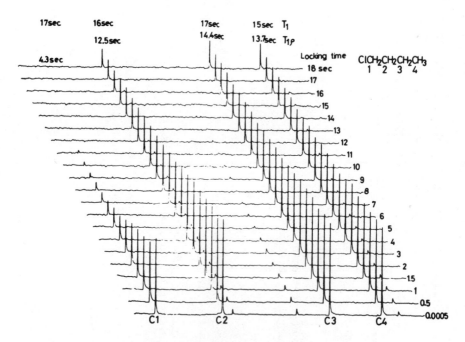

Figure 8.8.[34b]   *Carbon-13 $T_{1\rho}$ measurement spectra of n-butyl chloride with $B_{1\rho}$ = 2 G at 30°C. Values of T_1 are also given.*

10^{-8} sec) makes these studies uniquely attractive. Pulse T_2 measurements, for example, can be used to evaluate scalar relaxation of ^{13}C nuclei in compounds where spin-spin coupled quadrupolar nuclei shorten T_2 insufficiently to give broadened lines. An example is shown in Figure 8.8[34] where $T_{1\rho}$ for C-1 of n-butyl chloride is shortened by chlorine quadrupole-induced scalar T_2 relaxation.

REFERENCES

1. I. I. Rabi, N. F. Ramsey, and J. Schwinger, *Rev. Mod. Phys.* **26**, 167 (1964).
2. J. R. Lyerla, Jr. and G. C. Levy, *Topics in Carbon-13 NMR Spectroscopy*, Vol. 1, G. C. Levy, Ed., Wiley-Interscience, New York 1974, Chapter 3.
3. D. A. Wright, D. E. Axelson and G. C. Levy, *Topics in Carbon-13 NMR Spectroscopy*, Vol. 3, G. C. Levy, Ed., Wiley-Interscience, New York, 1979, Chapter 2.
4. F. W. Wehrli and T. Wirthlin, *Interpretation of Carbon-13 NMR Spectra*, Heyden, New York, 1976, Chapter 4.
5. G. C. Levy and U. Edlund, *J. Am. Chem. Soc.*, **97**, 5031 (1975).
6. G. C. Levy, D. M. White, and F. A. L. Anet, *J. Magn. Resonance*, **6**, 453 (1972).
7. R. R. Sharp, *J. Chem. Phys.*, **60**, 1149 (1974); R. R. Sharp and R. M. Hawk, *ibid.*, 1009.
8. G. C. Levy, *J. Chem. Soc., Chem. Commun.*, **1972**, 352.
9. A. Allerhand and K. Dill, *J. Am. Chem. Soc.*, **101**, 4376 (1979).
10. G. C. Levy, M. P. Cordes, J. S. Lewis, and D. E. Axelson, *J. Am. Chem. Soc.*, **99**, 5492 (1977).
11. R. A. Komoroski, I. R. Peat, and G. C. Levy, *Topics in Carbon-13 NMR Spectroscopy*, Vol. 2, G. C. Levy, Ed., Wiley-Interscience, New York, 1976, Chapter 4.
12. G. C. Levy, U. Edlund, and C. E. Holloway, *J. Magn. Resonance*, **24**, 375 (1976).
13. O. A. Gansow, A. R. Burke, and G. N. LaMar, *J. Chem. Soc., Chem. Commun.*, **1972**, 456; R. Freeman, K. G. R. Pachler, and G. N. LaMar, *J. Chem. Phys.*, **55**, 4586 (1971).
14. G. C. Levy, U. Edlund, and J. G. Hexem, *J. Magn. Resonance*, **19**, 259 (1975).
15. G. N. LaMar and J. W. Faller, *J. Am. Chem. Soc.*, **95**, 3817 (1973); J. W. Faller, M. A. Adams, and G. N. LaMar, *Tetrahedron. Lett.*, **1974**, 699.
16. P. M. Henrichs and S. Gross, *J. Magn. Resonance*, **17**, 399 (1975).
17. L. G. Werbelow and D. M. Grant, *J. Chem. Phys.*, **63**, 4742 (1976); see also work cited in L. G. Werbelow and D. M. Grant, *Adv. Magn. Resonance*, **9**, 189 (1977).
18. R. L. Vold, J. S. Waugh, M. P. Klein, and D. E. Phelps, *J. Chem. Phys.*, **48**, 3831 (1968).

19. D. Canet, G. C. Levy, and I. R. Peat, *J. Magn. Resonance*, **18**, 199 (1975).
20. R. Freeman and H. D. W. Hill, *J. Chem. Phys.*, **54**, 3367 (1971).
21. J. L. Markley, W. H. Horsley, and M. P. Klein, *J. Chem. Phys.*, **55**, 3604 (1971); G. G. McDonald and J. S. Leigh, Jr., *J. Magn. Resonance*, **9**, 358 (1973).
22. R. Freeman, H. D. W. Hill, and R. Kaptein, *J. Magn Resonance*, **7**, 82, (1972).
23. E. D. Becker, J. A. Ferretti, R. K. Gupta, and G. H. Weiss, *ibid.*, **37**, 381 (1980).
24. (a) K. A. Christensen, D. M. Grant, E. M. Schulman, and C. Walling, *J. Phys. Chem.*, **78**, 1971 (1974); (b) R. K. Gupta, *J. Magn. Resonance*, **25**, 231 (1977); R. K. Gupta, et al, *ibid.*, **35**, 301 (1979); (c) P. E. Fagerness, D. M. Grant and R. B. Parry, *ibid.*, **26**, 267 (1977); (d) D. M. Cantor and J. Jonas, *Anal. Chem.*, **48**, 1904 (1976).
25. G. C. Levy and I. R. Peat, *J. Magn. Resonance*, **18**, 500 (1975).
26. (a) C. L. Wilkins, T. R. Brunner, and D. J. Thoennes, *ibid.*, **17**, 373 (1975); (b) I. M. Armitage, H. Huber, D. W. Live, W. Pearson, and J. D. Roberts, *ibid.*, **15**, 142 (1974).
27. H. Hanssum, W. Maurer, and H. Ruterjans, *ibid.*, **31**, 231 (1978).
28. T. K. Leipert and D. W. Marquardt, *ibid.*, **24**, 181 (1976).
29. J. Kowalewski, G. C. Levy, L. F. Johnson, and L. Palmer, *ibid.*, **26**, 533 (1977).
30. M. Sass and D. Ziessow, *ibid.*, **25**, 263 (1977).
31. S. J. Opella, D. J. Nelson, and O. Jardetzky, *J. Chem. Phys.*, **64**, 2533 (1976); R. K. Harris and R. H. Newman, *J. Magn. Resonance*, **24**, 449 (1976).
32. J. Kowalewski, A. Ericsson, and R. Vestin, *ibid.*, **31**, 165 (1978).
33. (a) R. Freeman and H. D. W. Hill, in *Dynamic Nuclear Magnetic Resonance Spectroscopy*, L. M. Jackman and F. A. Cotton, Eds., Academic, New York, 1975; (b) D. G. Hughes, *J. Magn. Resonance*, **26**, 481 (1977).
34. (a) T. K. Leipert, J. H. Noggle, W. J. Freeman, and D. L. Dalrymple, *J. Magn. Resonance*, **19**, 208 (1975); (b) M. Ohuchi, T. Fujito, and M. Imanari, *ibid.*, **35**, 415 (1979).
35. T. L. James, G. B. Matson, I. D. Kuntz, R. W. Fischer and D. H. Buttlaire, *J. Magn. Resonance*, **28**, 417 (1977); T. L. James, G. B. Matson, and I. D. Kuntz, *J. Am. Chem. Soc.*, **100**, 3590 (1978).
36. G. C. Levy and G. L. Nelson, *J. Am. Chem. Soc.*, **34**, 4898 (1972).
37. G. C. Levy, J. D. Cargioli, and F. A. L. Anet, *J. Am. Chem. Soc.*, **35**, 699 (1972).
38. S. Berger, F. R. Kreissl, and J. D. Roberts, *J. Am. Chem. Soc.*, **96**, 4348 (1974).
39. T. C. Farrar, S. J. Druck, R. R. Shoup, and E. D. Becker, *J. Am. Chem. Soc.*, **94**, 699 (1972).

40. G. C. Levy, R. A. Komoroski, and J. A. Halstead, *J. Am. Chem. Soc.*, **96**, 5456 (1974).
41. E. Breitmaier and W. Voelter, ^{13}C *NMR Spectroscopy*, Verlag Chemie, Weinheim, West Germany, 1974, p. 114.
42. D. Doddrell and A. Allerhand, *J. Am. Chem. Soc.*, **93**, 1558 (1971).
43. N. Platzer, *Org. Magn. Resonance*, **11**, 350 (1978).
44. G. C. Levy, A. D. Godwin, J. M. Hewitt, and C. Sutcliffe, *J. Magn. Resonance*, **29**, 553 (1978).
45. M. F. Czarniecki and E. R. Thornton, *J. Am. Chem. Soc.*, **99**, 8273 (1977).
46. R. A. Komoroski and G. C. Levy, *J. Phys. Chem.*, **80**, 2410 (1976).
47. G. C. Levy, T. Holak, and A. Steigel, *J. Am. Chem. Soc.*, **98**, 495 (1976).
48. T. Holak and G. C. Levy, *J. Phys. Chem.*, **82**, 2595 (1978); G. C. Levy, T. A. Holak, and A. Steigel, *ibid.*, **79**, 2325 (1975).
49. U. Edlund, C. Holloway, and G. C. Levy, *J. Am. Chem. Soc.*, **98**, 5069 (1976).
50. See, for example: D. R. Bauer, E. R. Alms, J. I. Brauman, and R. Pecora, *J. Chem. Phys.*, **61**, 2255 (1974); C. M. Hu and R. Zwanzig, *ibid.*, **60**, 4354 (1974).
51. G. C. Levy, P. L. Rinaldi, J. J. Dechter, D. E. Axelson, and L. Mandelkern, in "Characterization of Macromolecules by NMR and ESR", A.C.S. Symposium Series, F. A. Bovey, Ed., in press.
52. H. Saito, H. H. Mantsch, and I. C. P. Smith, *J. Am. Chem. Soc.*, **95**, 8453 (1973).
53. J. B. Wooten, C. W. Jarvis, G. B. Savitsky, A. L. Beyerlin, and J. Jacobus, *J. Magn. Resonance*, **33**, 177 (1979) and earlier papers.
54. J. W. Apsimon, H. Beierbeck, and J. K. Saunders, *Can. J. Chem.*, **53**, 338 (1975).
55. (a) J. R. Lyerla and T. T. Horikawa, *J. Phys. Chem.*, **80**, 406 (1976); (b) R. C. Long, J. H. Goldstein, and C. J. Carman, *Macromolecules*, **11**, 574 (1978).
56. T. D. Alger, D. M. Grant, and R. K. Harris, *J. Phys. Chem.*, **76**, 281 (1972).
57. G. C. Levy, *Acc. Chem. Res.*, **6**, 161 (1973).
58. D. E. Axelson and C. E. Holloway, *Can. J. Chem.*, **54**, 2820 (1976).
59. A. P. Zens and P. D. Ellis, *J. Am. Chem. Soc.*, **97**, 5685 (1975).
60. J. W. Blunt and J. B. Stothers, *J. Magn. Resonance*, **27**, 515 (1977).
61. F. W. Wehrli, *Adv. Molec. Relaxation Processes*, **6**, 139 (1974).
62. R. E. Echols and G. C. Levy, *J. Org. Chem.*, **39**, 1321 (1974); M. J. Shapiro and A. D. Kahle, *Org. Magn. Resonance*, **12**, 235 (1979).

63. G. C. Levy, J. J. Dechter, and J. Kowalewski, *J. Am. Chem. Soc.*, **100**, 2308 (1978).
64. D. Doddrell, I. Burfitt, and N. V. Riggs, *Aust. J. Chem.*, **28**, 369 (1975).
65. J. G. Hexem, U. Edlund, and G. C. Levy, *J. Chem. Phys.*, **64**, 936 (1976).
66. R. E. Wasylishen and J. S. Cohen, *Nature (Lond.)*, **249**, 847 (1974).

Supplemental References

^{13}C NMR Relaxation Mechanisms in Methyl-Transition Metal Compounds, R. F. Jordan and J. R. Norton, *J. Am. Chem. Soc.*, **101**, 4853 (1979).

Internal Rotation of Methyl Groups in Terpenes. Variable Temperature Carbon-13 Spin-Lattice Relaxation Time Measurements and Force Field Constants, A. Ericsson, J. Kowalewski, T. Liljefors, and P. Stilbs, *J. Magn. Resonance*, **38**, 9 (1980).

^{13}C Magnetic Relaxation in Micellar Solutions. Influence of Aggregate Motion on T_1., H. Wennerstrom, B. Lindman, O. Soderman, T. Drakenberg, and J. B. Rosenholm, *J. Am. Chem. Soc.*, **101**, 6860 (1979).

Anisotropic Motion in 1-substituted Adamantanes From Carbon-13 NMR Relaxation Time Data, H. Beierbeck, R. Martino, and J. K. Saunders, *Can. J. Chem.*, **57**, 1224-8 (1979).

Use of Carbon-13 $T_{1\rho}$ Measurements to Study Dynamic Processes in Solution, D. M. Doddrell, M. E. Bendall, P. F. Barron, and R. T. Pegg, *J. Chem. Soc., Chem. Commun.*, 77 (1979).

Carbon-13 Spin-Lattice Relaxation Times as a Probe for the Study of Electron Donor-Acceptor Complexes, J. Blackburn and J. Friesen, *Org. Magn. Resonance*, **9**, 113 (1977).

Chapter 9

Biomolecules

INTRODUCTION

Applications of ^{13}C nmr spectroscopy to studies of bio-
molecular structure have expanded in an almost explosive manner
since the early 1970s. Although a comprehensive review is not
within the scope of this monograph, major developments are
summarized for the various classes of natural products as well as
biopolymer systems. In this chapter emphases are placed on the
unique methodologies developed for structural studies of these
complex molecules. The use of shielding effects, decoupling
techniques, spin-relaxation times, and chemical-substituent
effects for spectral assignments is covered. Discussion of ^{13}C
nmr studies of biopolymers is preceded by an introduction to the
unique characteristics of biopolymer spin relaxation. Proteins,
nucleic acids, polysaccharides, and lipids are then considered,
with emphasis on the unique contribution of ^{13}C nmr to understanding of details of molecular structure and structure/function
relationships.

TECHNIQUES FOR BIOMOLECULES

No special FT nmr techniques are necessary with low and
intermediate molecular weight natural products.[1] Solubility of
these molecules is generally sufficient for 0.05 to 0.5M samples.
In cases where sample availability is restricted, microcell probes
are used to optimize nmr sensitivity. Aqueous studies require
suitable, preferably inert, chemical-shift references. Dioxane
is often used, but DSS [sodium 3-(trimethylsilyl)-1-propanesul-
fonic acid] is preferable, if necessary, using the commercially
available partly deuterated compound.

For studies of biomolecules in buffered aqueous media, care
must be taken to avoid sample heating arising from proton
decoupling of the high dielectric sample. This problem can be
increasingly serious at higher magnetic fields (proton frequencies
above 150 MHz). Modern spectrometers have improved decoupling

modulation schemes that minimize this problem, but in some cases use of a two-level decoupling mode[2] may be necessary.

Biopolymer studies by ^{13}C nmr at natural isotopic abundance cannot generally be classified routine as a result of the limited concentrations generally obtainable ($\approx 10^{-3}$ M) and often the restricted availability of samples. Nevertheless, many applications are practical, especially for *smaller* proteins and nucleic acids, lipids, and so on.

Many of the experimental problems associated with the use of ^{13}C nmr in the study of biopolymers and other complex biomolecules can be eliminated by appropriate ^{13}C isotopic enrichment. A blanket enrichment to 10% ^{13}C would yield small ^{13}C-^{13}C satellites but increase sensitivity by almost an order of magnitude. Such an enrichment can be accomplished biosynthetically by growing an appropriate organism in a medium of enriched nutrient (e.g., ^{13}CO$_2$, [ul-^{13}C]glucose), yielding a large variety of ^{13}C enriched materials (e.g., proteins, nucleic acids, and membranes). In fact, items such as uniformly ^{13}C-labeled algal lyophilized cells, the whole hydrolysate of such cells, and mixtures of sugars or of amino acids are commercially available from several sources. Of course, although such an enrichment allows more dilute solutions to be studied (or greater S:N to be obtained at a given concentration), the problem of resolving the large number of resonances in the ^{13}C spectrum of the biopolymer still exists.

A considerably more useful but, in general, more difficult procedure is the incorporation of a residue, enriched at one or more carbons, at a specific point in the sequence of a biopolymer. Here the isolated resonance of the enriched carbon or carbons is superimposed on the natural-abundance ^{13}C biopolymer spectrum; for higher degrees of enrichment such background absorption poses no problem. This procedure allows one to monitor the behavior of the chemical shifts and all the relaxation parameters of a single carbon or group of resolved carbons at a specific site in a biopolymer.

TERPENOIDS AND STEROIDS

Assignment of ^{13}C resonances in acyclic or monocyclic terpenes is facilitated by ready comparison with model alkenes and cycloalkanes. Many assignment techniques and much earlier representative work have been described comprehensively by Wenkert et al.[3] Bohlmann and his group have reported the ^{13}C chemical shifts of nearly 100 cyclic and acyclic monoterpenes.[4] Examples are given in Table 9.1. Typical effects of *cis-trans* isomerism may be seen at the methyl carbon on C-3, and at C-4, of geraniol and nerol. Conversion of both to their respective acetates deshields C-1 slightly and shields C-2; these are β- and γ-effects. Effects of oxygen substitution are transmitted

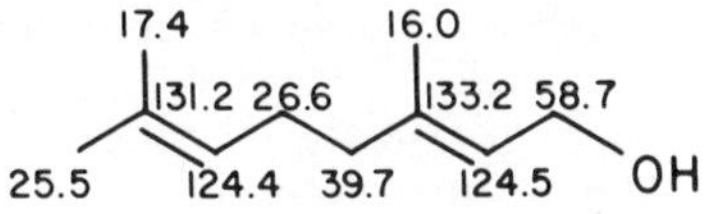

17.4   16.0
131.2 26.6   133.2 58.7
25.5   124.4 39.7 124.5 OH

Geraniol

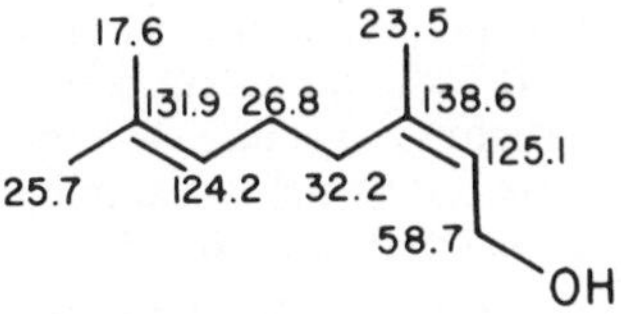

17.6   23.5
131.9 26.8   138.6
                    125.1
25.7 124.2 32.2
            58.7
                OH

Nerol

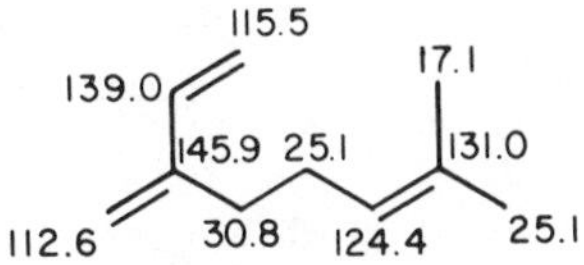

115.5
139.0   17.1
    145.9 25.1   131.0
112.6   30.8 124.4   25.1

Myrcene

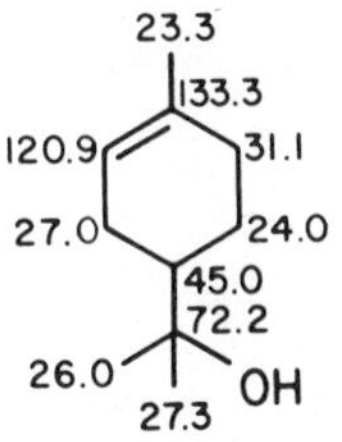

23.3
133.3
120.9   31.1
27.0   24.0
        45.0
        72.2
26.0   OH
    27.3

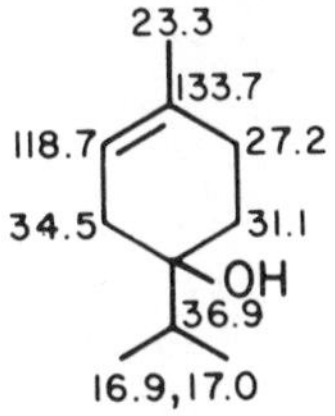

α-Terpineol

23.3
133.7
118.7   27.2
34.5   31.1
        OH
        36.9
16.9,17.0

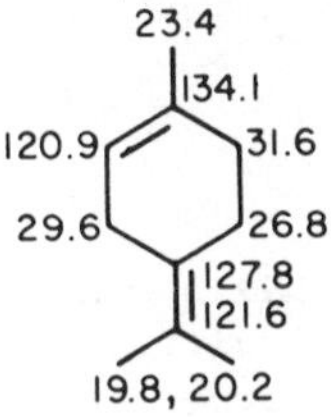

23.4
134.1
120.9   31.6
29.6   26.8
        127.8
        121.6
19.8, 20.2

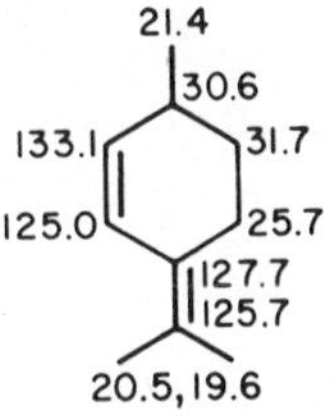

21.4
30.6
133.1   31.7
125.0   25.7
        127.7
        125.7
20.5,19.6

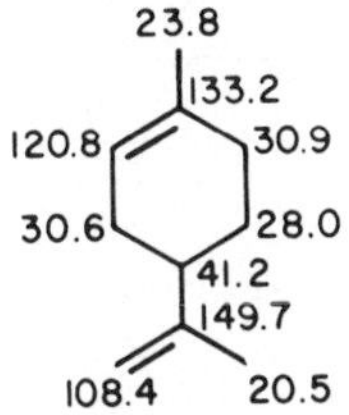

23.8
133.2
120.8   30.9
30.6   28.0
        41.2
        149.7
108.4   20.5

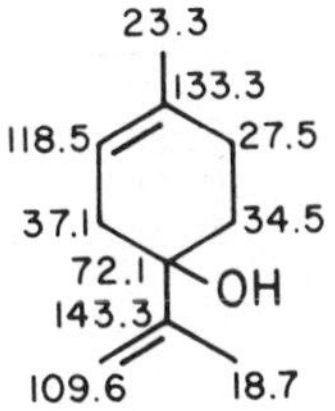

Limonene

23.3
133.3
118.5   27.5
37.1   34.5
72.1   OH
143.3
109.6   18.7

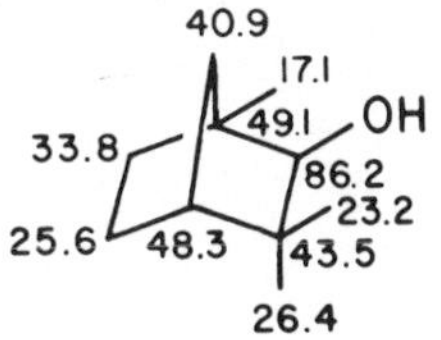

40.9
        17.1
        49.1 OH
33.8   86.2
            23.2
25.6 48.3 43.5
        26.4

β-Fenchol

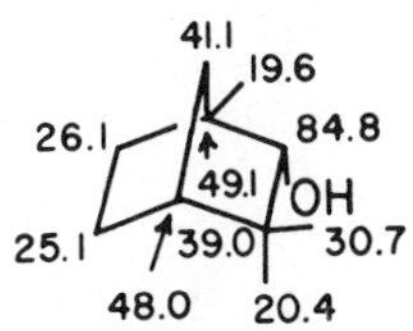

41.1
    19.6
26.1   84.8
    49.1 OH
25.1 39.0   30.7
48.0 20.4

α-Fenchol

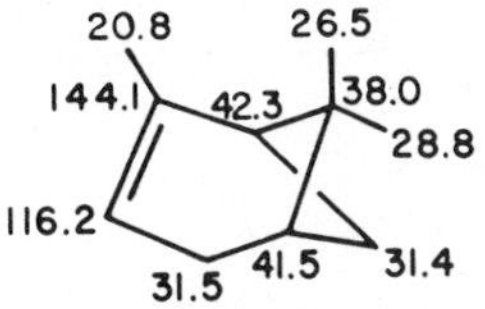

20.8   26.5
144.1   42.3 38.0
            28.8
116.2
    31.5 41.5 31.4

α-Pinene

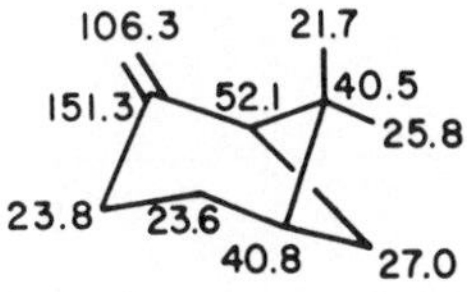

106.3   21.7
151.3   52.1 40.5
            25.8
23.8 23.6
        40.8   27.0

β-Pinene

249

similarly in the monocyclic terpenes (e.g., α-terpineol), and these compounds display the isopropyl methyl groups as magnetically nonequivalent. The influence of geometry can be seen by comparing α- and β-fenchol.

Larger terpenes--sesqui-, di-, and triterpenes--benefit from the large body of data available for the monoterpenes. This is seen in comparison of β-bisabolene with myrcene and limonene (Table 9.1).[3] Good examples of this are found with β-carotene[5]

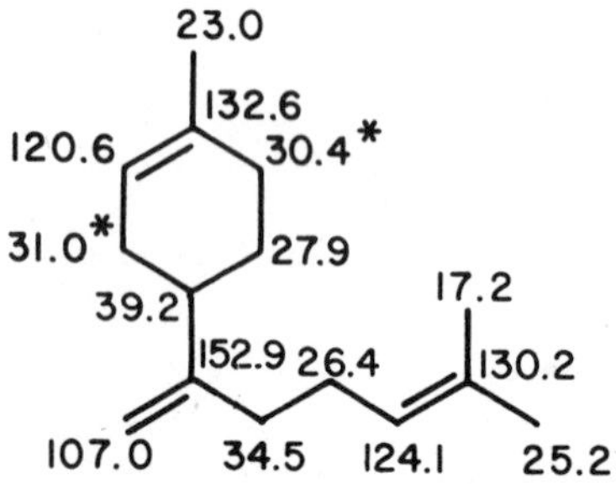

β-Bisabolene (chemical shifts in ppm;

* = assignments interchangeable)

PROBLEM 9.1.   *On the basis of the following ^{13}C chemical shifts and multiplicities in an off-resonance decoupled spectrum, derive structures for the monoterpene $C_{10}H_{18}O$ and its corresponding acetate $C_{12}H_{20}O_2$:*

$C_{10}H_{18}O$	$C_{12}H_{20}O_2$[a]
17.5 (q)	17.5 (q)
22.6 (t)	22.5 (t)
25.3 (q)	23.8 (q)
27.2 (q)	25.7 (q)
41.2 (t)	39.7 (t)
72.7 (s)	82.8 (s)
111.3 (t)	113.0 (t)
124.6 (d)	124.1 (s)
130.3 (s)	131.5 (s)
145.0 (d)	142.1 (s)

[a]Acetyl resonances not given.

and the retinals.[6] Various techniques were used for the latter compounds: single-frequency and off-resonance decoupling, T_1 measurements, addition of relaxation reagents, and solvent

effects. The 11-*cis* isomer was found to exhibit distinct differences from the others, and from these observations and from
T_1 results it was inferred to be partly nonplanar. Representative data are as follows along with chemical shifts for the
model β-ionone:

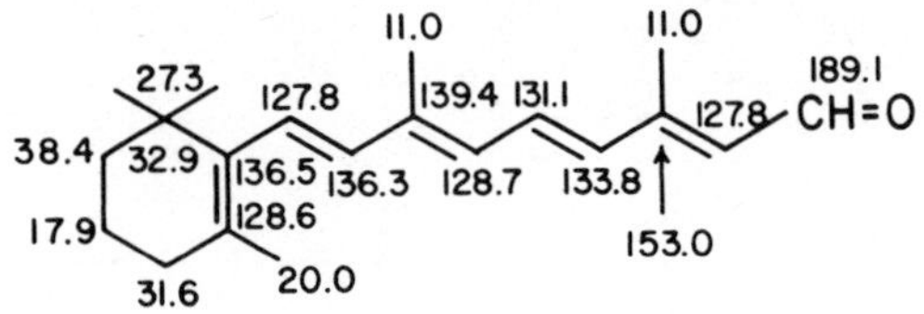

All-*trans*

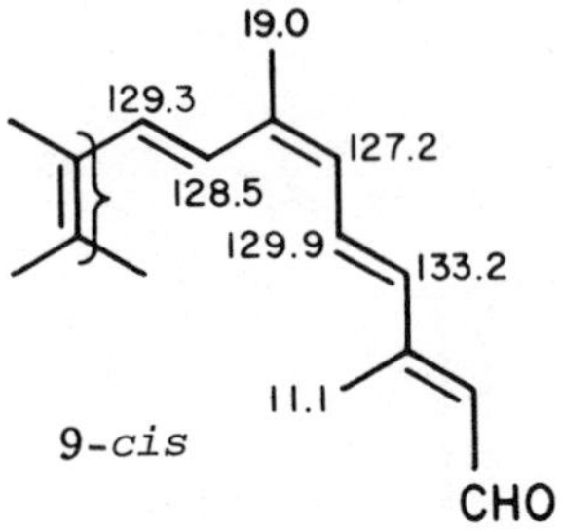

9-*cis*

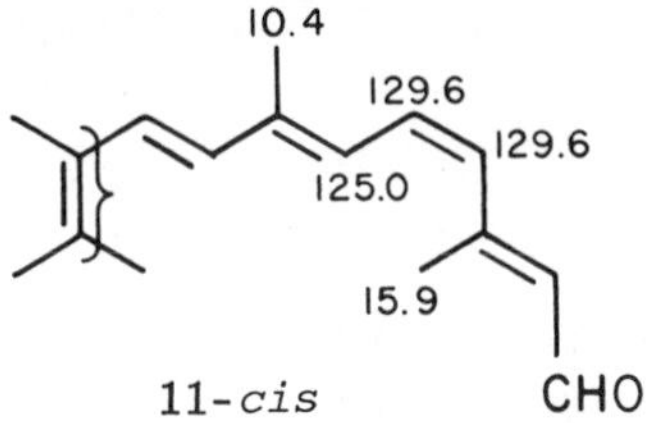

11-*cis*

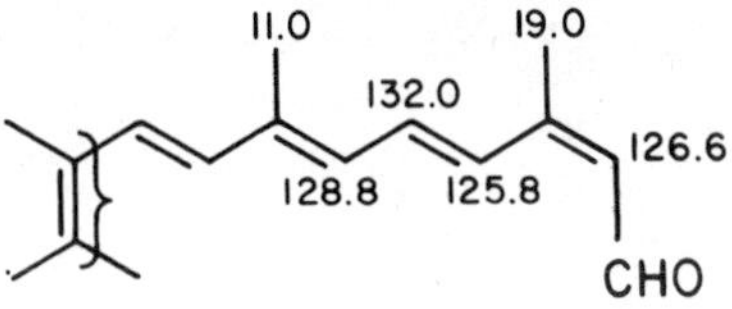

13-*cis*

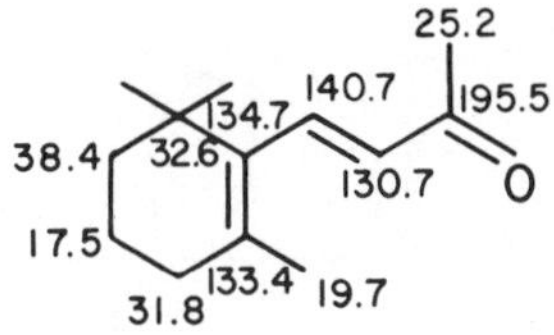

β-Ionone

Among the terpenoids the steroids represent one of the
largest classes of compounds.[7] Very comprehensive critical
reviews reporting data for over 400 examples have been published
by Blunt and Stothers,[7a] as well as by Smith.[7c] Success in
assignments of resonances arises from the additivity of substituent effects (with some exceptions; see ref. 7) and from a
reasonable understanding of relaxation properties. Off-resonance
decoupling is also useful, as shown in Figure 9.1.[7a] Quartets
at 14.4 and 18.3 ppm in the off-resonance spectrum identify the
methyl groups, and the sharp singlets at 39.9 and 44.1 ppm
characterize the quaternary carbons. Similarly, doublets at
58.8, 63.0, and 68.5 ppm are assignable to the methine carbons
at C-9, C-17, and C-11, respectively. The triplet at 50.2 ppm

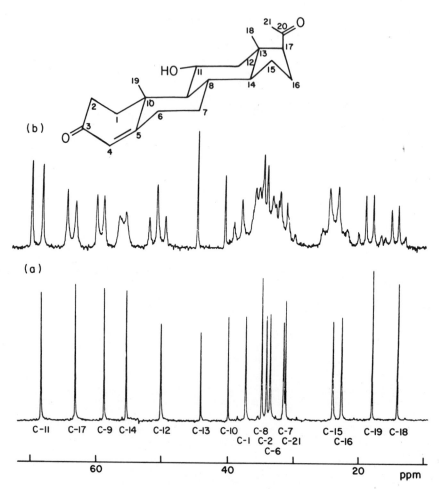

FIGURE 9.1.[7a]   *High-field portions of the* ^{13}C *spectra of 11α-hydroxypregn-4-ene-3,20-dione, with (a) proton noise-decoupling, and (b) with off-resonance decoupling.*

is assigned to C-12. The remaining signals in the off-resonance spectrum are broad and ill-defined doublets and triplets from overlapping methylene and methine carbons. In addition, the signals display some second-order character because the chemical shifts of the directly attached protons are similar to those of protons on adjacent carbons.

The rigid geometry of steroids allows them to be used as monitors for structural and stereochemical influences on ^{13}C nmr parameters,[7,8] thus paralleling their role in studies of organic reactions. The sensitivity of shifts to geometry is shown with 3α- and 3β-androstanol.[7b] Geometry of the *A/B* ring fusion is

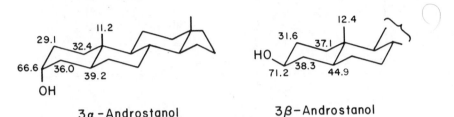

3α-Androstanol   **3β-Androstanol**

also reflected in the resonance position of C-19: as a result of additional γ-alkyl effects present in a   *cis*-fused (5β) system, C-19 methyls are shielded more than are those in *trans*-fused (5α) steroids.   This is exemplified with 5α- and 5β-androstane:

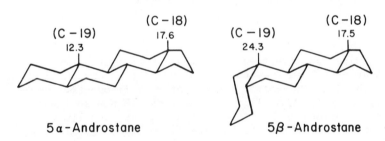

5α-Androstane   **5β-Androstane**

Correlations of this type among large numbers of related compounds are highly useful for assignments of resonances and are frequently summarized in the form of correlation diagrams.   These allow ready assessment of structural influences on chemical shifts.

An example is shown in Figure 9.2 for pregnane derivatives.[10] Compared to cholestane, the substituted C-5 carbons are deshielded (α-effect), as are C-4 and C-6 (β effects).   C-3, C-7 and C-9-- which are γ-*gauche* to the 5-substituent--are shielded substantially.   The remaining carbons remain relatively unaffected.

Other representative studies on terpenes have appeared. These include humulones,[11] germacranolides,[12] squalene derivatives,[13] and pentacyclic triterpene glycyrrhetic and liquiritic acids.[14] Biosynthetic studies using ^{13}C nmr, including that of cholesterol,[15] have also been reported.[16,17]

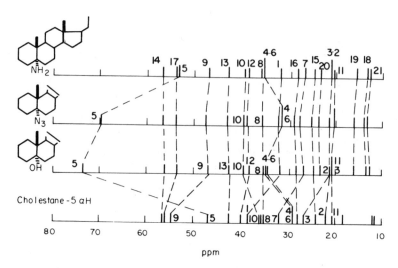

ppm

FIGURE 9.2.[10]   *Chemical-shift correlation diagram for*
5α-substituted pregnanes.

ALKALOIDS

The size and structural complexity of alkaloids can make
assignment of ^{13}C resonances a substantial challenge. Added to
this is the fact that compounds of interest frequently are
available only in submilligram quantities. Hence stringent
demands are placed on instrument capabilities, and limitations
may exist on the ability to chemically modify materials for
comparison.

Some perspective on the area may be obtained from reviews
by Crabb[18a] and van Binst.[18b] In his review, Wenkert provides an
illuminating discussion of techniques for assignment of ^{13}C
resonances in isoquinoline alkaloids,[3] which are broadly
applicable to other classes as well. Single-frequency proton
decoupling is particularly useful because longer-range couplings
are often revealed that may allow distinction between carbons with
similar chemical shifts. For example, C-11 in isocorydine is
distinguishable from C-1, C-2, and C-10 by irradiation of the
aromatic protons. Carbon 11 appears as a sharp singlet, whereas
the other three give rise to unresolved signals because of
vicinal coupling with methyl hydrogens.

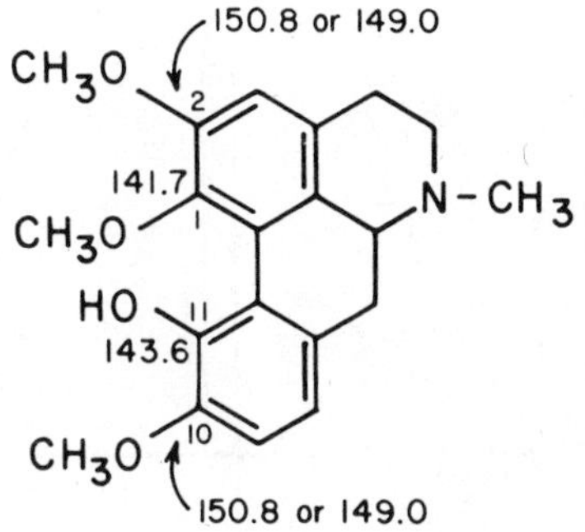

Isocorydine

Gribble et al.[19] used both specific deuteration and T_1 measurements to assign chemical shifts in the indoloquinolizideine shown in 1. This compound serves as a model for the general classes of *Rauwolfia* and *Corynanthe-Yohimbe* alkaloids. The spectrum shows 15 lines. Individual incorporations of deuterium at C-1, C-3, C-4, C-6, C-7, C-9, and C-12b caused the respective individual lines to disappear or to be replaced by very weak low-intensity multiplets. An example of this is shown in Figure 9.3, which also shows the usefulness of varying the pulse repetition rate for assigning resonances. The rapid repetition rate used for the upper spectrum results in reduced intensities for all quaternary carbons and no signal for the deuterated C-6. At the longer repetition rate employed for the lower trace, the quaternary carbons are more extensively relaxed and the signals are comparable with those of the protonated carbons. In addition, a weak broad multiplet for C-6 is now discernable (note also the increased intensity for the $CDCl_3$ multiplet). The assigned chemical shifts are given with structure 1. Values for C-9 and C-10 are reversed from those originally reported for indole.[20]

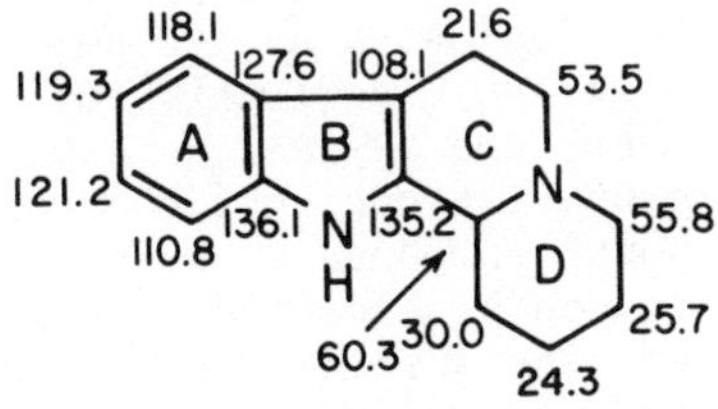

1

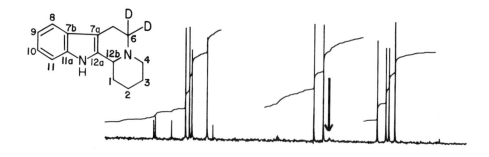

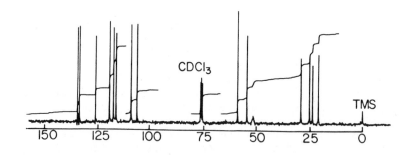

FIGURE 9.3.[19]   *Carbon-13 67.9-MHz FT nmr spectrum of 1-6-d₂:
(top) 300 90° (34-µsec) pulses, repetition time
1 sec; (bottom) 120 90° pulses, repetition
time 20 sec. Each spectrum is 15 kHz with 16K
transform. The arrow in the top spectrum
indicates the C-6 resonance position in the
nondeuterated material.*

Gribble et al.[19] also demonstrated that the geometry of the
c/D ring fusion may be elucidated by ^{13}C nmr. The chemical
shifts of the *cis-2-tert*-butyl derivative of 1, which has a *trans
c/D* ring fusion, are the same as in 1, except for normal α and
β effects of the *tert*-butyl group. By contrast, the ^{13}C spectrum
of the *trans-2-tert*-butyl compound shows some striking differences,
mostly in carbons that can be expected to experience γ-*gauche*
interactions. However, the shielding experienced by C-6 and C-12b
cannot be rationalized in these terms, and differences in the
manner in which the nitrogen lone pair interacts with adjacent

C-H orbitals was suggested to account for this.   Chemical shifts
are shown in the following partial structures:

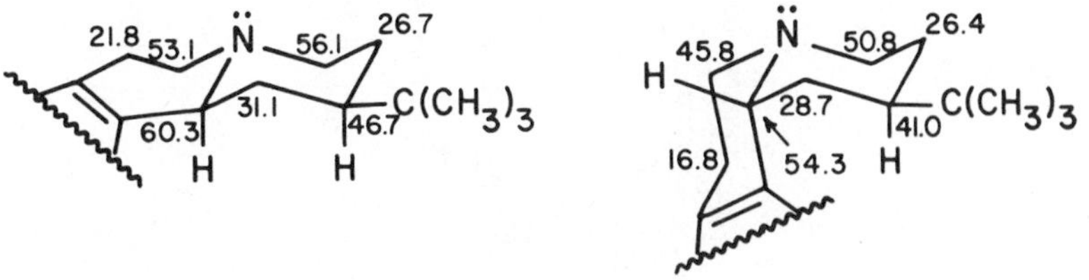

Bohlmann and Zeisberg[21] have assigned the chemical shifts of
a large number of lupine alkaloids and model compounds containing
quinolizidine rings.   Specific deuteration and off-resonance
decoupling were required to confirm some of the assignments.
Effects of geometry can be seen by comparing the bridging C-8
carbons in sparteine and isosparteine.

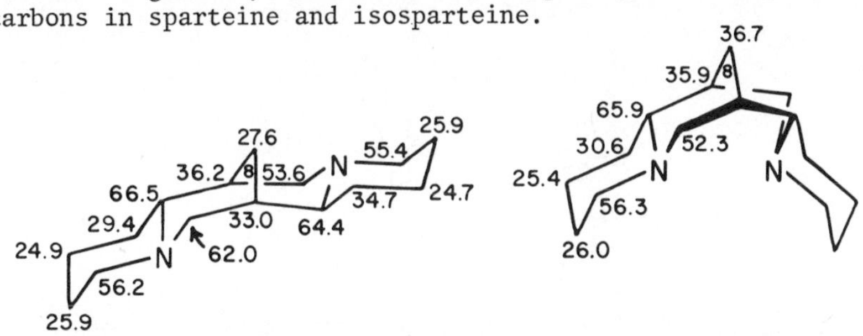

A potentially useful procedure for assigning alkaloid
resonances is to take advantage of the effects of nitrogen
protonation or quaternization on nearby carbons (see Chapter 3).
However, such studies must be carried out with caution.   In a
study of *Strychnos* alkaloids, Leung and Jones[22] showed that
earlier assignments[23] of some of the resonances in strychnine
and brucine, which were based on protonation shifts, must be
reassigned.   Because protonation shifts may be unexpectedly
small, they suggest comparison with corresponding quaternary
N-methyl and *N*-oxide derivatives as a means of assignment.   For
the quaternary salts, carbons α to nitrogen are generally
deshielded (β effect), whereas β-carbons are slightly shielded
(γ effect).   Chemical shifts for strychnine, its hydrochloride
and methiodide, are given in Table 9.2.

Carbon-13 shifts of several ergot alkaloids have been
reported,[24] and assignments for agroclavine, a useful model for
this series, are given with the structure.   The shifts of C-12
and C-13 were assigned by analogy with corresponding carbons in
indole and methylindoles.[20]   In view of the results due to
Gribble, et al.[19] these may have to be reversed.

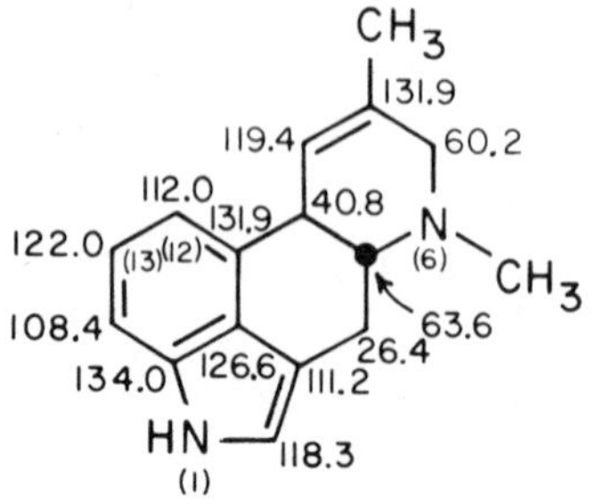

CH₃ 131.9

Agroclavine

Atropine serves as a model for the tropane alkaloids. Of special note is the effect of the chiral center on the remote ring carbons.[25]

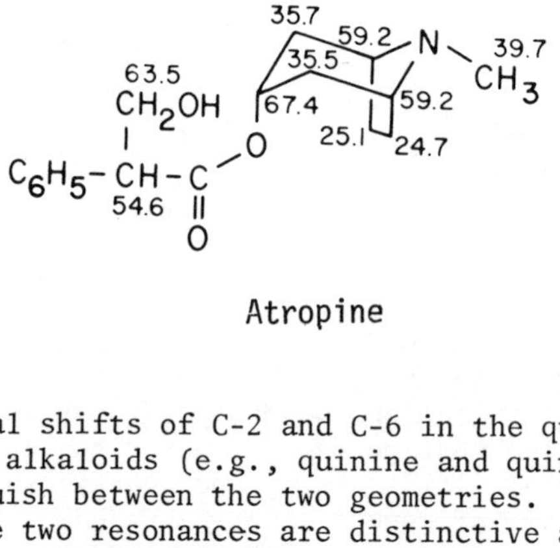

Atropine

The chemical shifts of C-2 and C-6 in the quinuclidine ring of the *Cinchona* alkaloids (e.g., quinine and quinidine) may be used to distinguish between the two geometries. As shown in Table 9.3, these two resonances are distinctive among the aliphatic carbons. Whereas C-6 is γ-*gauche* to C-9 in quinine, C-2 displays this relationship in quinidine. Hence, C-2 is shielded in quinidine.

Using very similar techniques for assignment, ^{13}C chemical-shift assignments have been reported for morphine,[27] *Aconitum* and *Delphinium*,[28] *Aspidosperma*,[29] *Rauwolfia*,[30] *Vinca*,[31] Amaryllidaceae,[32] and *Lycopodium*[33] alkaloids.

OTHER NATURAL PRODUCTS

Carbon-13 nmr has been used to characterize lignin components derived from different types of tree.[34,35] For example, two peaks at 105 and 107 ppm, assigned to C-2,6 of the syringyl

TABLE 9.2. ^{13}C CHEMICAL SHIFTS (IN PARTS PER MILLION) OF STRYCHNINE AND DERIVATIVES[22]

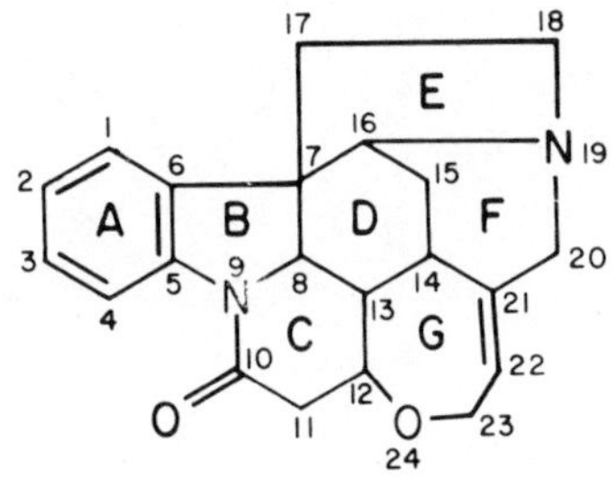

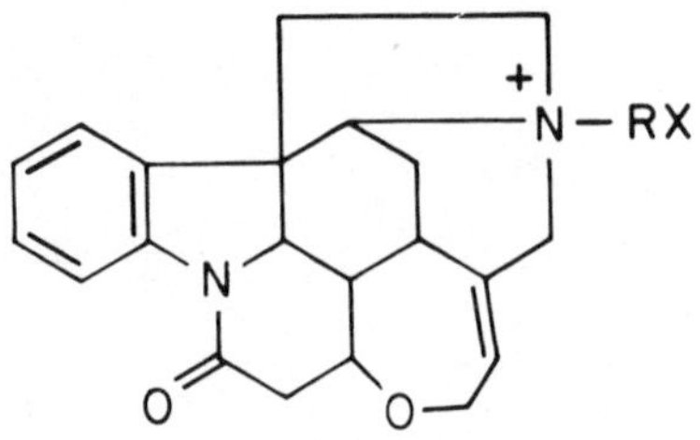

CARBON	PARENT COMPOUND	RX = HCl	RX = CH$_3$I
C-1	124.99	125.53	125.22
C-2	122.67	122.94	124.89
C-3	128.61	130.33	130.72
C-4	116.52	116.79	116.42
C-5	142.21	142.20	142.80
C-6	132.91	129.15	130.50
C-7	52.33	52.38	53.84
C-8	60.42	62.31	59.40
C-10	170.31	170.04	170.21
C-11	42.46	42.24	41.27
C-12	77.73	77.31	76.72
C-13	48.44	47.42	47.42
C-14	31.83	30.86	29.89
C-15	26.92	25.35	25.47
C-16	60.42	59.45	75.31
C-17	43.05	41.32	39.71
C-18	50.49	51.14	62.74
C-20	52.76	52.54	64.58
C-21	40.04	132.65	133.85
C-22	128.99	136.54	136.70
C-23	64.84	64.41	64.90

residue, appear in beech but not in spruce lignin and thus allow a distinction between hardwood and softwood lignins.

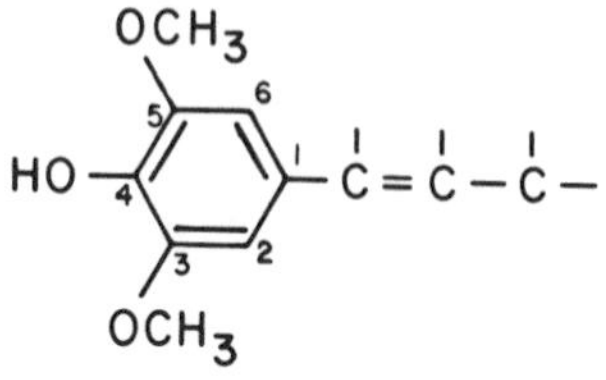

Syringyl residue

In assigning the ^{13}C spectrum of the nebramycin factor apramycin, an antibiotic produced by *Streptomyces tenebrarius*, Wenkert and Hagaman took advantage of amine protonation effects on ^{13}C chemical shifts.[36] Of the three anomeric carbons

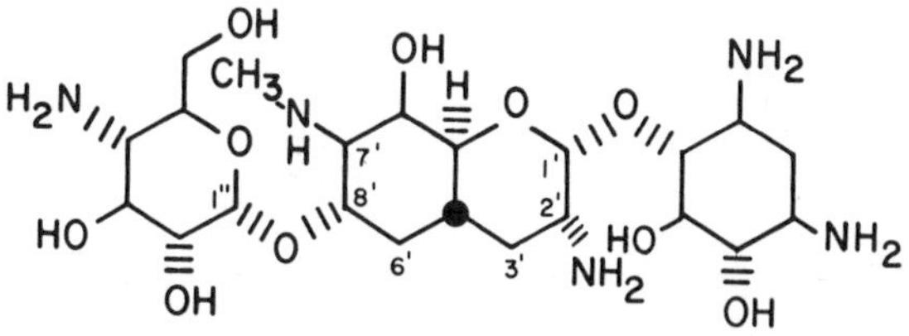

Apramycin

the two close to the amino groups, C-1' and C-8', are shielded by 3 to 5 ppm when the spectrum is determined at pH < 1 compared with the spectrum at pH > 11. Carbon 1" remains unchanged and thus may be distinguished from the other two. In the central saccharide unit the presence of both a primary and a secondary amine is evidenced by the different characteristic effects of protonation on the β carbons. Carbons β to a primary amine (C-3' and C-1') are shielded to 4 to 6 ppm, whereas those β to a secondary amine (C-6', C-8') are shielded by 2 to 3 ppm. These results serve to characterize the corresponding carbons. The complete structural assignment, based also on comparison with smaller residues derived from apramycin, identifies all 21 carbons.[36] Similar studies were carried out in a very elegant manner on the related but more complex nucleoside disaccharide antibiotic anthelmycin.[37]

Other antibiotics characterized by ^{13}C nmr include nogala-mycin,[38] actinomycin,[39] ezomycin complexes,[40] erythromycins,[41] and tetracycline.[42] In the last case microscopic dissociation

TABLE 9.3. ^{13}C CHEMICAL SHIFTS (IN PARTS PER MILLION) OF QUININE AND QUINIDINE[26]

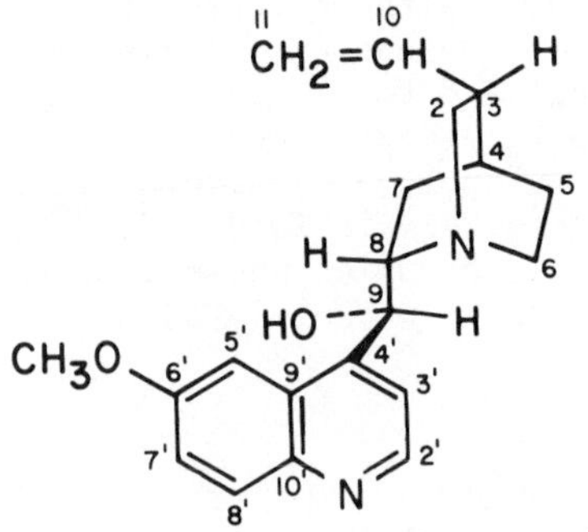

Quinine

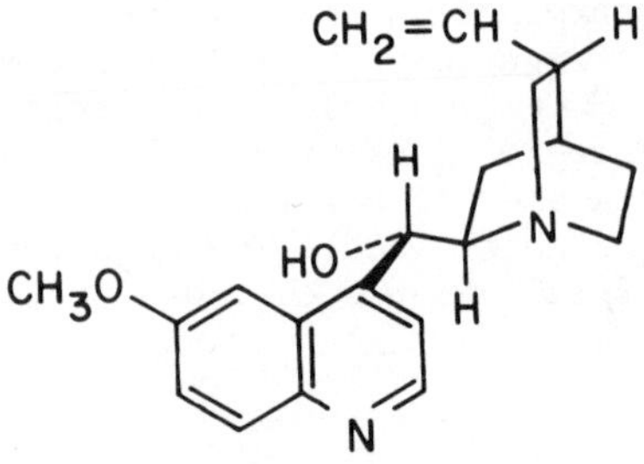

Quinidine

CARBON		
2	56.9	49.9
3	39.8	40.0
4	27.7	28.1
5	27.5	26.2
6	43.0	49.4
7	21.4	20.8
8	59.9	59.6
9	71.5	71.5
10	141.6	140.5
11	114.1	114.2
CH$_3$O	55.4	55.3
2'	147.0	147.1
3'	121.1	121.1
4'	148.3	148.2
5'	101.4	101.3
6'	157.4	157.3
7'	118.3	118.3
8'	130.9	130.9
9'	126.4	126.3
10'	143.7	143.6

constants were determined from the titration behavior of the ^{13}C resonances. Substituent and conformational influences of macro-lide antibiotics have also been investigated.[43-47] Early assign-ments,[43a] which had relied largely on single-frequency and off-resonance decoupling data and had been done without access to degradation products for comparison, were subsequently revised.[44] Determination of T_1 values has been suggested to be useful for distinguishing between carbohydrate and aglycone carbons.[46]

Cox and Cole[48] have reported the ^{13}C spectra of several *Aspergillus* aflatoxins, which occur extensively in foodstuffs.[48] Data are shown for aflatoxin B$_1$, one of 12 examples whose spectra were determined. The data are expected to be useful in iden-tifying other such fungal metabolites.

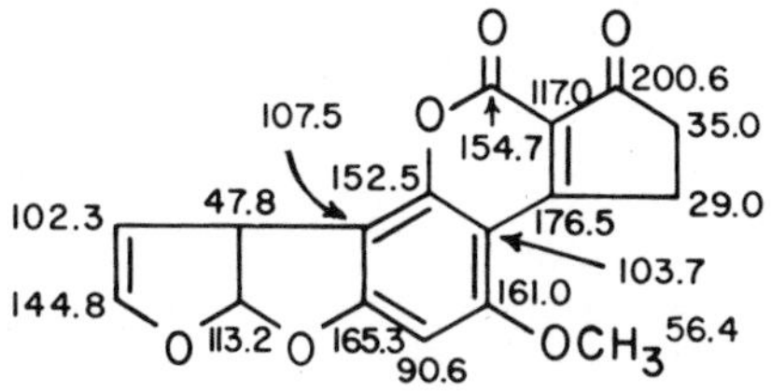

Aflatoxin B$_1$

The coumarin ring of aflatoxins is also a structural compo-nent of flavones and related plant components, and in part of the widely used anticoagulant warfarin. The ^{13}C spectra of these compounds have been extensively characterized.[49-56] The resonances of flavone itself were assigned with the aid of deuterated analogues and by use of shift reagents.[50] Naturally occurring flavonoids,

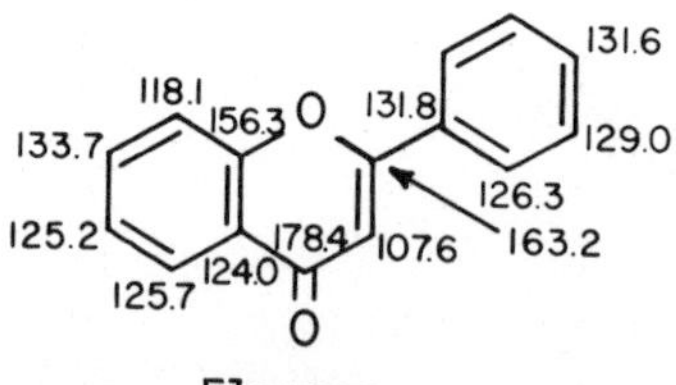

Flavone

in which the parent flavone structure bears a wide variety of terpenoid and glycoside substituents, have been assigned using the techniques discussed in earlier chapters.[51,55b] Of interest is the use of long-range C-H coupling constants, obtained from gated decoupled spectra. A detailed discussion of the technique applied to flavonoids has been presented by Chang.[53] Indeed, on the basis of the coupling between C-4a and the hydroxyl proton of naringenin in dry DMSO, the existence of the intramolecularly hydrogen-bonded species could be demonstrated. In the presence

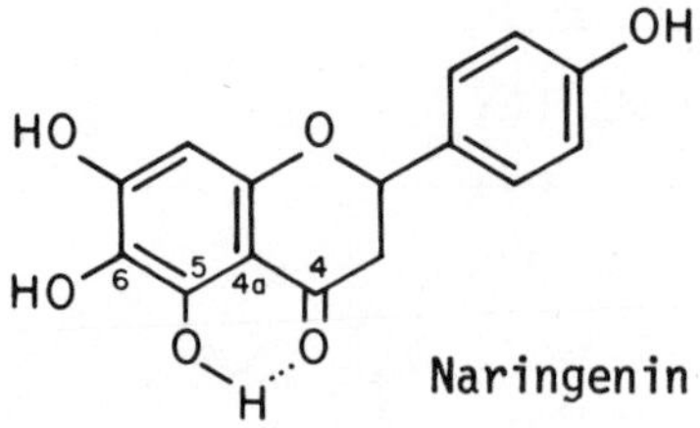

Naringenin

of a small (0.2 to 0.5%) amount of water, the signals from C-4, C-4a, and C-5 are split into two peaks each, differing by 0.1 to 0.2 ppm. This was attributed to the presence of both intra-molecularly hydrogen-bonded and free species. Coalescence of the peaks was observed at ~100°.

Using ^{13}C nmr, the structure of warfarin in solution was shown to exist as a mixture of cyclic hemiketals rather than as the acyclic ketoenol. Each numbered carbon appears as a pair of

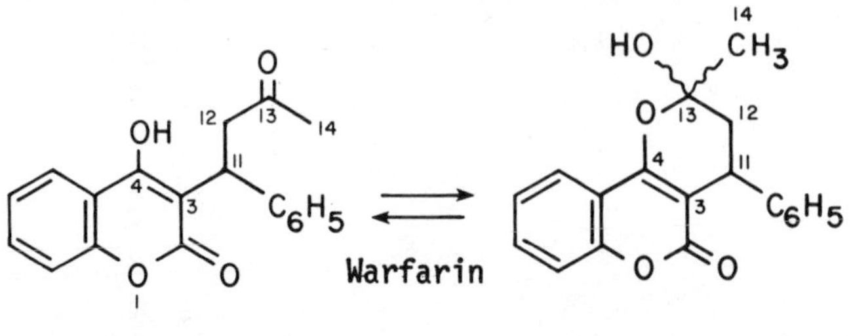

Warfarin

peaks (of unequal intensity), and no ketone carbonyl resonance is present. By contrast, the sodium salt of warfarin, in which hemiketal formation is not possible, displays only single resonances for the indicated carbons, as well as a ketone carbonyl at 216.5 ppm.

Other oxygen-containing natural products whose ^{13}C spectra have been assigned include the antifungal agent griseofulvin[57] and several gibberellins.[58] Assignments for griseofulvin and its C-6' epimer are given in Table 9.4. Note that C-2' in

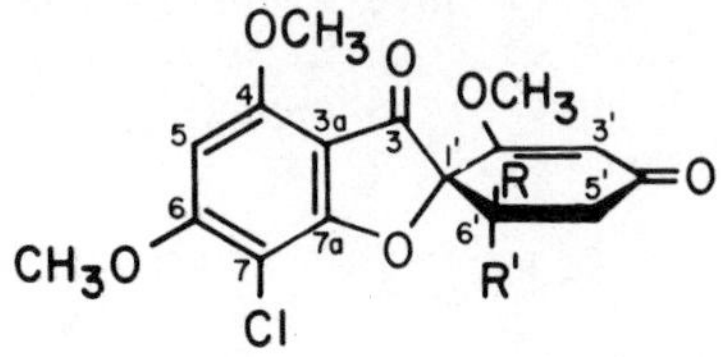

R = CH₃, R' = H: griseofulvin
R = H, R' = CH₃: episgriseofulvin

TABLE 9.4. ^{13}C CHEMICAL SHIFTS OF GRISEOFULVIN AND
EPIGRISEOFULVIN[57]

	GRISEOFULVIN	EPIGRISEOFULVIN
C-3	191.2	192.3
C-3a	104.1	105.1
C-4	157.6	157.3
C-5	91.2	91.2
C-6	169.6	169.6
C-7	95.3	95.6
C-7a	164.5	164.2
C-1'	90.1	89.4
C-2'	170.3	167.9
C-3'	104.7	105.8
C-4'	195.6	195.7
C-5'	39.5	40.1
C-6'	35.4	34.2
CH_3	13.8	13.2
4-OCH_3	57.5	57.4
6-OCH_3	56.6	56.4
2'-OCH_3	57.1	56.8

griseofulvin, in which the γ-situated methyl group at C-6' is
equatorial, is less shielded than C-2' in the epigriseofulvin.
    The ^{13}C resonances of several prostaglandins and some of
their synthetic analogues have been assigned.[59]   Relaxation
behavior as a function of concentration was used in confirming
the assignments.   Figure 9.4 shows a correlation diagram for a
number of these compounds.[59a]

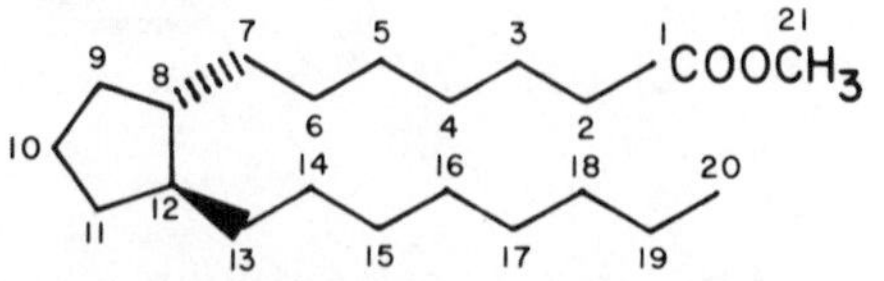

Prostaglandin numbering

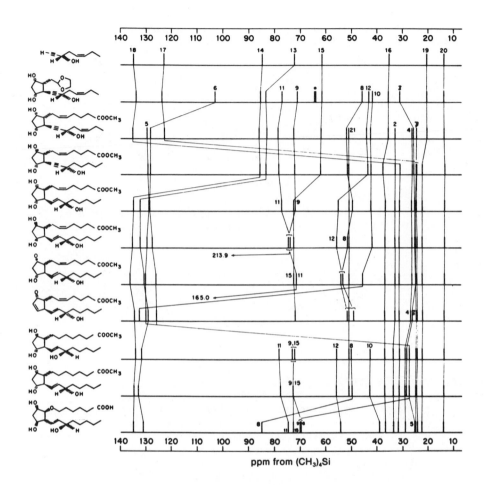

FIGURE 9.4.[59a]   *Correlation diagram of ^{13}C nmr chemical shifts of some prostaglandins, prostaglandin analogues, and synthetic intermediates.*

Carbon-13 chemical shifts for derivatives of acetylcholine, atropine, and other cholinergic neurotransmitters have been determined.[60] Not surprisingly, there is no apparent correlation between the shifts and the potencies of these compounds.

The ^{13}C chemical shifts of thiamine and several related structures have been assigned.[61] There is no simple correlation with electron density; indeed, the pyrimidine and the thiazolium-ring carbon shifts display rough correlations with approximate π densities, which have opposite slopes. Both the kinetic acidity and ^{13}C resonance position of C-2 appear to depend on the electronegativity of the N-3 substituent.[61]

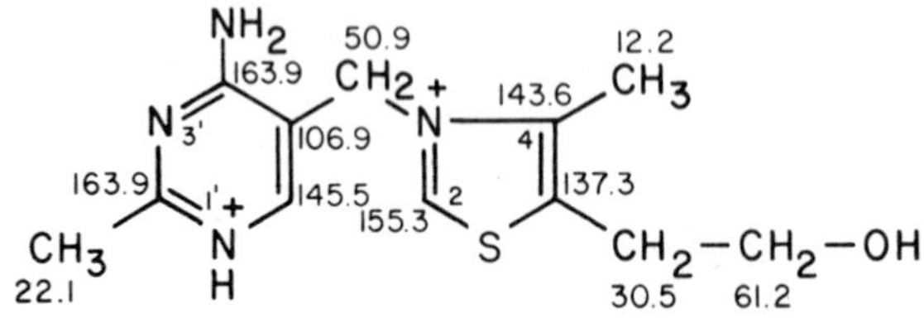

Thiamine (pH 0.6)

Carbon-13 resonance positions of pyridoxal (vitamin B-6) and several of its derivatives depend in a complex way on pH.[62]

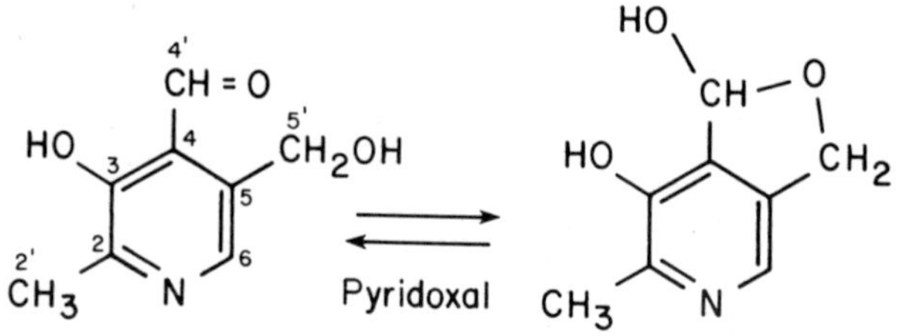

The resonance at 99.6 ppm for C-4' of pyridoxal itself is consistent with the presence of the hemiacetal form. Carbons C-2, C-4, and C-6 display pH dependencies similar to that of pyridine (see Chapter 4), but the behavior of C-3 depends very much on which compound is under examination. However, the three sites display distinctive behavior at different pH values. Figure 9.5 displays the pH dependence of pyridoxal phosphate.

The pyridoxal group represent essential cofactors in enzymatic metabolism of amino acids. Schiff base formation has been suggested as the initial step in the process. This reaction has been investigated by ^{13}C nmr[63]; several pH-dependent species of the Schiff base and the carbinolamine could be detected on the basis of chemical shifts and intensities.

Both ^{13}C chemical shifts[64] and T_1 values[64a] have been reported for several series of corrinoid coenzymes. The shifts of both methyl and cyano groups bound to the coordinated cobalt of these compounds are influenced by the remaining ligand. Resonance positions of the methine carbons of the corrin ring are shielded by ~2 ppm when a weak ligand (e.g., H_2O) is displaced by a stronger one (pyridine). Substitution of a weak ligand by a strong one was suggested to reduce electron demand by the cobalt from the corrin ring and hence to increase the charge density in the ring and increase the shielding.

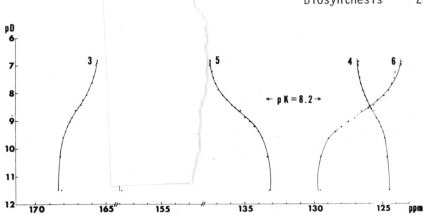

FIGURE 9.5.[62a]   *The pD dependence of chemical shifts for the*
^{13}C resonances of pyridoxal phosphate.   The
reference is Me₄Si.   Arrows refer to reported
pK values, determined in H₂O.

In a very elegant study employing partially enriched material
and INDOR techniques, Boxer et al. assigned all 55 resonances
of chlorophyll a and its magnesium- and phytyl-free precursor
pheophytin a.[65]   The effect of the magnesium in general is to
deshield carbons bonded to pyrrolelike nitrogens and shield
those bonded to pyridinelike ones. This parallels the behavior
of pyrrole and pyridine on deprotonation and protonation,
respectively.

Carbon-13 resonance assignments for porphin and chlorin
degradation products from chlorophyll have also been reported.[66]
The ^{13}C-^{57}Fe coupling into the carbonyl group of sperm whale
myoglobin has been measured. The value 27.1 Hz is essentially
identical to that found in simple models and in the range
exhibited by small inorganic molecules (23 to 28 Hz). Thus the
C=O group appears not to be "tilted" from the heme plane.[67]

BIOSYNTHESIS

Elucidation of biosynthetic pathways using ^{13}C labels and ^{13}C
nmr analyses has a number of advantages in comparison to tradi-
tional experiments that use ^{14}C radiolabel incorporation. In particular, the ^{13}C nmr method allows identification of incorporation
levels and *sites of incorporation* (without need for chemical
degradation). With bis-labeled precursors, ^{13}C nmr methods even
detail *patterns* of biosynthetic incorporation. The sole disadvan-
tage of ^{13}C methodology is low "sensitivity," resulting from
the need for at least ∿0.5 to 1% incorporation levels at least
for mono-^{13}C-labeled incorporation, as contrasted with ^{14}C-
labeling studies that may operate at far lower incorporation

PROBLEM 9.2.   *Determine the structure of this "natural product" and list assignments.*

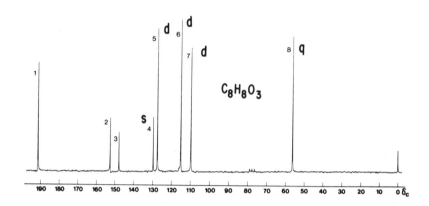

PROBLEM 9.3.   *Identify the compound whose ^{13}C spectrum (in D_2O) appears below. Assign each resonance line to the appropriate carbon or carbons. The molecular formula is $C_{11}H_{17}NO_3 \cdot HCl$.*

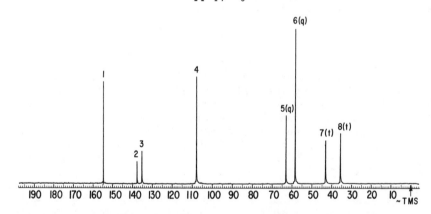

levels. In fact, with bis-^{13}C labels, incorporation levels may be considerably lower.

Several excellent reviews have appeared, detailing versatile approaches for ^{13}C nmr biosynthetic experiments.[68-72]

Sometimes ^{13}C biosynthesis studies use a simple mono- or bis-labeled precursor and simply follow its incorporation into the compound of interest. Sometimes a number of ^{13}C enriched precursor molecules are used in separate experiments to unambiguously

map the entire biosynthesis of the compound.   In these cases
precursor ^{13}C labels can be usefully supplemented with other
isotopes (e.g., ^{15}N and ^{2}H).   One example is the [13]C- and [15]N-
labeled biosynthetic analysis reported for tenellin (2).[73]

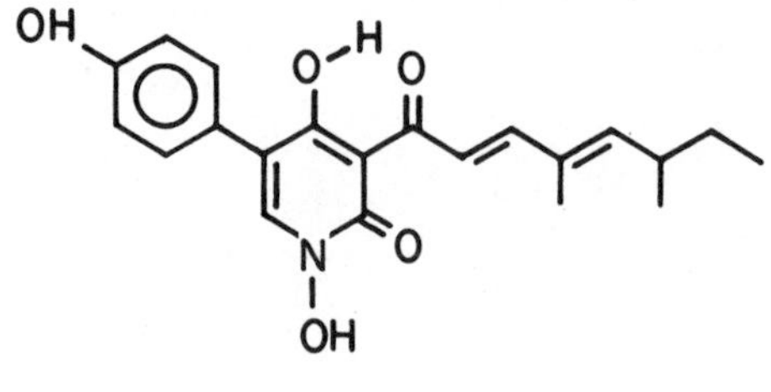

2

The ^{13}C and ^{15}N nmr experiments of this study established
patterns of biosynthetic incorporation and also confirmed the
structure of the heterocyclic ring.   Figure 9.6[73] shows ^{13}C nmr
spectra for 2 at natural abundance and with several isotopic
incorporations.

Initial radiotracer experiments showed that acetate,
methionine, and phenylalanine were efficiently incorporated
into 2 produced by fungus *Beauveria bassiana*, but that tyrosine
was poorly incorporated.   The ^{13}C nmr study showed that 2 is
formed from five acetate units (or the biogenetic equivalent)
condensed with a unit derived from phenylalanine.   A possible
scheme for the biosynthesis of tenellin is:

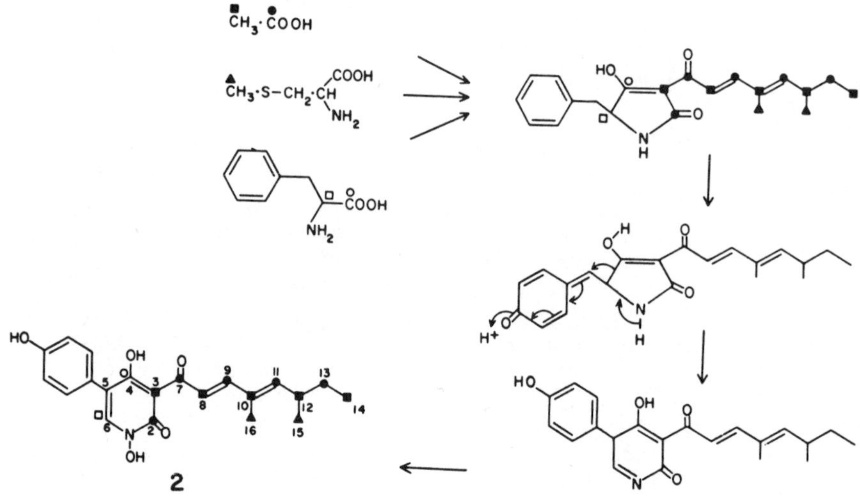

2

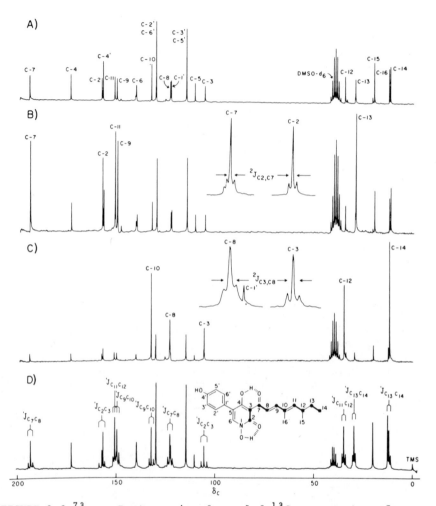

FIGURE 9.6.[73]   Proton-noise-decoupled ^{13}C nmr spectra of
tenellin in dimethyl sulfoxide-d_6: (A) at
natural abundance ^{13}C concentration; (B) biosynthesized from 54.2% enriched sodium $[1-^{13}C]$
acetate (insert shows satellites due to spin-
spin coupling ($^2J_{C2C7}$)); (C) biosynthesized from
90% enriched sodium $[2-^{13}C]$ acetate (insert
shows satellites due to spin-spin coupling
($^2J_{C3C8}$)); and (D) biosynthesized from 90%
enriched sodium $[1,2-^{13}C]$ acetate, showing
satellites due to spin-spin coupling ($^1J_{CC}$)
within acetate units incorporated intact.

A second example, investigation of the biosynthesis of cephalosporins and penicillins, demonstrated biosynthetic stereoselectivity by using chirally labeled valine precursors:[74]

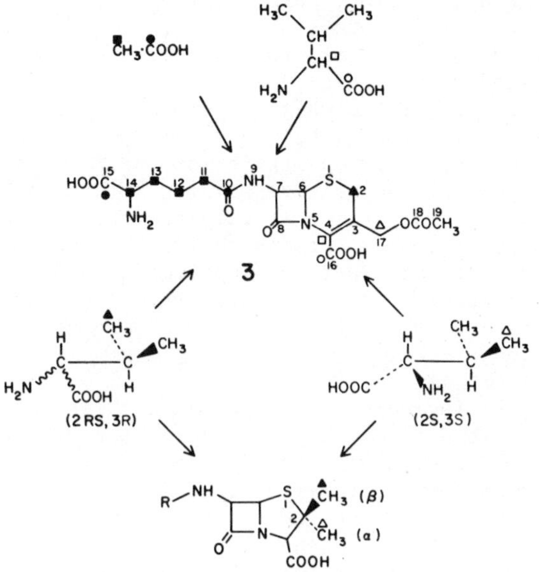

SPIN RELAXATION IN BIOPOLYMERS

The major difficulty encountered in applying natural-abundance ^{13}C nmr to biopolymers and to biological problems involving small molecules is that of sensitivity. Several factors render the direct observation of ^{13}C spectra of biopolymers more difficult than obtaining spectra of typical organic molecules. First, as the overall reorientation of the molecule (assuming no internal motions) becomes slower and the extreme narrowing condition ceases to be fulfilled, the NOE decreases from a value of 2 to about 0.1, thus raising the minimum concentration necessary for observation of a given ^{13}C resonance. Even for small, native biopolymers the situation is particularly unfavorable for protonated carbons because of the broad natural linewidths ($\gtrsim 30$ Hz) encountered. Nonprotonated carbons *may* have linewidths on the order of several hertz and hence do not suffer as much from this cause of sensitivity loss. Of course, for observation of the gross ^{13}C nmr spectral features of complex biological macromolecules, the situation is relatively favorable since the presence of "equivalent" carbons with roughly identical shifts and the overlap of resonances from various types of carbon raises the effective concentration, particularly in the carbonyl and aliphatic regions of ^{13}C protein spectra.

In "normal" pulse FT nmr, the sample is excited by a series of equally spaced 90° rf pulses[†] with a waiting period comparable to or exceeding T_1. If the waiting period is much longer than the time for data acquisition (i.e., if $T_1 \gg T_2^*$, where $1/\pi T_2^*$ is the total linewidth), the sensitivity of the FT experiment is considerably reduced. Repetitive pulsing before equilibration of all the spins with the lattice is complete results in attenuation or disappearance of long T_1 resonance lines. The condition $T_1 \gg T_2^*$ holds for large native biopolymers, and especially for nonprotonated carbons in such biopolymers. Thus, areas of the ^{13}C spectra of biopolymers that arise predominantly from non-protonated carbons may appear attenuated relative to the pro-tonated carbon regions.

Although the increase in resolution obtained by employing large magnetic fields is highly desirable, the expected improve-ment of basic nmr sensitivity to observe the ^{13}C FT nmr spectra of biopolymers may not be realized in full.[75] For molecular systems where the extreme narrowing condition is not fulfilled, T_1, T_2, and the NOE are frequency dependent. For rotational correlation times typical of most proteins and other large, native biopolymers ($\tau_c \gtrsim 10^{-8}$ sec), the NOEF has essentially reached its minimum value of 0.1 at fields of 1.4 T and an increase of field causes no further loss. However, for systems undergoing reorientation at an intermediate rate ($\tau_c \sim 10^{-10}$ to 10^{-9} sec), such as the cyclic peptide gramicidin S, the NOE may be considerably reduced at superconducting fields. Similar behavior may occur for carbons on side chains of large rigid biopolymers, if the correlation time(s) for internal rotation of the sidechain is (are) in the "intermediate" region. Figure 9.7 demonstrates the dependence of T_1 and T_2 on magnetic field for isotropic overall reorientation. For carbons in macromolecules outside the extreme narrowing region, T_1 increases substantially as one goes to a higher field. For example, the methine carbons in tRNA that have $T_1 \approx 50$ msec at 1.4 T are predicted to have T_1s exceeding 1 sec above 6.3 T. Proportional increases in non-protonated carbon T_1s (which are roughly on the order of 500 msec at low field) can be expected. Thus, nonprotonated carbon resonances obtained at high field on biopolymers such as tRNA may be subject to significant saturation effects at typical pulse repetition rates. On the other hand, chemical-shift anisotropy relaxation will begin to dominate sp^2 or sp nonprotonated carbon

---

[†]Of course, narrower flip angles may be utilized when the necessary pulse interval is significantly shorter than T_1.

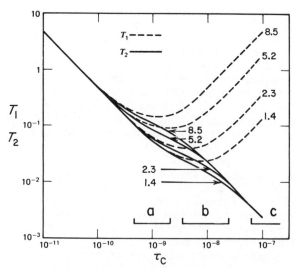

FIGURE 9.7.   *Calculated dependence of T_1 and T_2 on the correlation time τ_C for various magnetic fields (in Tesla) for a ^{13}C nucleus in a CH group (proton decoupled, $r_{CH} = 1.09$ Å), assuming isotropic rotational reorientation. Regions: (a) peptides; (b) small proteins, tRNA; (c) larger nucleic acids and proteins*

T_1s above 4 or 5 T, leading to some shortening of observed T_1 (and T_2).

A few detailed treatments of biopolymer spin relaxation have appeared.[75-77] Recently, authors have questioned the applicability of several usual assumptions, such as single-correlation time models[78] or constant C-H bond lengths.[79]

AMINO ACIDS, PEPTIDES, AND PROTEINS

Carbon-13 nmr applications with peptide and protein biopolymers have proved particularly successful. Following initial measurements on individual amino acids (deriving appropriate shielding additivity constants[80]) studies of small and biologically active peptide molecules were reported.[81] Initial cmr protein measurements were completed in the early 1970s. However, versatile applications became possible only with the improved instrumentation of the late 1970s.

Amino Acids and Peptides.   Individual amino acid shielding
trends parallel those of aliphatic compounds, with additional
complications arising from functional group titrations. In
short peptides, amino acid shieldings are found to reflect position in the peptide, but they are not sensitive to substitution
of neighboring residues (except for neighboring proline).[82]

Identification of individual amino acid titrations in small
peptides is straightforward. Selected titrations may be
assignable even in spectra of complex proteins[83] (Figures 9.8
and 9.9[83a]). Deuterium isotope effects on amino acid ^{13}C
shieldings have been measured to use the effect to aid spectral
assignments in complex peptides.[84]

Spin-spin couplings to carbon in amino acids and peptides
may be quite useful for analysis of peptide and side-chain
conformations.[85,86] Conformations of the ionophore peptide
antibiotic valinomycin in several solvent systems and with K^+ and
Tl^+, were refined using ^{13}C-^{1}H, ^{1}H-^{1}H, ^{1}H-^{15}N, and ^{13}C-metal
couplings (with the spin 1/2 metal ^{203,205}Tl).[87]

Carbon-13 conformational analyses of peptides containing
proline (or hydroxyproline) require introduction of cis-trans
isomerism around the X-Pro bond.[88] Recent studies by ^{13}C nmr
especially at high magnetic fields,[89] have shown that cis-proline
is more prevalent than anticipated and that even hydroxyproline
can have measureable cis isomers.[89a]

Carbon-13 studies have proved very useful for analysis
of conformational flexibility of small- and intermediate-sized
peptides.[89b,90] These studies give information about the degree
of segmental motion observed along the peptide chains. They
also reflect restricted group rotations of amino acid side chains.
However, a note of caution should be sounded. Amino acid
carboxyl T_1s and basic amine sites (e.g., the imidazole ring in
histidine) are especially susceptible to paramagnetic relaxation
from small amounts of paramagnetic metal impurities. An initial
literature report[91a] of complex pH and concentration dependences
for ^{13}C carboxyls in three amino acids was found to be in
error.[91b]

In large polypeptides ^{13}C T_1s and linewidths (proportional
to $1/T_2$ for simple systems) monitor random coil to helix transitions.[92,93] In a particularly elegant study Torchia et al.[93]
studied a 36 amino acid linear peptide fragment of rat skin
collagen, α1-CB2. The sequence of this peptide is known and
prediction of the ^{13}C nmr spectrum from empirical rules was
successful; agreement for the α-carbon region is shown in Figure
9.10. The ^{13}C spectrum taken at 2°C dramatically shows the effect
of helix formation, with greatly restricted flexibility producing
very broad resonances. Figure 9.11 compared the random coil

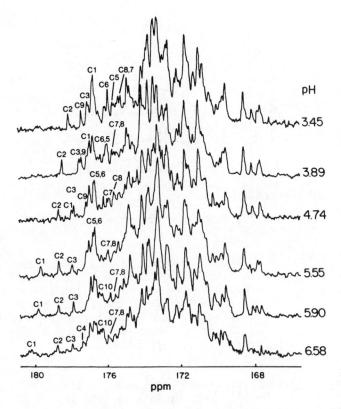

FIGURE 9.8.[83a]   Carboxyl and carbonyl carbon resonances of
natural-abundance ^{13}C FT nmr spectra of hen
egg-white lysozyme (10 mM) in H_2O containing
D_2O (4:1) and 0.1 M NaCl. Each spectrum was
recorded at 67.9 MHz with 8000 real points,
15 kHz sweep width, and 30,000 to 40,000
transients (total time 17 to 22 hr). The
peaks labeled C-1 to C-10 were observed
to titrate on change of pH.

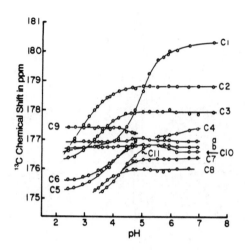

FIGURE 9.9.[83a]   *Carbon-13 chemical shift as a function of pH*
for the resolved titrating resonances corres-
ponding to the carboxyl carbon atoms of hen
egg-white lysozyme. Two nontitrating
resonances a and b, which probably derive from
glutamine and asparagine carbonyl groups,
respectively, are also shown.

spectrum (30°C) with the triple helix form of α1-CB2 at 2°C.
Other details of the structure of this peptide were also
discussed.[93]

A number of research groups have recently attempted to reach
more detailed understanding of dynamic processes in peptides and
proteins. Theoretical and experimental problems make exact
descriptions elusive.[94]

Proteins.   No application of ^{13}C nmr has benefited more from
instrumentation improvements than has the elucidation of protein
structure. In 1973 Allerhand showed significant sensitivity
improvement with his 20-mm sample probe operating at 15.1 MHz;[95]
subsequently, large sample probes became available at 2.3 T and
for superconducting spectrometers operating to 8.4 T (90 MHz for
[13]C). Large-volume samples necessarily limit ^{13}C studies to a
few easily available proteins. Recently, side-spinning superconducting probe designs have shown greatly improved sensitivities
with small to moderate-sized samples.[96] Additional improvements
in ^{13}C spectroscopy of proteins have come from the inherent
advantages of higher-field spectrometers, although it is not yet
clear whether the highest fields will be advantageous for very

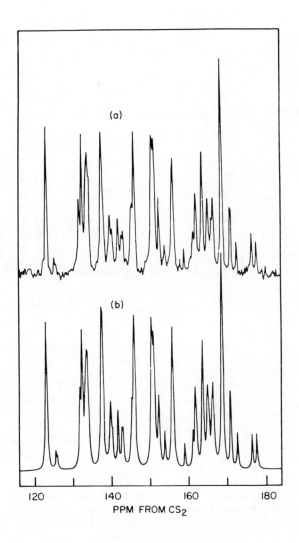

FIGURE 9.10.[93]   High field region of ^{13}C spectrum of α1–CB2:
(a) experimental; (b) calculated using the
known sequence of the peptide.

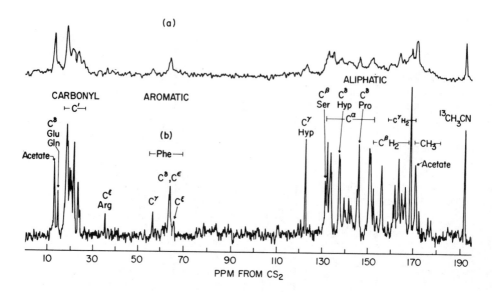

FIGURE 9.11.[93]   *Comparison of αl-CB2 spectra obtained at two*
temperatures using 90°-t-90° pulse sequences:
(a) 2°C, t = 1.5 sec, 32K scans of the FID
accumulated; (b) 30°, t = 4.5 sec, 16K scan FID
accumulated with vertical scale multiplied by
2 to facilitate comparison with the 2° spectrum.
Protein concentration, 25mg/ml in 0.15 M
NaOAc buffer at pH 4.8. Chemical-shift scale
in parts per million from external CS₂.

large, native proteins. Relaxation considerations may become a
significant factor for some studies of these molecules.

Hen egg-white lysozyme is the protein most studied to date
by ^{13}C nmr.[95,97] This low-molecular-weight enzyme (mw ≈ 14000)
shows significant detail in the low-field region of the [13]C
spectrum. Indeed, *individual amino acid* residues are resolved
for nonprotonated carbons of several residues. For example, for
the six tryptophan residues in lysozyme, four Cγ carbons are
resolved (Figure 9.12).

The situation for protonated carbons of proteins is less
favorable since efficient dipolar relaxation leads to broad lines,
even with small systems such as lysozyme. Use of higher magnetic
fields is particularly important under these circumstances

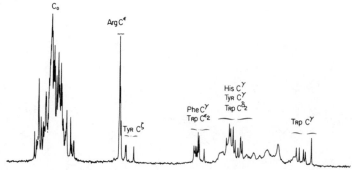

FIGURE 9.12.   *The unsaturated carbon region of the 45 MHz ^{13}C
spectrum of native HEW lysozyme (0.015 M in a
25 mm sample tube, 9000 scans). Numerous
single nonprotonated carbon resonances are
observed with a high ratio of signal-to-noise
(spectrum provided by J. D. Roberts and C. H.
Bradley).*

(Figure 9.13[97a]), although *single* C-H carbon resonances remain
elusive at 6.3 T (limited success has been achieved).

An interesting class of techniques has been developed for
spectral isolation of sharp lines in biopolymer spectra. The
most used technique, convolution difference spectroscopy,[98] has
been applied to lysozyme and other proteins. Examples are given
in Figure 9.14.[99] Other techniques have been developed with
similar goals.[100]

Another small protein that is well studied is the basic
pancreatic trypsin inhibitor (BPTI).[101] The 90.5-MHz ^{13}C spectrum of BPTI showed large improvement over the spectrum recorded
at 25.2 MHz.[101a]

Other systems investigated include the histones of chromatin
core protein[102] and several structural proteins[103] observed with
special techniques used in solid-state ^{13}C nmr (see Chapter 10).
Carbon-13 labeling has been used to monitor enzyme active-site
chemistry and protein binding and enzyme kinetics.[105] Chemical
modifications of proteins have been followed by ^{13}C nmr.[106]

CARBOHYDRATES

Carbohydrates were among the first natural products to be
examined by ^{13}C nmr; their fixed geometries allowed geometrical
influences on ^{13}C chemical shifts and coupling constants to be
examined in some detail.[107-110] At the same time, chemical-shift

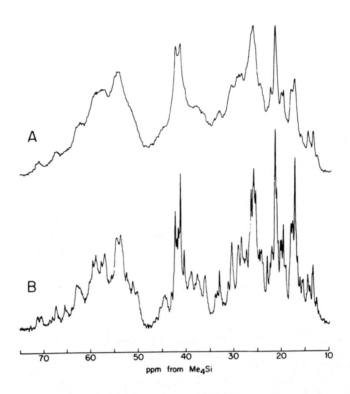

FIGURE 9.13[97a]   *Aliphatic region ^{13}C nmr spectra of HEW*
lysozyme in H_2O and 0.1M NaCl: (a) at 15.2 MHz,
14.1 mM lysozyme in 10.5 ml (2.1g protein),
32K scans (20 hr total time), spectrum
recorded with 0.9 Hz digital filter; (b) at
67.9 MHz, 15.4 mM lysozyme in 2 ml (0.44g of
protein), 32K scans (27 hr total time), 2.8 Hz
digital broadening.

assignments frequently were based on expected effects of substituents and geometry as discussed in Chapter 3. Subsequent
work,[111] especially using ^{13}C and deuterium-labeled hexopyranoses, confirmed many of the earlier assignments[109] but
revealed some uncertainties in others.[107,108] Table 9.5 gives
chemical-shift values for several hexopyranoses and their methyl
pyranosides as determined by Walker et al.[111] These assignments
derive from measurements for the ^{13}C-1-enriched compounds and
rely on observation of direct and geminal ^{13}C-^{13}C couplings to
C-2, C-3, and C-5. Thus $^1J_{12}$ = 42 to 47 Hz in all the compounds
in Table 9.5, and hence C-2 could be readily distinguished from

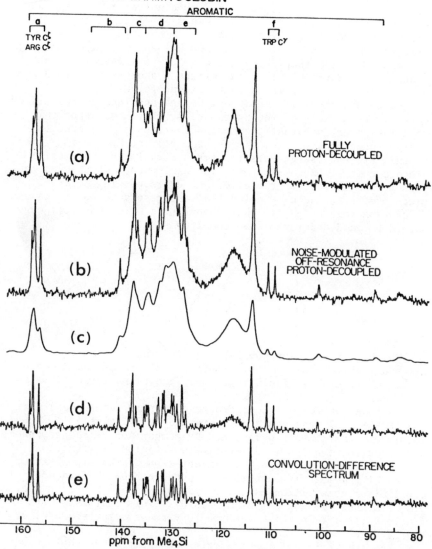

WHALE CYANOFERRIMYOGLOBIN

FIGURE 9.14.[99]   15.1 MHz ^{13}C spectra of whale cyanoferrimyoglobin. (a) Broadband decoupling; (b) off-
resonance broadband decoupling; (c) same data
as in (b), but with a stronger exponential
multiplication of the FID; (d) difference
spectrum equal to (b) —0.9(c); (e) baseline
adjustment of spectrum (d).

closely lying C-5 resonances. Similarly, C-3 was differentiated from C-5 by the presence of 2 to 5 Hz coupling to C-1 in the β-anomer and its absence in the α-anomer. Interestingly, vicinal C-C coupling exists between C-1 and C-6, which are oriented anti to each other, but not between the *gauche*-oriented C-1 and C-4. The coupling constant $^3J_{16}$ is larger for the β-anomers (4.0 to 4.3 Hz) than for the α-anomers (3.0 to 3.5 Hz).

Table 9.5 demonstrates that, except for C-1, the ^{13}C resonance positions of the methyl hexopyranosides do not differ markedly from those of the pyranoses. It is because of this relative invariance that the monosaccharides may be used to assign resonances of di-, oligo-, and polysaccharides. Typical of this are the assignments for amylose, a linear polymer of glucose units linked α-1→4.[112] The assignments given with the structure are based on those for the corresponding methyl glucoside. It should be noted that the resonance positions can depend somewhat on pH, so that comparisons with model compounds should be made under the same conditions.

$$\left[\begin{array}{c} \overset{62.0}{\underset{80.6}{\text{CH}_2\text{OH}}} \\ \text{HO} \quad \overset{72.6\ \ 73.8}{\underset{75.4}{}} \\ \text{OH} \quad 102.9 \end{array}\right]_n$$

Wilbur et al. have published a 67.9-MHz ^{13}C spectrum[113] of D-mannose that allowed a quantitative analysis of the mannose solution structures, including the pyranose and furanose pairs of anomers (Figure 9.15[113]).

Resonances for all four fructose tautomers have been assigned (Table 9.6),[114a] and earlier literature ambiguities[115-117] were resolved. Assignments were aided by comparison of the resonance positions for the 3-O-methyl-D-fructose tautomers with those of D-fructofuranose-6-phosphate. Integration of the signal intensities of the methylfructosides in water established the equilibrium composition as 18% α-pyranoside, 37% β-pyranoside, 11% α-furanoside, and 34% β-furanoside. This contrasts with fructose itself, in which β-D-fructopyranose comprises >70% of the mixture.[114b]

The complexity of carbohydrate ^{13}C spectra has challenged the inventiveness of chemists in developing assignment techniques. In addition to the methods of specific isotopic labeling and of chemical-shift comparison with model compounds, other techniques include correlation with T_1 values,[118] use of relaxation[119] and shift[120] reagents, and deuterium isotope effects.[121] The last technique has been used to differentiate hydroxy-bearing carbons from others. For example, the ^{13}C spectrum of *partially*

TABLE 9.5.   [13]C CHEMICAL SHIFTS (IN PARTS PER MILLION) OF SOME HEXOPYRANOSES AND PYRANOSIDES IN H_2O.[111]

	C-1	C-2	C-3	C-4	C-5	C-6	OCH3
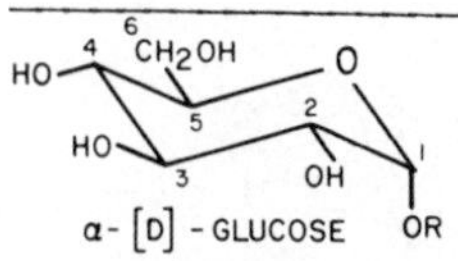 **α-[D]-GLUCOSE** R = H	93.6	73.2	74.5	71.4	73.0	62.3	
R = CH_3	100.6	72.7	74.7	71.2	73.0	62.2	56.5
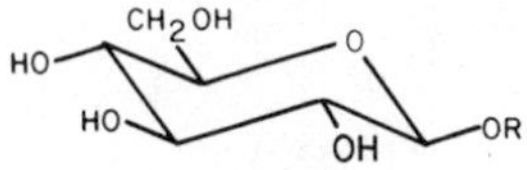 **β-[D]-GLUCOSE** R = H	97.4	75.9	77.5	71.3	77.4	62.5	
R = CH_3	104.6	74.6	77.4	71.2	77.3	62.4	58.5
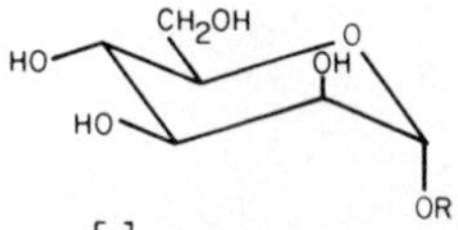 **α-[D]-MANNOSE** R = H	95.5	72.3	71.9	68.5	73.9	62.6	
R = CH_3	102.2	71.4	72.1	68.3	73.9	62.5	56.1
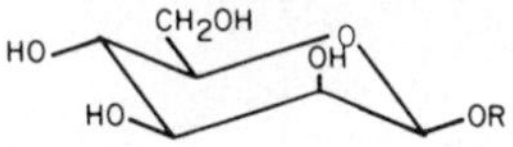 **β-[D]-MANNOSE** R = H	95.2	72.8	74.8	68.3	77.6	62.6	
R = CH_3	102.3	71.7	74.5	68.4	77.6	62.6	
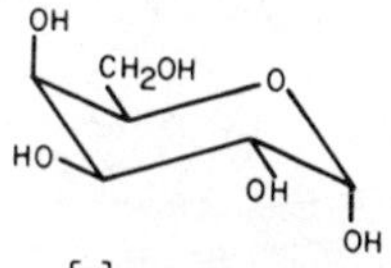 **α-[D]-GALACTOSE**	93.8	70.0	70.8	70.9	72.0	62.8	
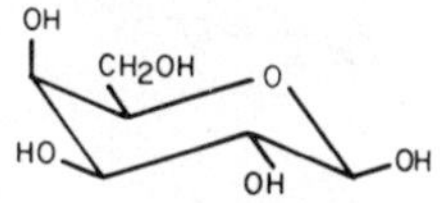 **β-[D]-GALACTOSE**	98.0	73.6	74.4	70.4	76.6	62.6	

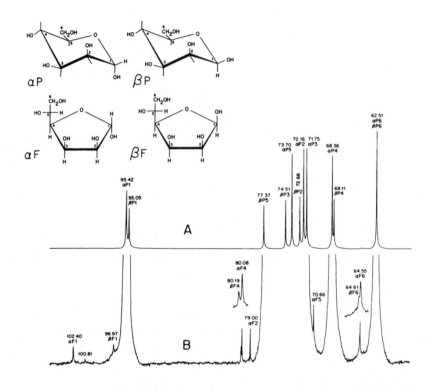

FIGURE 9.15.[113]   *Proton decoupled 67.9 MHz ^{13}C FT nmr spectrum*
of equilibrated 4M D-Mannose in H_2O at 36°,
5 hr accumulation (B shows 36 fold vertical
expansion; insets show additional fourfold
horizontal expansion).

O-deuterated methyl α-D-glucopyranoside, under conditions of slow
exchange (DMSO solution) displays narrowly separated doublets
for C-2, C-3, C-4, and C-6; thus C-5 is readily identifiable.
In more complex molecules, for example, methyl β-D-cellobioside,
the hydroxy-bearing carbons occasionally appear as broadened,
hence less intense singlets, so that the nonhydroxylated carbons
may still be identified.

METHYL β-D-CELLOBIOSIDE

TABLE 9.6.   [13]C CHEMICAL SHIFTS (IN PARTS PER MILLION) OF
D-FRUCTOSE TAUTOMERS IN H_2O[114a]

	C-1	C-2	C-3	C-4	C-5	C-6
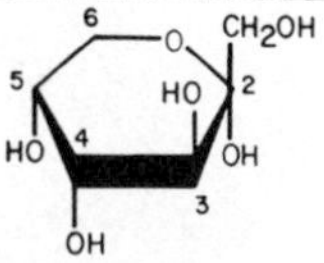	65.9	99.1	70.9	71.3	70.0	61.9
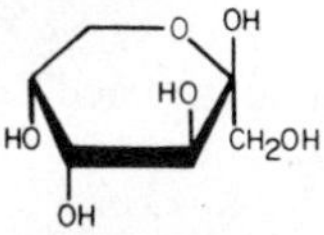	64.7	99.1	68.4	70.5	70.0	64.1
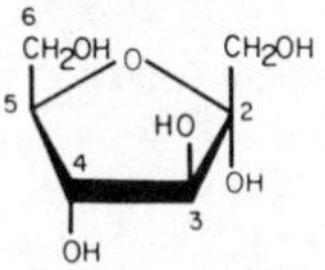	63.8	105.5	82.9	77.0	82.2	61.9
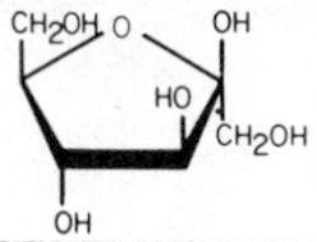	63.6	102.6	76.4	75.4	81.6	63.2

Extension of mono- and disaccharide studies to carbohydrate
polymers has been relatively successful, in good part because
the shifts remain essentially unperturbed when the monomer units
are incorporated into the polymers. Much of this work has been
reviewed by Perlin and Hamer.[122] The blood anticoagulant heparin,
for example, has been shown to exist in two main classes,
according to its source.[122,123] Type A is characterized by that
from hog mucosa, and type B derives from beef lung. The latter
consists essentially completely of the repeat biose structure
shown, whose ^{13}C spectrum displays twelve signals. Type A heparin

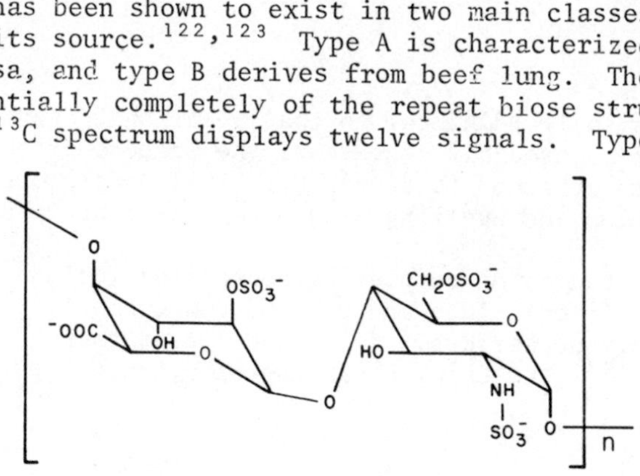

consists of the biose units copolymerized with 15-30% of
2-deoxy-2-acetamido-α-D-glucopyranose and β-D-glucopyranuronic
acid, as shown. A telling feature of the spectrum was the
presence of the N-methyl group at ~25 ppm.

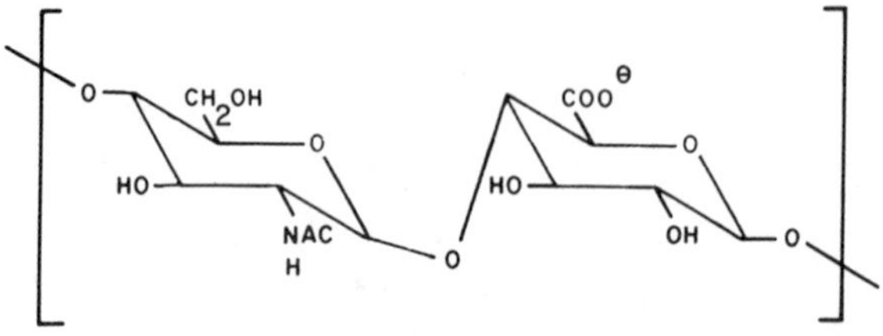

Conformations of carbohydrate polymers have been examined
with the aid of vicinal ^{13}C-H couplings, in particular, the
coupling across the glycosidic (C-O-C-H) bond.[122,124] Refine-
ment of a Karplus type of curve is hampered by the difficulty
in obtaining molecules that provide appropriate ^3J values for
the 0° to 60° region. Cyclohexaamylose has been used as a
suitable model, and spectral analysis has been simplified by
selective deuteration. Representative values are $^3J_{C_1H_4}$ = 4.8 Hz,
and $^3J_{C_4H_1}$= 5.2 Hz. Although the dihedral angle relating C_1 and

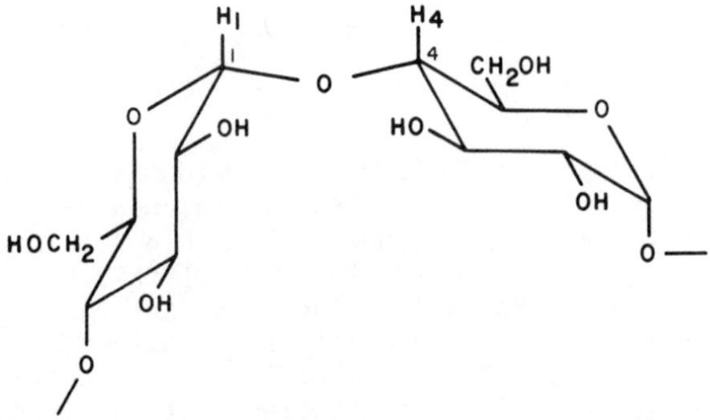

H_4 is essentially identical to that relating C_4 and H_1, the
values differ, probably because of the effects of the different
substituents attached to C_1 and C_4. This underscores some of
the limitations on applying a Karplus treatment to carbohydrate
geometry.
Other studies on complex carbohydrates include aldopyranosyl
phosphates,[125] cellulose and related compounds,[121,126] menin-
gococcal polysaccharides,[109,127] and intact tissue-bound pro-
teoglycans.[128] The chemical shifts of vitamin C (L-ascorbic

acid) have been determined; values are shown with the structure.[129]

L-ASCORBIC ACID

NUCLEOTIDES AND NUCLEIC ACIDS

Carbon-13 nmr has been used in pioneer studies of nucleic acid structure; however, applications to date have not been as promising as in the case of peptides and proteins. This results from a number of factors, and especially from the very large molecular weights of nucleic acids (tRNAs $\simeq$ 25,000, mRNAs $\approx 10^5$, and DNAs $> 10^7$) that restrict motion of these biopolymers and also lower practical concentrations. Furthermore, biochemical preparation of nucleic acid samples is not easily performed on a scale useful for ^{13}C nmr at natural abundance. Despite these problems, significant progress on nucleic acid structure has been realized by use of modern instrumentation. Additionally, many studies have used ^{13}C nmr to evaluate mono- and dinucleotide structure and conformation; ^{13}C spin-relaxation measurements have examined molecular dynamics of polynucleotides and tRNA. Reviews have appeared dealing largely with polynucleotide and tRNA structure and conformational analysis.[130]

Several parameters are available for ^{13}C nmr analysis of simple mono- and polynucleotides.[131] Chemical shifts due to ring currents, which are dramatic in ^1H nmr of base-stacked nucleotides, are not as easily applied in ^{13}C nmr where the ^{13}C resonances undergo relatively small (but still usable) changes as a function of concentration and base stacking (Figure 9.16[130a]). Three-bond $^{13}C-^{31}P$ scalar couplings have been used to probe backbone structure of oligonucleotides. Despite the complexities associated with these analyses, several systems have yielded useful results.[130a,132,133] Figure 9.17[132] shows the ^{13}C spectrum of a dinucleotide monophosphate, A-A(3',5'). Syn-anti conformation of the base ring can be evaluated from ^{13}C shieldings and ^{13}C relaxation parameters (T_1s and NOEs). Several authors have viewed nucleotides as a function of pH.[133-135] Chenon et al.[136] developed a set of shielding

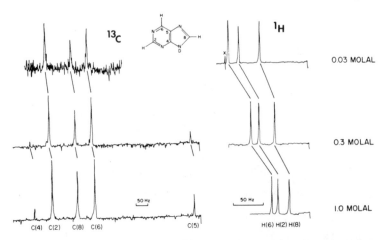

FIGURE 9.16.[130a]   Proton and ^{13}C nmr spectra of purine in
D_2O displaying the differential concentration
dependence of the ^1H resonances and the
ion-differential concentration dependence of
the ^{13}C resonances. The C-4 and C-5
resonances at low purine concentration are
not visible because of long spin-lattice
(T_1) relaxation times. The peak marked X is
a spectrometer artifact.

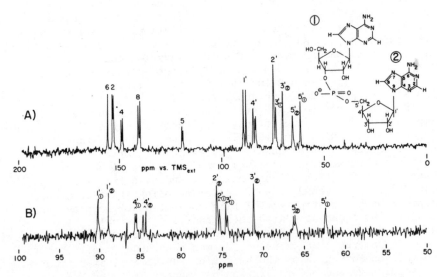

FIGURE 9.17.[132]   (A) Carbon-13 nmr spectrum (25 MHz) of A-A
(3',5'); (b) detail of the ribose region of
spectrum A.

parameters to facilitate determination of tautomeric populations for nucleotide bases (purines and pyrrolopyrimidines).[136]

The original reports of natural abundance ^{13}C nmr of unfractionated yeast tRNA came from Komoroski and Allerhand.[137] The initial report[137a] showed that thermally denatured tRNA ($\approx$ 80°C) gave a well-resolved spectrum similar to that of poly-A, whereas the spectrum showed significant broadening for the folded biopolymer at lower temperatures. In a subsequent study[137b] a large volume sample probe was used to improve performance, allowing detection of rare base carbons from dihydrouridine residues (Figure 9.18).

Grant and co-workers[138] incorporated ^{13}C-labeled uracil (labeled 90% at C-4) into tRNA by using a mutant bacterium. The high degree of sensitivity of the ^{13}C spectrum of 55 mg of bulk tRNA with ^{13}CO uridine (and a few other molecules synthesized *in vivo* in the experiment) is shown in Figure 9.19. The native tRNA shows a band of resonances near 165 ppm due to uridines in different chemical environments, plus two other sharp peaks. Heat denaturation results in a collapse of the uridine signal, as expected when secondary and tertiary structure is broken. Grant tentatively assigned line 1 to C-4 of dihydrouridine and line 3 to C-4 of pseudouridine. In a later study 5S ribosomal RNA was separated after ^{13}C-uracil incorporation.[139] Figure 9.20 shows the ^{13}C spectra of native and denatured labeled 5S RNA (This molecule contains 20 uridines but probably does not contain bases modified from uracils).

Other authors examined tRNA modification from alkylation[140] and bisulfite addition[141] to yeast RNA.

Deoxyribonucleic Acid.   Nuclear magnetic resonance studies of native DNA must contend with the extremely restricted solution dynamics of this very large, "rigid" molecule and resulting unfavorable relaxation behavior. High-resolution ^{13}C nmr techniques have only restricted application for DNA; newer solids nmr technology has not yet been used for DNA studies. One original report of a natural-abundance ^{13}C spectrum of DNA appeared in 1973,[130b] but this was later ascribed to largely denatured molecules. Recently improved instrumentation has been used to obtain natural-abundance ^{13}C spectra of native DNAs isolated from calf thymus and prepared for nmr studies.[142]

Small, relatively homogeneous (140 ± 20 nucleotide pairs long) double-stranded DNA was prepared from chromatin digested with micrococcal nuclease and then chromatographed. Carbon-13 spectra show resolved lines for sugar and base carbons (Figures 9.21[142a] and 9.22[142b]). Assigned resonances agreed well with assignments reported for mononucleotides, with variations in shieldings of about ±3 ppm. Intercalation of ethidium into the

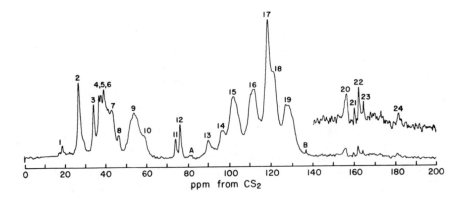

FIGURE 9.18.[137b]   15.2 MHz ^{13}C nmr spectrum of unfractionated baker's yeast tRNA in water (150 mg/ml) at pH 7.1 and 41°C. Note minor base resonances 20-24.

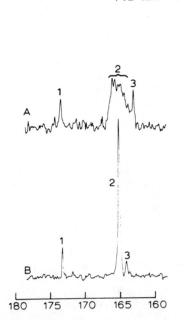

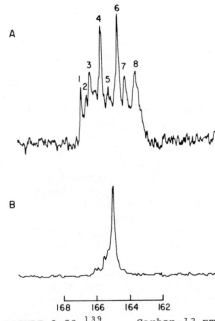

FIGURE 9.19.[138]   Carbon-13 nmr spectra of ^{13}C-enriched tRNA. Spectrum A was taken at 37°C. Spectrum B was taken at 82°C. Chemical shifts were referenced to TMS.

FIGURE 9.20.[139]   Carbon-13 nmr spectra of the C-4 uridine carbons in 5S RNA at (A) 37°C with 18,000 transients, 0.25-Hz digital resolution, 4-sec cycle time, 60° pulse width and (B) 75°, 4,000 transients, 2-sec cycle time (TMS shift scale).

290

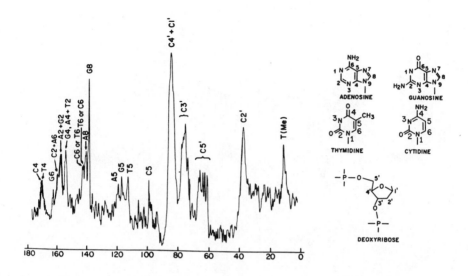

FIGURE 9.21.[142a]   *Natural abundance 67.9 MHz* [13]*C nmr spectrum of native DNA (140 ± 30 nucleotide-pair length, 171 mg in 3.5 ml of phosphate buffer, pH 7.35); 43,000 scans were accumulated, with a cycle repeat time of 1.5 sec, 18.8 hrs total time. Chemical shifts relative to TMS; base carbon assignments are shown above.*

DNA broadened resonances markedly, whereas heat denaturation lowered linewidths by a factor of 2. ^{13}C T_1s have also been determined for these DNA samples (Table 9.7 and Figure 9.22). The relaxation times and linewidths reflect conformational mobility for the sugar carbons and particularly exocyclic C-5'. Base-ring carbon T_1s indicate a reduced degree of motional flexibility, but a greater degree of motion than anticipated from rigid-rod models. Interestingly, preliminary data indicate little effect on ribose carbon T_1s from ethidium intercalation into DNA.

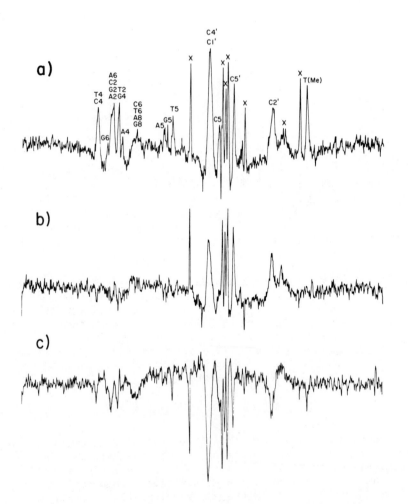

FIGURE 9.22.[142b]   Natural abundance 37.7 MHz ^{13}C FIRFT nmr
spectra of native DNA (120 ± 30 nucleotide
pair length, 250 mg in 2.6 ml of phosphate
buffer, pH 7.35); special side-spinning
solenoid coil probe utilized; 10,000
scans were accumulated with a cycle
repeat time of 1.2 sec, 65 hrs total
time for 9 spectra; τ values: (a) 3.6,
(b) 0.3, and (c) 0.005 sec (spectra have been
baseline-flattened to minimize probe background signal).

TABLE 9.7. ^{13}C T_1s (IN SECONDS) OF DOUBLE-STRANDED DNA[a,142b]

Sugar Carbons	37.7 MHz[b]	67.9 MHz[b]	67.9 MHz[c]
2'	0.15	0.15	0.28
3'	0.18	0.31	0.46
1',4'	0.21	0.38	0.57
5'	0.065	0.11	0.22
Base Carbons			
T(Me)	0.79	0.43	----
T5	1.56	----	----
T4,C4	1.53	----	----

[a]Preliminary data; values determined at 37.7 and 67.9 MHz and 32 ± 3°C.

[b]120 Nucleotide-pair DNA, 250 mg in 2.6 ml of phosphate buffer, pH 7.35.

[c]160 Nucleotide-pair DNA, 175 mg in 3.4 ml of phosphate buffer, pH 7.35.

LIPIDS

The structure and the chemical function of biological membranes have received increasing attention with the availability of modern spectroscopic methods. Carbon-13 nmr, ^{2}H nmr, and electron spin resonance (esr)-labeling studies have each contributed to our understanding of this complex organelle. Recent models derive from the fluid mosaic model,[143] consisting of an oriented lipid bilayer with imbedded proteins (some traversing the structure), channels, and so on. Diffraction methods (X-ray and neutron) have proved valuable to understanding the static structure of membranes, but magnetic resonance methods are better suited for studying membrane dynamic structure. Electron spin resonance and ^{2}H nmr have been used to evaluate the *order parameter* that describes cooperativity of organization of the lipids, and ^{13}C nmr has been used to examine the degree of segmental group mobility along lipid chains. The original ^{13}C nmr studies of lipid structure examined sonicated dipalmitoyllecithin micelles,[144] reporting strong segmental motion for the alkyl chain, with the last four or five carbons undergoing relatively free solution dynamics. Subsequent papers dealt with more quantitative (although still approximate) interpretation of ^{13}C T_1s in phospholipid bilayers.[145] Smith has used biosynthetic incorporation of $^{13}CH_3CO_2^-$ and $CH_3^{13}CO_2^-$ to produce alternately labeled lipid chains, thus increasing ^{13}C nmr sensitivity without introducing the complexities of extensive ^{13}C-^{13}C couplings.[147] Fatty acid carboxyl shieldings[146] and aqueous lanthanide (paramagnetic) shift reagents[147] have been incorporated to probe lipid bilayer structures. Other workers have incorporated chemicals into lipid bilayers and observed structural changes in the vesicles.[148] In one study[148a] differential effects were observed for cholesterol and lanosterol incorporation.

Lipids other than phosphatidyl cholines (serines, etc.) have been investigated, including lipoproteins from human plasma,[149] the lipids in intact rod outer segments,[150] and bovine white matter.[151] Carbon-13 PRFT spectra of purified gangliosides confirmed that lipid chain motions were anisotropic.[152]

Several workers[153] have shown practical application for the generally high mobility of lipids in cells. It is possible to *nondestructively* examine the lipid content of *viable* seeds to determine polyunsaturated lipid ratios, thus enhancing prospects for genetically increasing polyunsaturated fatty acid content.

REFERENCES

1. For a comprehensive review of the ^{13}C spectra of natural products, see F. W. Wehrli and T. Nishida, in *Progress in the Chemistry of Organic Natural Products*, Vol 36, (W. Herz, H. Grisebach, and G. W. Kirby, Eds., Springer-Verlag, Vienna, 1979, p. 1.

2. G. C. Levy, I. R. Peat, R. Rosanske, and S. Parks, *J. Magn. Resonance*, **18**, 205 (1975).

3. E. Wenkert, B. L. Buckwalter, I. R. Burfitt, M. J. Gasic, H. E. Gottlieb, E. W. Hagaman, F. M. Schell, and P. M. Wovkulich, *Topics in Carbon-13 NMR Spectroscopy*, Vol. 2, G. C. Levy, Ed., Wiley-Interscience, New York, 1975, Chapter 2.

4. F. Bohlmann, R. Zeisberg, and E. Klein, *Org. Magn. Resonance*, **7**, 426 (1975).

5. W. Bremser and J. Paust, *Org. Magn. Resonance*, **6**, 433 (1974).

6. (a) R. Rowan, III and B. D. Sykes, *J. Am. Chem. Soc.*, **96**, 7000 (1974); (b) R. S. Becker, S. Berger, D. K. Dalling, D. M. Grant, and R. J. Pugmire, *ibid.*, **6**, 7008 (1974).

7. (a) J. W. Blunt and J. B. Stothers, *Org. Magn. Resonance*, **9**, 439 (1977); (b) H. Eggert, C. L. VanAntwerp, N. S. Bhacca, and C. Djerassi, *J. Org. Chem.*, **41**, 71 (1976); (c) W. B. Smith, *Annu. Rep. NMR Spectrosc.*, **8**, 199 (1979).

8. W. B. Smith, *J. Phys. Chem.*, **82**, 234 (1978).

9. J. L. Gough, J. P. Guthrie, and J. B. Stothers, *J. Chem. Soc., Chem. Commun.*, **1972**, 979.

10. Q. Khuong-Huu, A. Pancrazi, and I. Kabore, *Tetrahedron*, **30**, 2579 (1974).

11. F. Borreman, M. de Potter, and D. D. Keukeleire, *Org. Magn. Resonance*, **7**, 415 (1975).

12. N. S. Bhacca, F. W. Wehrli, and N. H. Fischer, *J. Org. Chem.*, **38**, 3618 (1973).

13. L. Crombie, R. W. King, and D. A. Whiting, *J. Chem. Soc., Perkin Transact. I*, **1975**, 913.

14. (a) H. Duddeck, M. H. A. Elgamal, G. Severini Ricca, B. Danieli, and G. Palmisano, *Org. Magn. Resonance*, **11**, 130 (1978); (b) *ibid.*, **11**, 163 (1978).

15. G. Popjak, J. Edmond, F. A. L. Anet, and N. Easton, *J. Am. Chem. Soc.*, **99**, 931 (1977).

16. D. E. Cane and R. H. Levin, *J. Am. Chem. Soc.*, **98**, 1183 (1976).

17. S. Seo, Y. Tomita, and K. Tori, *J. Chem. Soc., Chem. Commun.*, **1975**, 270.

18. (a) T. A. Crabb, *Annu. Rep. NMR Spectrosc.*, **6A**, 249 (1975); (b) D. Tourwe and G. van Binst, *Heterocycles*, **9**, 507 (1978).

19. G. W. Gribble, R. B. Nelson, J. L. Johnson, and G. C. Levy, *J. Org. Chem.*, **40**, 3720 (1975).

20. R. G. Parker and J. D. Roberts, *J. Org. Chem.*, **35**, 996 (1970).

21. F. Bohlmann and R. Zeisberg, *Chem. Ber.*, **108**, 1043 (1975).

22. J. Leung and A. J. Jones, *Org. Magn. Resonance*, **9**, 333 (1977).

23. P. R. Srinivasan and R. L. Lichter, *Org. Magn. Resonance*, **8**, 198 (1976).

24. (a) N. J. Bach, H. E. Boaz, E. C. Kornfeld, C.-J. Chang, H. G. Floss, E. W. Hagaman, and E. Wenkert, *J. Org. Chem.*, **39**, 1272 (1974); (b) L. Zetta and G. Gatti, *Tetrahedron*, **31**, 1403 (1974); (c) *Org. Magn. Resonance*, **9**, 218 (1977).

25. (a) V. I. Stenberg, N. K. Narain, and S. P. Singh, *J. Heterocyclic Chem.*, **14**, 225 (1977); (b) A. M. Taha and G. Rucker, *J. Pharm. Sci.*, **67**, 775 (1978); (c) M. Lounasmaa, P. M. Wovkulich, and E. Wenkert, *J. Org. Chem.*, **40**, 3694 (1975).

26. C. G. Moreland, A. Philip, and F. I. Carroll, *J. Org. Chem.*, **39**, 2413 (1974).

27. (a) F. I. Carroll, C. G. Moreland, G. A. Brine, and J. A. Kepler, *J. Org. Chem.*, **41**, 996 (1976); (b) G. A. Brine, D. Prakash, C. K. Hart, D. J. Kotchmar, C. G. Moreland, and F. I. Carroll, *ibid.*, **41**, 3445 (1976).

28. (a) S. W. Pelletier and Z. Djarmati, *J. Am. Chem. Soc.*, **98**, 2626 (1976); (b) A. J. Jones and M. H. Benn, *Can. J. Chem.*, **51**, 486 (1973).

29. E. Wenkert et al., *J. Am. Chem. Soc.*, **95**, 4990 (1973).

30. (a) R. H. Levin, J.-Y. Lallemand, and J. D. Roberts, *J. Org. Chem.*, **38**, 1983 (1973); (b) E. Wenkert et al., *J. Am. Chem. Soc.*, **98**, 3645 (1976).

31. (a) E. Wenkert et al., *Helv. Chim. Acta*, **58**, 1560 (1975); (b) D. E. Dorman and J. W. Paschal, *Org. Magn. Resonance*, **8**, 413 (1976).

32. L. Zetta, G. Gatti and C. Fuganti, *J. Chem. Soc., Perkin Transact. II.*, 1180 (1973).

33. (a) T. T. Nakashima, P. P. Singer, L. M. Browne, and W. A. Ayer, *Can. J. Chem.* **53**, 1936 (1975); (b) E. Wenkert et al., *J. Am. Chem. Soc.*, **95**, 8427 (1973).

34. H. D. Ludemann and H. Nimz, *Biochem. Biophys. Res. Commun.*, **52**, 1162 (1973).

35. (a) H. D. Ludemann and H. Nimz, *Makromol. Chem.*, **175**, 2393, 2409, 2577 (1974); (b) H. Nimz, I. Mogharab, and H. D. Ludemann, *ibid.*, **175**, 2563 (1974); (c) H. Nimz, H. D. Ludemann, and H. Becker, *Z. Pflanenphys.*, **73**, 226 (1974).

36. (a) E. Wenkert and E. W. Hagaman, *J. Org. Chem.*, **41**, 701 (1976); (b) K. F. Koch, J. A. Rhoades, E. W. Hagaman, and E. Wenkert, *J. Am. Chem. Soc.*, **96**, 3300 (1974).

37. M. Vuilhorgne, S. Ennifar, B. C. Das, J. W. Paschal, R. Nagarajan, E. W. Hagaman, and E. Wenkert, *J. Org. Chem.*, **42**, 3289 (1977).

38. P. F. Wiley, R. B. Kelly, E. L. Caron, V. H. Wiley, J. H. Johnson, F. A. MacKellar, and S. A. Mizsak, *J. Am. Chem. Soc.*, **99**, 3736 (1977).

39. U. Hollstein, E. Breitmaier, and G. Jung, *J. Am. Chem. Soc.*, **96**, 8036 (1974).
40. K. Sakata, J. Uzawa, and A. Sakurai, *Org. Magn. Resonance*, **10**, 230 (1977).
41. Y. Terui, K. Tori, K. Nagashima, and N. Tsuji, *Tetrahedron Lett.*, **1975**, 2583.
42. G. L. Asleson and C. W. Frank, *J. Am. Chem. Soc.*, **98**, 4745 (1976).
43. (a) J. G. Nourse and J. D. Roberts, *J. Am. Chem. Soc.*, **97**, 4584 (1975); (b) G. Lukacs et al., *ibid.*, **97**, 4001 (1975).
44. G. Lukacs et al., *J. Am. Chem. Soc.*, **100**, 663 (1978).
45. S. Omura, A. Neszmelyi, M. Sangare, and G. Lukacs, *Tetrahedron Lett.*, **1975**, 2939.
46. (a) A. Neszmelyi, S. Omura, and G. Lukacs, *J. Chem. Soc., Chem. Commun.*, **1976**, 97; (b) A. Neszmelyi, S. Omura, T. T. Thang, and G. Lukacs, *Tetrahedron Lett.*, **1977**, 725.
47. S. Omura, A. Nakagawa, H. Takeshima, J. Miyazawa, C. Kitao, F. Piriou, and G. Lukacs, *Tetrahedron Lett.*, **1975**, 4503.
48. R. H. Cox and R. J. Cole, *J. Org. Chem.*, **42**, 112 (1977).
49. (a) D. D. Giannini, K. K. Chan, and J. D. Roberts, *Proc. Nat. Acad. Sci. (USA)*, **71**, 4221 (1974); (b) K. K. Chan, D. D. Giannini, A. H. Cain, J. D. Roberts, W. R. Porter, and W. F. Trager, *Tetrahedron*, **33**, 899 (1977).
50. P. Joseph-Nathan, J. Mares, M. C. Hernandez, and J. N. Shoolery, *J. Magn. Resonance*, **16**, 447 (1974).
51. E. Wenkert and H. E. Gottlieb, *Phytochemistry*, **16**, 1811 (1977).
52. K. Tori, T. Hirata, O. Koshitani, and T. Suga, *Tetrahedron Lett.*, **16**, 1311 (1976).
53. C. J. Chang, *Lloydia*, **41**, 17 (1978).
54. A. Pelter, R. S. Ward, and R. J. Bass, *J. Chem. Soc., Perkin Transact. I.*, **1978**, 666.
55. (a) B. Ternai and K. R. Markham, *Tetrahedron*, **32**, 565 (1976); (b) J. Y. Lallemand and M. Duteil, *Org. Magn. Resonance*, **9**, 179 (1977).
56. D. Bergenthel, K. Szendrei, and J. Reisch, *Arch. Pharm.*, **310**, 390 (1977).
57. S. G. Levine, R. E. Hicks, H. E. Gottleib, and E. Wenkert, *J. Org. Chem.*, **40**, 2540 (1975).
58. I. Yamaguchi, N. Takahashi, and K. Fujita, *J. Chem. Soc., Perkin Transact I*, **1975**, 992; I. Yamaguchi, M. Migamoto, H. Yamani, N. Hurofushi, N. Takahashi, and K. Fujita, *ibid.*, **1975**, 996.
59. (a) G. F. Cooper and J. Fried, *Proc. Nat. Acad. Sci. (USA)*, **70**, 1579 (1973); (b) *Proc. First Internat. Conf. Stable Isotope Chem., Biol. Med.*, **1973**, 72; (c) W. W. Conover and J. Fried, *Proc. Nat. Acad. Sci. (USA)*, **71**, 2157 (1974).
60. L. Simeral and G. E. Maciel, *Org. Magn. Resonance*, **6**, 226 (1974).

61.   A. A. Gallo and H. Z. Sable, *J. Biol. Chem.*, **249**, 1382 (1974);
      R. E. Echols and G. C. Levy, *J. Org. Chem.*, **39**, 1321 (1974).
62.   (a) T. H. Witherup and E. H. Abbott, *J. Org. Chem*, **40**, 2229
      (1975); (b) R. C. Harruff and W. T. Jenkins, *Org. Magn. Reso-
      nance*, **8**, 1548 (1976).
63.   B. H. Jo, V. Nair, and L. Davis, *J. Am. Chem. Soc.*, **99**, 4467
      (1977).
64.   (a) T. E. Needham, N. A. Matwiyoff, T. E. Walker, and H. P. C.
      Hogenkamp, *J. Am. Chem. Soc.*, **95**, 5019 (1973); H. P. C. Hogen-
      kamp, R. D. Tkachuck, M. E. Grant, R. Fuentes, and N. A. Mat-
      wiyoff, *Biochemistry*, **14**, 3707 (1975).
65.   S. G. Boxer, G. L. Closs, and J. J. Katz, *J. Am. Chem. Soc.*,
      96, 7058 (1974).
66.   (a) K. M. Smith and J. F. Unsworth, *Tetrahedron*, **31**, 367 (1975);
      (b) R. J. Abraham, H. Pearson, K. H. Smith, P. Loftus, and J.
      D. Roberts, *Tetrahedron Lett.*, **1976**, 877.
67.   G. N. LaMar, D. B. Viscio, D. L. Budd, and K. Gersonde, *Bio-
      chem. Biophys. Res. Commun.*, **82**, 19 (1978).
68.   A. G. McInnes, J. A. Walter, J. L. C. Wright, and L. C. Vining,
      in *Topics in Carbon-13 NMR Spectroscopy,* Vol. 2, G. C. Levy,
      Ed., Wiley-Interscience, New York, 1976, Chapter 3.
69.   A. G. McInnes and J. L. C. Wright, *Acc. Chem. Res.*, **8**, 313
      (1975).
70.   M. Tanabe, in *Specialist Periodical Reports, Biosynthesis*,
      Vol. 4, Chemical Society, London, 1976, p. 204.
71.   G. Kunesch and C. Poupat, in *Isotopes in Organic Chemistry*,
      Vol. 3, E. Buncel and C. C. Lee, Eds., Elsevier, Amsterdam,
      1977.
72.   T. J. Simpson, *Chem. Soc. Rev.*, **4**, 497 (1975).
73.   J. L. C. Wright, L. C. Vining, A. G. McInnes, D. G. Smith, and
      J. A. Walter, *Can. J. Biochem.*, **55**, 678 (1977); A. G. McInnes,
      D. G. Smith, J. A. Walter, L. C. Vining, and J. L. C. Wright,
      J. Chem. Soc., Chem. Commun., **1974**, 282; A. G. McInnes, D. G.
      Smith, C.-K. Wat, L. C. Vining, and J. L. C. Wright, *ibid.*,
      1974, 281.
74.   N. Neuss, C. H. Nash, J. E. Baldwin, P. A. Lemke, and J. B.
      Grutzner, *J. Am. Chem. Soc.*, **95**, 3797 (1973).
75.   R. A. Komoroski, I. R. Peat, and G. C. Levy, *Topics in Carbon-
      13 NMR Spectroscopy*, Vol. 2, G. C. Levy, Ed., Wiley-Inter-
      science, New York, 1976, Chapter 4.
76.   E. Oldfield, R. S. Norton, and A. Allerhand, *J. Biol. Chem.*,
      250, 6368 (1975).
77.   W. Egan, H. Shindo, and J. S. Cohen, in *Annu. Rev. Biophys.
      Bioeng.*, **3**, 383 (1977).
78.   R. A. Komoroski, I. R. Peat, and G. C. Levy, *Biochem. Biophys.
      Res. Commun.*, **65**, 272 (1975); O. W. Howarth, *J. Chem. Soc.
      Farad. Transact. 2*, **75**, 863 (1979); R. E. London, in *Magnetic
      Resonance in Biology*, J. S. Cohen, Ed., Wiley-Interscience,

New York, in press; R. E. London and J. Avitabile, *J. Am. Chem. Soc.*, **100**, 7159 (1978) and previous papers.

79. M. Llinás, W. Meier, and K. Wüthrich, *Biochem. Biophys. Acta*, **492**, 1 (1977); K. Dill and A. Allerhand, *J. Am. Chem. Soc.*, **101**, 4376 (1979).

80. W. Horsley, H. Sternlicht, and J. S. Cohen, *J. Am. Chem. Soc.*, **92**, 680 (1970) and preliminary reports; see also J. C. MacDonald, G. G. Bishop, and M. Mazurek, *Can. J. Chem.*, **54**, 1226 (1976).

81. Early references cited in R. Deslauriers and I. C. P. Smith, *Topics in Carbon-13 NMR Spectroscopy*, Vol. 2, G. C. Levy, Ed., Wiley-Interscience, New York, 1976, Chapter 1.

82. G. Grathwohl and K. Wüthrich, *J. Magn. Resonance*, **13**, 217 (1974); P. Keim, R. A. Vigna, J. S. Morrow, R. C. Marshall, and F. R. N. Gurd, *J. Biol. Chem.*, **248**, 6104 (1973); M. Christl and J. D. Roberts, *J. Am. Chem. Soc.*, **94**, 4565 (1972).

83. (a) H. Shindo and J. S. Cohen, *Proc. Nat. Acad. Sci. (USA)*, **73**, 1979 (1976); (b) D. J. Wilbur and A. Allerhand, *FEBS Lett.*, **79**, 144 (1977); (c) *ibid.*, *J. Biol. Chem.*, **251**, 5187 (1976).

84. K. H. Ladner, J. J. Led, and D. M. Grant, *J. Magn. Resonance*, **20**, 530 (1975).

85. See review by V. F. Bystrov, *Progr. NMR Spectrosc.*, **10**, 41 (1976).

86. P. E. Hansen, J. Feeney, and G. C. K. Roberts,*J. Magn. Resonance*, **17**, 249 (1975); J. Feeney, *ibid.*, **21**, 473 (1976); R. E. London, T. E. Walker, V. H. Kollman, and N. A. Matwiyoff, *J. Am. Chem. Soc.*, **100**, 3723 (1978); W. G. Espersen and R. B. Martin, *J. Phys. Chem.*, **80**, 741 (1976).

87. V. Bystrov, Y. D. Favrilov, V. T. Ivanov, and Y. A. Ovchinnikov, *FEBS Eur. J. Biochem.*, 1978.

88. W. A. Thomas and M. Kevin Williams, *J. Chem. Soc., Chem. Commun.*, **197.**, 994.

89. (a) D. S. Clark, J. J. Dechter, and L. Mandelkern, *Macromolecules* in press; (b) R. Deslauriers, I. C. P. Smith, G. C. Levy, R. Orlowski, and R. Walter, *J. Am. Chem. Soc.*, **100**, 3912 (1978).

90. R. Deslauriers, E. Ralston, and R. L. Somorjai, *J. Molec. Biol.*, **113**, 697 (1977); R. Deslauriers and R. Somorjai, *J. Am. Chem. Soc.*, **98**, 1931 (1976); R. A. Komoroski, I. R. Peat, and G. C. Levy, *Biochem. Biophys. Res. Commun.*, **65**, 272 (1975); H. E. Belich, J. D. Cutnell, and J. A. Glasel, *Biochemistry*, **15**, 2455 (1976).

91. (a) I. A. Armitage, H. Huber, H. Pearson, and J. D. Roberts, *Proc. Natl. Acad. Sci. (USA)*, **71**, 2096 (1974); (b) H. Pearson, D. Gust, I. M. Armitage, H. Huber, J. D. Roberts, R. R. Vold, and R. L. Vold, *ibid.*, **72**, 1599 (1975).

92. (a) D. A. Torchia and J. R. Lyerla, Jr., *Biopolymers*, **13**, 97
    (1974); (b) H. Saito and I. C. P. Smith, *Archives Biochem.
    Biophys.*, **158**, 154 (1973); (c) H. J. Lader, R. A. Komoroski,
    and L. Mandelkern, *Biopolymers*, **16**, 895 (1977).
93. D. A. Torchia, J. R. Lyerla, Jr. and A. J. Quattrone, *Biochem-
    istry*, **14**, 887 (1975).
94. R. E. London, *J. Am. Chem. Soc.*, **100**, 2678 (1978); R. E. Lon-
    don and J. Avitabile, *ibid.*, **100**, 7159 (1978); D. J. Wilbur,
    R. S. Norton, A. O. Clouse, R. Addleman, and A. Allerhand,
    J. Am. Chem. Soc., **98**, 8250 (1976); K. Dill and A. Allerhand,
    ibid., **101**, 4376 (1979); M. Llinás, W. Meier, and K. Wüthrich,
    Biochem. Biophys. Acta, **492**, 1 (1977).
95. A. Allerhand, R. F. Childers, and E. Oldfield, *Ann. N. Y. Acad.
    Sci.*, **222**, 764 (1973).
96. (a) E. Oldfield and M. Meadows, *J. Magn. Resonance*, **31**, 327
    (1978); (b) G. C. Levy and J. Terry Bailey, 20th Experimental
    NMR Conference, Asilomar, Calif. 1979.
97. (a) R. S. Norton, A. O. Clouse, R. Addleman, and A. Allerhand,
    J. Am. Chem. Soc., **99**, 79 (1977); (b) G. Van Binst, M. Biese-
    mans, and A. O. Barel, *Bull. Soc. Chim. Belg.*, **84**, 1 (1975);
    (c) K. Dill and A. Allerhand, *J. Am. Chem. Soc.*, **99**, 4508
    (1977).
98. I. D. Campbell, C. M. Dobson, R. J. P. Williams, and A.
    Xavier, *J. Magn. Resonance*, **11**, 172 (1973).
99. E. Oldfield, R. S. Norton, and A. Allerhand, *J. Biol. Chem.*,
    250, 6381 (1975); *ibid.*, **250**, 6368 (1975).
100. R. R. Ernst, *Advance Magn. Resonance*, **2**, 1 (1966); A. DeMarco
    and K. Wüthrich, *J. Magn. Resonance*, **24**, 201 (1976); M. Guèron,
    ibid., **30**, 515 (1978); J. A. B. Lohman, *ibid.*, **38**, 163 (1980);
    D. L. Rabenstein, *Anal. Chem.*, A**50**, 1265 (1978); K. Roth, *J.
    Magn. Resonance*, **38**, 65 (1980).
101. (a) R. Richarz and K. Wüthrich, *FEBS Lett.*, **79**, 64 (1977); (b)
    K. Wüthrich and R. Baumann, *Org. Magn. Resonance*, **8**, 532
    (1976); (c) W. Maurer, W. Harr, and H. Ruterjans, *Z. Phys.
    Chemie Neue Folge*, **93**, 119 (1974).
102. D. M. Lilley, J. F. Pardon, and B. M. Richards, *Biochemistry*,
    16, 2853 (1977).
103. (a) D. A. Torchia and D. L. Vanderhart, in *Topics in Carbon-13
    NMR Spectrsocopy*, Vol. 3, G. C. Levy, Ed., Wiley-Interscience,
    New York, 1979, Chapter 8; (b) J. Schaefer, E. O. Stejskal,
    C. F. Brewer, H. D. Keiser, and H. Sternlicht, *Arch. Biochem.
    Biophys.*, **19** , 657 (1978); D. W. Urry and L. W. Mitchell,
    Biochem. Biophys. Res. Commun., **68**, 1153 (1976).
104. C. F. Brewer, D. M. Marcus, A. P. Grollman, and H. Sternlicht,
    J. Biol. Chem., **249**, 4614 (1974); D. C. Harris, G. A. Gray,
    and P. Aisen, *ibid.*, **249**, 5261 (1974); C. M. Deber, M. A.
    Moscarello, and D. D. Wood, *Biochemistry*, **17**, 898 (1978); A.
    Schejter, A. Lamir, I. Vig, and J. S. Cohen, *J. Biol. Chem.*,

253, 3768 (1978); J. S. Morrow, P. Keim, and F. R. N. Gurd, *J. Biol. Chem.*, **249**, 7484 (1974); R. B. Moon and J. H. Richards, *Biochemistry*, **13**, 3437 (1974); R. E. London, C. T. Gregg, and N. A. Matwiyoff, *Science*, **188**, 266 (1975).

105. H. L. Johnson, D. W. Thomas, M. Ellis, L. Cary, and J. I. DeGraw, *J. Pharm. Sci.*, **66**, 1660 (1977).

106. R. S. Norton and A. Allerhand, *J. Biol. Chem.*, **251**, 6522 (1976); K. Dill and A. Allerhand, *Biochemistry*, **16**, 5711 (1977).

107. D. E. Dorman, S. J. Angyal, and J. D. Roberts, *J. Am. Chem. Soc.*, **92**, 135 (1970).

108. D. E. Dorman and J. D. Roberts, *J. Am. Chem. Soc.*, **92**, 1355 (1970).

109. H. J. Koch and A. S. Perlin, *Carbohydrate Res.*, **15**, 403 (1970).

110. A. Allerhand and D. Doddrell, *J. Am. Chem. Soc.*, **93**, 2777 (1971).

111. T. E. Walker, R. E. London, T. W. Whaley, R. Barker, and N. A. Matwiyoff, *J. Am. Chem. Soc.*, **98**, 5807 (1976).

112. P. Colson, H. J. Jennings, and I. C. P. Smith, *J. Am. Chem. Soc.*, **96**, 8081 (1974).

113. D. J. Wilbur, C. Williams, and A. Allerhand, *J. Am. Chem. Soc.*, **99**, 5450 (1977).

114. (a) T. A. W. Koerner, Jr., R. J. Voll, L. W. Cary, and E. S. Younathan, *Biochem. Biophys. Res. Commun.*, **82**, 1273 (1978); (b) L. Hyronen, P. Varo, and P. Koivistoinen, *J. Food Sci.*, **42**, 652, 657 (1977).

115. (a) D. Doddrell and A. Allerhand, *J. Am. Chem. Soc.*, **93**, 2779 (1971); (b) T. A. W. Koerner, Jr., L. W. Cary, N. S. Bhacca, and E. S. Younathan, *Biochem. Biophys. Res. Commun.*, **51**, 543 (1973).

116. L. Que and G. R. Gray, *Biochem.*, **13**, 146 (1974).

117. S. J. Angyal and G. S. Bethell, *Aust. J. Chem.*, **29**, 1249 (1976).

118. (a) K. Bock and L. D. Hall, *Carbohydrate Res.*, **40**, C3 (1975); (b) G. Williams and A. Allerhand, *ibid.*, **56**, 173 (1977).

119. B. Casu, G. Gatti, N. Cyr, and A. S. Perlin, *Carbohydrate Res.*, **41**, C6 (1975).

120. P. A. J. Gorin and M. Mazwek, *Can. J. Chem.*, **52**, 3070 (1974).

121. D. Gagnaire and M. Vincendon, *J. Chem. Soc., Chem. Commun.*, **1977**, 509; D. Gagnaire, O. Mancier, and M. Vincendon, *Org. Magn. Resonance*, **11**, 344 (1978).

122. A. S. Perlin and G. K. Hamer, in *Carbon-13 NMR in Polymer Science*, W. M. Pasika, Ed., ACS Symposium Series No. 103, 1979, p. 123.

123. (a) A. S. Perlin, in *Heparin: Structure, Cellular Functions, and Clinical Applications*, Academic, New York, 1979, p. 25; (b) A. S. Perlin, *Fed. Proc.*, **36**, 106 (1977).

124. (a) N. Cyr, G. K. Hamer, and A. S. Perlin, *Can. J. Chem.*, **56**, 297 (1978); (b) G. K. Hauser, F. Balza, N. Cyr, and A. S. Perlin, *ibid.*, 3109 (1978).

125. J. V. O'Connor, H. A. Nunez, and R. Barber, *Biochemistry*, **18**, 500 (1979).

126. A. Parfondry and A. S. Perlin, *Carbohydrate Res.*, **57**, 39 (1977).

127. D. R. Bundle, I. C. P. Smith, and H. J. Jennings, *J. Biol. Chem.*, **249**, 2275 (1974); A. K. Bhattacharjee, H. J. Jennings, C. P. Kenny, A. Martin, and I. C. P. Smith, *ibid.*, **250**, 1926 (1979).

128. D. A. Torchia, M. A. Hasson, and V. C. Hascall, *J. Biol. Chem.*, **252**, 3617 (1977).

129. T. Ogawa, J. Uzawa, and M. Matsui, *Carbohydrate Res.*, **59**, C32 (1977).

130. (a) T. Schleich, B. P. Cross, B. J. Blackburn, and I. C. P. Smith, in *Structure and Conformation of Nucleic Acids and Protein-Nucleic Acid Interactions*, M. Sundaralingam and S. T. Rao, Eds., University Park Press, Baltimore, 1975, pp. 223-251; (b) I. C. P. Smith, H. H. Mantsch, R. D. Lapper, R. Deslauriers, and T. Schleich, "The Jerusalem Symposium on Quantum Chemistry and Biochemistry V," *Israel Acad. Sci. Humanities*, **1973**, 381-402.

131. B. Birdsall and J. Feeney, *J. Chem. Soc., Perkin Transact. II*, **1972**, 1654; D. E. Dorman and J. D. Roberts, *Proc. Nat. Acad. Sci. (USA)*, **65**, 19 (1970).

132. T. Schleich, B. P. Cross, and I. C. P. Smith, *Nuc. Acids Research*, **3**, 355 (1976).

133. M. Blumenstein and M. A. Raftery, *Biochemistry*, **12**, 3585 (1973).

134. R. M. Riddle, R. J. Williams, T. A. Bryson, R. B. Dunlap, R. R. Fisher, and P. D. Ellis, *J. Am. Chem. Soc.*, **98**, 4286 (1976); S. Uesugi and M. Ikehara, *ibid.*, **99**, 3260 (1977); M. P. Schweitzer, E. B. Banta, J. T. Witkowski, and R. K. Robins, *ibid.*, **95**, 3770 (1975).

135. (a) T. J. Williams, A. P. Zens, J. C. Wisowaty, R. R. Fisher, R. B. Dunlap, T. A. Bryson, and P. D. Ellis, *Arch. Biochem. Biophys.*, **172**, 490 (1976); (b) P. D. Ellis, R. R. Fisher, R. B. Dunlap, A. P. Zens, T. A. Bryson, and T. J. Williams, *J. Biol. Chem.*, **248**, 7677 (1973).

136. M. T. Chenon, R. J. Pugmire, D. M. Grant, R. P. Panzica, and L. B. Townsend, *J. Am. Chem. Soc.*, **97**, 4627 (1977); *ibid.*, 4626, 4736.

137. (a) R. A. Komoroski and A. Allerhand, *Proc. Nat. Acad. Sci. (USA)*, **69**, 1804 (1972); (b) R. A. Komoroski and A. Allerhand, *Biochemistry*, **13**, 369 (1974).

138. W. D. Hamill, Jr., D. M. Grant, W. J. Horton, R. Lundquist, and S. Dickman, *J. Am. Chem. Soc.*, **98**, 1276 (1976).

139. W. D. Hamill, Jr., D. M. Grant, R. B. Cooper, and S. A. Harmon, *J. Am. Chem. Soc.*, **100**, 633 (1978).
140. J. W. Triplett, N. W. Chow, S. L. Smith, and G. C. Digenis, *Biochem. Biophys. Res. Commun.*, **77**, 1170 (1977).
141. C.-J. Chang and C.-G. Lee, *Can. Res.*, **38**, 3734 (1978).
142. (a) R. L. Rill, P. R. Hilliard, J. T. Bailey, and G. C. Levy, *J. Am. Chem. Soc.*, **102**, 418 (1980); (b) R. L. Rill, P. R. Hilliard, and G. C. Levy, to be submitted; (c) a recent report has appeared with ^{13}C nmr data on less-well characterized DNA: P. H. Bolton and T. L. James, *Biochemistry*, **19**, 1388 (1980).
143. S. J. Singer and G. L. Nicolson, *Science*, **175**, 720 (1972).
144. Y. K. Levine, N. J. M. Birdsall, A. G. Lee, and J. C. Metcalfe, *Biochemistry*, **11**, 1416 (1972).
145. A. G. Lee, N. J. M. Birdsall, J. C. Metcalfe, G. B. Warren, and G. C. K. Roberts, *Proc. Roy. Soc. (Lond.) B*, **193**, 253 (1976); (b) I. C. P. Smith, *Can. J. Chem.*, **57**, 1 (1979): (c) I. C. P. Smith, A. P. Tulloch, G. W. Stockton, S. Schreier, A. Joyce, K. W. Butler, Y. Boulanger, B. Blackwell, L. G. Bennett, *Ann. N. Y. Acad. Sci.*, **308**, 8 (1978).
146. C. F. Schmidt, Y. Barenholz, C. Huang, and T. E. Thompson, *Biochemistry*, **6**, 3948 (1977).
147. Yu. E. Shapiro, A. V. Viktorov, V. I. Volkova, L. I. Barsukov, V. F. Bystrov, and L. D. Bergelson, *Chem. Phys. Lipids*, **14**, 227 (1975).
148. (a) P. L. Yeagle, R. B. Martin, A. K. Lala, A.-K. Lin, and K. Bloch, *Proc. Nat. Acad. Sci. (USA)*, **74**, 4924 (1977); (b) P. E. Godici and F. R. Landsberger, *Biochemistry*, **14**, 3927 (1975); (c) G. C. Levy, U. S. Report EPA-600/1-77-045 (1977).
149. (a) J. A. Hamilton and E. H. Cordes, *J. Biol. Chem.*, **253**, 5193 (1978); (b) W. Stoffel, P. Metz, and B. Tunggal, *Hoppe-Seyler's Z. Physiol. Chem.*, **359**, 465 (1978); (c) B. Sears, R. J. Deckelbaum, M. J. Janiak, G. G. Shipley, and D. M. Small, *Biochemistry*, **15**, 4151 (1976).
150. F. Miller, P. A. Hagrave, and M. A. Raftery, *Biochemistry*, **12**, 3591 (1973).
151. E. C. Williams and E. H. Cordes, *Biochemistry*, **15**, 5792 (1976).
152. P. L. Harris and E. R. Thornton, *J. Am. Chem. Soc.*, **100**, 6738 (1978).
153. (a) J. Schaefer and E. O. Stejskal, *J. Amer. Oil Chem. Soc.*, **51**, 210 (1974); (b) *ibid.*, **51**, 562 (1974); (c) *ibid.*, **52**, 366 (1975); (d) V. Rutar, M. Burgar, R. Blino, and L. Ehrenberg, *J. Magn. Resonance*, **27**, 83 (1977); (e) J. N. Shoolery and W. C. Jankowski, Varian Application Notes NMR-73-4 (1973).

Chapter 10

Special Techniques and Applications

This chapter outlines some specialized techniques and additional applications of ^{13}C nmr in organic chemistry and allied
fields. Included are discussions of reaction mechanisms, dynamic
cmr, specialized pulse sequences and other excitation methods,
two-dimensional FT nmr, and studies of organic molecules in motionally restricted environments.

REACTION MECHANISMS

The first reported ^{13}C nmr work directed toward the elucidation
of organic reaction mechanisms predated commercial availability
of FT cmr.[1,2] Carbon-13 nmr has been used--often in conjunction
with ^{13}C, ^{15}N, or ^2H isotopic incorporation--to evaluate reaction
pathways and to determine structures of transient intermediates.
In 1972 to 1974, Stothers and co-workers used ^2H exchange to
probe carbanion and enolate formation in a variety of systems.[3]
More recently ^{13}C nmr has been used to study Friedel-Crafts
cyclialkylations of halogenopropiophenones to indanones.[4] The
^{13}C measurements gave clear evidence that cyclization proceeds
through the methacrylophenone by the following scheme:

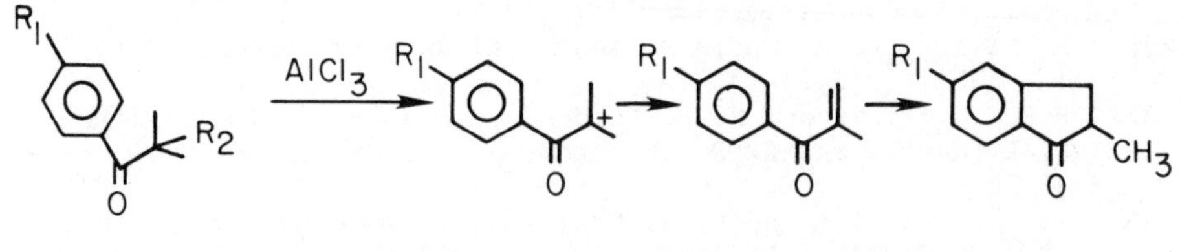

R_1 = H, F
R_2 = Br, Cl

R. G. Shulman and co-workers have shown the applicability of
^{31}P and ^{13}C to metabolic analyses in living organs or cells. In
one paper[5] 1-^{13}C glucose was fed to suspensions of anaerobic

Escherichia coli cells. Glycolysis was demonstrated in which the α- and β-glucose anomers disappeared at different rates, showing lactate, succinate, acetate, alanine, and valine accumulating as end products. Additionally, fructose bisphosphate was detected as an intermediate. The sites of ^{13}C labels in these products facilitate analysis of the metabolic pathways involved. Another example is the elegant study of Scott and his co-workers[6] shown in Figure 10.1. This set of spectra shows the time course of por- phobilinogen (PBG) metabolism by a strain of live cells in a glu- cose-rich medium. The course of the metabolic processes is followed by ^{13}C nmr. Loss of PBG labeled at C-11 and appearance of coproporphyrinogen I and III was observed. Since nonenriched glucose was periodically added to the medium during the first 24 h, its metabolism to propionate, acetate, and succinate was demon- strated (Figure 10.1). In addition, αα-trehalose was observed to accumulate in the glucose-rich medium, later to be consumed by the starved cells.

Carbon-13 chemically induced dynamic nuclear polarization (CIDNP) mechanistic studies have examined a variety of organic reactions.[7-9] Descriptions of CIDNP experimental methods set out requirements for accurate, quantitative evaluations of reacting systems.[7,10] Spectra are obtained as a function of time after initiation of photolysis. The observed CIDNP effects reflect the reaction mechanism (cage effects, etc.).

DYNAMIC CMR

Carbon-13 nmr has been shown to be particularly useful in studies of rapid chemical processes.[11] Several factors cause dy- namic cmr studies to be distinctly advantageous relative to ana- logous proton nmr studies:

1. Frequency differences between "exchanging" sites are gen- erally far greater in ^{13}C nmr.
2. Carbon resonance bandshapes that are not undergoing ex- change can be easily controlled; most resonances are singlets.
3. The relative simplicity of ^{13}C spectra facilitates inter- pretation of complex situations, for example, sorting out two or more frozen configurations.

The first lineshape kinetic analyses by cmr were reported in 1971.[12-14] In one study[12] rotation around the carbon-nitrogen amide linkage in *N,N*-dimethyltrichloroacetamide was observed as a function of temperature; in the second paper [10]-annulene con- formational equilibria were studied.[13] Dalling, Schneider and co- workers evaluated conformational dynamics in dimethylcyclohexanes and *cis*-decalins.[14] Professor F. A. L. Anet has extensively used ^{13}C nmr to evaluate conformational dynamics of alicyclic sys- tems.[11,15] Anet's recent work has combined ^{13}C and ^{1}H nmr results

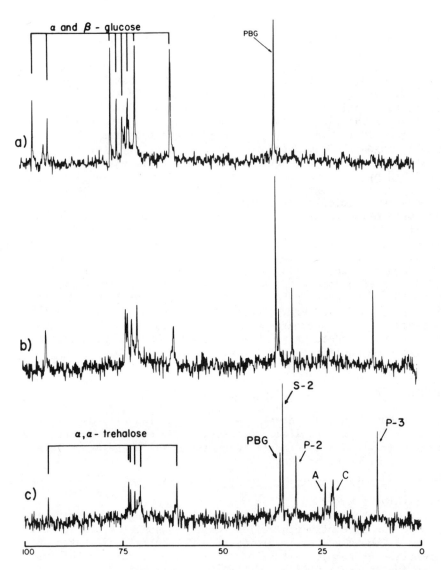

FIGURE 10.1.[6]   *Metabolism of $[11-^{13}C]$ porphobilinogen (PBG) by cells of Propionibacterium Shermanii (ATCC 9614). Cells were grown anaerobically for 17 days, harvested by centrifugation and incubated at 28°C in phosphate buffer-glucose (pH 7.6). 20 MHz spectra (15000 scans) recorded at (a) 4 hrs; (b) 31 hrs; (c) 82 hrs, after addition of PBG (P-2, P-3 are carbons 2 and 3 of propionate; A is acetate; S-2 is C-2 of succinate; C arises from labeled coproporphyrinogen).*

with molecular mechanics calculations for middle-sized cycles containing one or more double bonds. Other groups have also discussed the methods and advantages of dynamic [16a] ^{13}C nmr. In dynamic
nmr studies, determination of ΔH and ΔS is particularly sensitive
to errors from inaccurate temperature measurement[16b] and other
sources. Unfortunately, sample temperature measurement and control
in ^{13}C nmr is not as facile as in ^{1}H nmr spectroscopy.[17,18]

One particularly dramatic spectral example of dynamic ^{13}C nmr
is shown in Figure 10.2.[14b] Here the ^{13}C spectra of cis-1,2-

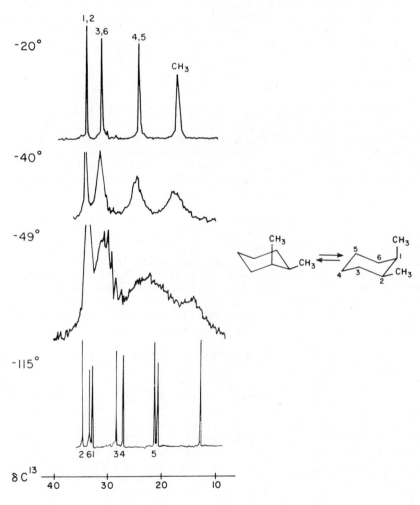

FIGURE 10.2.[14b]   *Temperature-dependent 25.2 MHz* ^{13}C *spectra of cis-*
1,2-dimethylcyclohexane.

dimethylcyclohexane are shown, near the high-temperature (fast exchange) limit, under conditions of intermediate exchange, and at
-115°C, where the ring-flipping process is slowed enough to display a frozen conformation. In this case four sets of equilibrating peaks are present. Derivation of kinetic parameters follows
from four separate two-site exchange processes that have equivalent
kinetics. Values for ΔH and ΔS are obtained from least-squares regression analysis.

Fluxional organometallic compounds have also been widely investigated by dynamic cmr (some of this work is cited in Chapter 6).
Cotton's group has been particularly active, examining a number of
metal cluster compounds.[19] Gansow and others have also studied
the complex dynamic processes that occur in these molecules.[20]

ALTERNATIVE PULSE SEQUENCES AND EXCITATION MODES

A number of alternative and specialized pulse sequences have
been developed for FT nmr spectroscopy. Additionally, nonpulse
rapid-scan (correlation) FT nmr has been used, largely in [1]H
nmr where effective suppression is necessary for large solvent
resonances. These methods all require a level of implementation
higher than that for single-pulse FT nmr but offer new possibilities for selective excitation or, in one case, for increased
sensitivity.

Rapid-scan or correlation FT nmr spectroscopy was developed[21]
primarily to allow selective excitation of one region of an nmr
spectrum. In this technique the response from a rapid spectral
scan (typically several hundred milliseconds) is correlated with
the response from a similar scan of a single sharp line (e.g.,
TMS). Fourier transformation then yields the usual absorption
spectrum. Suppression of peaks just outside of the scan range by
a factor of several hundred is typical. An additional advantage
of this technique is that a high-power pulse amplifier is not
needed.

It was realized quite early that stochastic (i.e., random) modulation of low-power rf excitation could produce effective FT nmr
experiments.[22] This form of excitation can be usefully paired with
a Hadamard transform prior to FT, where the final result is the
usual frequency/absorption nmr spectrum.[22,23] Stochastic excitation itself, however, offers few advantages relative to single-
pulse FT nmr, and its widespread use is not anticipated. On the
other hand, a related scheme has considerable advantages. If,
instead of stochastic modulation of the rf, the time sequence of
modulation is *calculated* to produce *selective* or *tailored* excitation, especially useful experiments may be performed.[24] In particular, the effective excitation window may be chosen to exclude one
or more regions, or to apply different degrees of excitation.
(Figure 10.3)

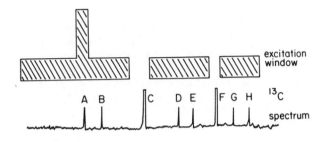

FIGURE 10.3.   *This excitation window would effectively produce full transverse magnetization (equivalent to a 90° pulse) of nuclei B, D, E, G, and H, while not appreciably exciting solvent nuclei C and F. The magnetization of nucleus A would be largely saturated.*

Morris and Freeman developed a form of *tailored detection* for ^{13}C nmr[25] that may be used in conjunction with various pulse sequences. There are a number of other strong-peak-suppression pulse sequences available. Most of these have been developed primarily for ^1H nmr and are based on selective saturation of the solvent lines.

Cross-polarization pulse methods have been shown to significantly improve sensitivity of ^{13}C spectra in solids by pumping net ^{13}C magnetization and short circuiting long ^{13}C spin-lattice relaxation times (see text that follows). In an interesting extention of this technique to liquids, Bertrand et al.[26a] have shown that cross polarization may be effected via J (scalar) coupling in liquid samples. Large improvements in ^{13}C sensitivity have been noted, although the advantage is expected to be diminished somewhat for nonprotonated carbons and in larger molecules where ^{13}C T_1s are short and comparable with ^1H T_1s. These experiments require high rf homogeneity and careful experimental design. In another type of cross polarization experiment[26b,c] Morris and Freeman showed that enhanced sensitivity could be achieved without the stringent requirements of J cross polarization. This experiment was appropriately named *INEPT* (*I*nsensitive *N*uclei *E*nhanced by *P*olarization *T*ransfer).

Jakobsen and others have promoted selective population transfer experiments for use in spin-spin-coupled ^{13}C spectra.[27a] Non-selective population transfer may also be used for cross-polarization sensitivity enhancement.[27,26a]

TWO-DIMENSIONAL FT ^{13}C NMR

The various techniques of two-dimensional FT nmr have already had impact on ^{13}C spectroscopy.[28,29] In a 2DFT nmr experiment spectral changes that result from two time variables are monitored. Fourier transformation along both axes yields spectra that are functions of two frequency variables. Stacked plots that result from spin-relaxation measurements (e.g., Figures 8.5, 8.8) may be thought of in a sense as two-dimensional FT experiments (sorting T_1 and shielding parameters) but they are actually two-dimensional displays of multiple one-dimensional FT experiments--that is, there is no FT calculated along the second axis. Likewise, graphical displays (Figure 1.7) of sequential single frequency decoupling experiments give a two-dimensional view of *a set of one-dimensional FT nmr experiments.*

In two-dimensional FT nmr experiments the general time-frame division is as shown in Figure 10.4. The exact experimental conditions present during the three periods determine the two-dimensional information content. In several J spectroscopy variations of two-dimensional FT nmr, spin-spin coupled ^{13}C multiplets are sorted on the second frequency axis by chemical shift. Thus coupled ^{13}C spectra of complex compounds are obtainable with no interferences due to multiplet overlaps. In one experiment, chemical shifts from heteronuclear spin-spin coupled nuclei can be correlated (Figure 10.5).

Two-dimensional FT ^{13}C nmr is not yet routine, largely as a result of lower effective sensitivity (experimental time is increased by the number of experiments performed to achieve desired resolution along the second axis) and the need for new software. Phase-corrected two-dimensional FT nmr spectra are not yet generally available; most groups show absolute-value spectra to avoid problems of phasing.[28,29] Two-dimensional FT ^{13}C nmr is also useful for nmr imaging (or zeugmatography).[30,31]

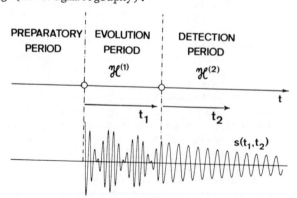

FIGURE 10.4.   *Generalized 2D FT nmr experiment showing two time variables.*

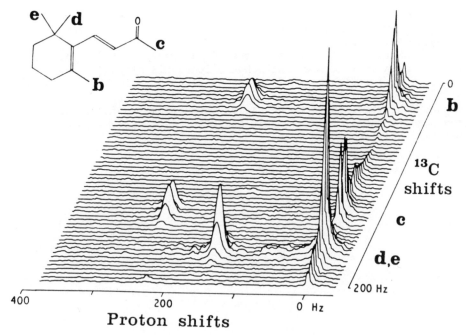

FIGURE 10.5.[28b]   *Two-dimensional FT ^{13}C nmr spectral set. This
type of experiment correlates ^{13}C and ^{1}H shieldings, as demonstrated here for the three types
of CH$_3$ groups.*

^{13}C NMR IN LIQUID CRYSTALS

Carbon-13 nmr spectroscopy in liquid crystalline (particularly
nematic) phases enjoys certain advantages. As a result of the low
natural abundance of ^{13}C and the ability to dipolar decouple* all
proton spins, these spectra may be observed as composites of singlet resonances (in the absence of abundant nuclides with nonzero
spin). In cases where dipolar couplings are desired, they can be
observed. These factors enable direct observation of nematic
phases by natural abundance ^{13}C nmr. In 1974 Pines and Chang[32]
observed ^{13}C spectra of *p*-azoxydianisole in isotropic and nematic
phases. Grant and co-workers (in Schwartz et al.[33]) observed that
the spin-relaxation behavior of CH$_3$I was modified by motional restrictions in a nematic phase. Carbon-13 nmr spectra of liquid
crystals may also be obtained on standard high-resolution ^{13}C
spectrometers with little or no modification.[34] Figure 10.6 com-

---

*Note that dipolar decoupling requires 10 to 100 times the power
of ordinary wide-band scalar decoupling.

pares ^{13}C spectra for isotropic and nematic p,p'-di-n-hexyldi-phenyldiacetylene.

The characteristics of spectrum 10.6a are typical for this type of nematic molecule, which has a long axis aligned along the magnetic field (with rapid rotation possible around this molecular axis). The shielding changes in the nematic phase reflect the emphasis of shielding components for the molecular orientation of the nematic phase. The aromatic and acetylenic carbons are especially shifted.

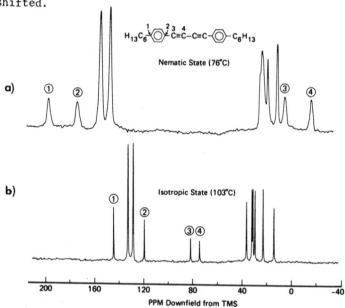

FIGURE 10.6.[34]   Proton-decoupled ^{13}C spectra of p,p'-di-n-hexyl-diphenyldiacetylene in the isotropic and nematic state.

^{13}C NMR OF ADSORBED MOLECULES

Carbon-13 nmr spectra of adsorbed molecules can give information on site geometry, adsorption chemistry, degree of immobilization of adsorbed species, and so forth. Early studies examined simple hydrocarbons or CO_2 on surfaces of alumina, silica, and chrysotile asbestos (the latter actually reacted with simple organic molecules).[35] Catalytic activity of surfaces can be evaluated, for example, by monitoring olefin isomerizations.

Sefcik, Schaefer, and Stejskal,[36a,b] and Nagy et al.[36c] have examined zeolite surfaces. In one study zeolite catalyzed alkylation of toluene (with CH_3OH) to styrene monomer was followed by ^{13}C nmr. Cesium ion zeolite catalysis was found to give enhanced yield of the desired monomer.

^{13}C NMR OF SOLIDS

High-resolution ^{13}C solids nmr techniques as applied to synthetic high-molecular-weight polymers were briefly discussed in Chapter 7. Spectroscopists are now able to observe high-resolution ^{13}C spectra from a wide variety of solid materials.[37] Three techniques are available that, when used in concert, remove line broadening from ^{13}C spectra of motionally restricted samples and circumvent inefficient ^{13}C spin-lattice relaxation in those samples that have long T_1s: (1) *dipolar decoupling,* (2) *magic-angle spinning,* and (3) *cross polarization.* These techniques are useful for ^{13}C nmr as a result of the low abundance of spin-1/2 ^{13}C nuclei. Thus the only dipolar interactions present in most organic compounds are those between a ^{13}C nucleus and nearby protons, which can *exceed* 10 kHz.

Line broadening in nmr spectra of solid powders arises jointly from static dipolar interactions and from anisotropy in ^{13}C shielding tensors. These effects are averaged to zero for rapidly reorienting molecules, as in isotropic liquids. Dipolar decoupling of protons effectively removes ^{13}C-^{1}H static dipolar broadening, but anisotropy of shielding from variable orientations of the molecule with respect to the magnetic field (in a powder sample*) are, of course, not removed. This cause of line broadening may be eliminated or greatly attenuated by spinning the sample at an angle Θ with respect to the static magnetic field, thus reducing the term $(3 \cos^2 \Theta - 1)$ to zero. This angle, referred to as the *magic angle,* is $54.7°$. Sample spinning at the magic angle also removes residual static dipolar interactions since those reflect direction cosine terms that will be averaged out. Generally, it is necessary to spin the sample at a rate comparable with the linewidth observed in the absence of spinning. For ^{13}C at low magnetic fields, 2- to 3-kHz spinning rates are adequate, whereas 5-to 10-kHz rates are required at higher fields. There are techniques that use slower spinning, but these require special experimental conditions.

Both dipolar decoupling and magic angle spinning are available on modern commercial FT nmr spectrometers, although these experiments are still not entirely routine. Manufacturers often show high-resolution ^{13}C solids spectra of adamantane, bicyclic camphor derivatives, or hexamethylbenzene. These molecules are especially favorable as a result of significant site reorientational mobility in the solid state (i.e., adamantane is a so-called plastic crystal).

One additional problem exists for ^{13}C nmr of solids. In cases where molecular motion is absent ^{13}C T_1s may become extremely long (sometimes measured in *hours*). In these cases it is necessary

---

*In single-crystal dipolar-decoupled ^{13}C studies sharp lines are observed, with the line position a function of the crystal orientation.

to utilize a technique that will allow spectral accumulation at rates greater than those allowed by the ^{13}C T_1s. Such a method is cross polarization. The ^1H and ^{13}C spins are placed in contact in a joint rotating frame of reference (using adjusted, simultaneous ^1H and ^{13}C rf fields). This is known as the *Hartmann-Hahn condition,* first achieved in 1962,[38] and first applied to ^{13}C nmr by Pines et al.[39] In this process, polarization of the ^{13}C nuclei is achieved from the abundant proton spins. This polarization occurs in a time closer to T_2 than to T_1, typically on the order of 100 µsec. Furthermore, it is often possible to make multiple contacts, giving additional ^{13}C signal strength.

High-resolution ^{13}C nmr experiments utilizing both cross polarization (with dipolar decoupling) and magic-angle spinning are called CPMAS experiments. The state of the art is such that one group has reported CPMAS spectra of frozen liquids[40] including the first high-resolution ^{13}C spectra of carbocations in the solid state.[41] One example of a CPMAS spectrum is given in Figure 10.7.[42] Note that achieved resolution in this spectrum is excellent.

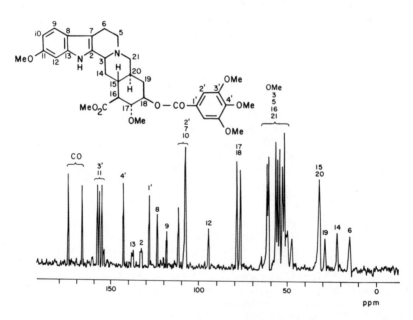

FIGURE 10.7.   *75.46 MHz carbon-13 solid-state CPMAS spectrum of reserpine obtained on a Bruker CXP-300. Sideband removal was accomplished by a block-multiplication technique (300 scans plus 364 scans). (Spectrum courtesy of Dr. D. Müller, Bruker Analytische Messtechnik GmbH.)*

Alla and Lippmaa[43] reported T_1, $T_{1\rho}$ and T_2 solid-state ^{13}C pulse experiments for norbornadiene. Using a modification of the usual experiment, they showed that it was possible to *reconstruct* the *coupled* lineshapes of individual resonances selectively. Figure 10.8 shows one such reconstruction.[43]

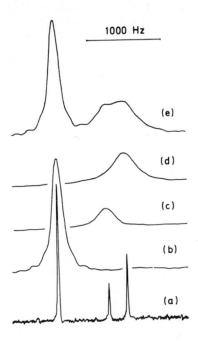

FIGURE 10.8.[43]   *Separated lines of groups of nonequivalent carbon atoms in solid norbornadiene at 210 K: (a) high-resolution solid-state spectrum; (b) unsaturated carbons; (c) bridge carbon; (d) bridgehead carbons; (e) reconstituted single resonance spectrum. Lines (b), (c), and (d) are formed by FT of 12 to 20 data points of the transverse magnetization decay.*

There have been frequent reports[44] of ^{13}C shielding-tensor studies that use dipolar decoupled ^{13}C spectroscopy without magic-angle spinning. It has also been shown that ^{13}C shielding tensors may be evaluated by using "slow" magic-angle spinning.[45]

^{13}C NMR OF FOSSIL FUELS

Carbon-13 nmr analyses of natural and synthetic fuels have be-come of increasing interest. In fact, high-resolution studies by both liquid and solid-state techniques have appeared. Materials as intractable as raw coal,[46] coal tar pitches,[47] and oil shales[48] have been examined. Crude oils[49] and coal-derived liquids[50] have also been examined. Generally, these studies have evaluated com-ponent distributions that in some cases are limited to degree of aromaticity. In more favorable cases individual types of compounds were quantified. Figure 10.9[48a] shows ^{13}C cross-polarization spec-tra of oil-shale samples from Colorado and Kentucky. The Kentucky

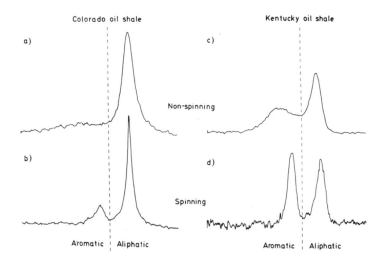

FIGURE 10.9.[48a]   Carbon-13 nmr spectra of oil shales obtained by high-power ^{1}H decoupling and ^{13}C-^{1}H cross-polar-ization, (a) Colorado oil shale, without magic-angle spinning; (b) Colorado oil shale, with magic-angle spinning; (c) Kentucky oil shale, without magic-angle spinning; (d) Kentucky oil shale, with magic-angle spinning.

oil shale clearly demonstrates much higher aromaticity (note that without magic-angle spinning there is considerable overlap be-tween aromatic and aliphatic resonances; see Figs. 10.9a and 10.9c). CPMAS spectra of coal samples show similar differentiation, as in Figure 10.10.[46b] Figure 10.11[50a] shows a normal ^{13}C FT nmr spec-trum of a light oil obtained by catalytic hydrogenation of Utah bituminous coal.

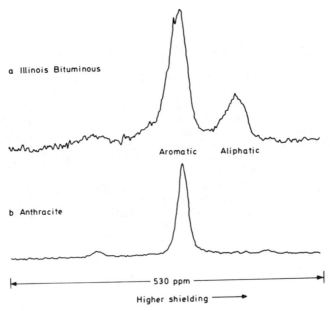

a Illinois Bituminous

Aromatic   Aliphatic

b Anthracite

|← 530 ppm →|

Higher shielding ⟶

FIGURE 10.10[46b]   Carbon-13 nmr spectra of (a) an Illinois bituminous coal and (b) anthracite.

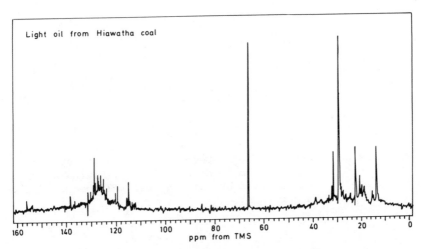

Light oil from Hiawatha coal

160   140   120   100   80   60   40   20   0
ppm from TMS

FIGURE 10.11.[50a]   Proton-decoupled cmr spectra of light-oil
fraction of catalytically hydrogenated Hiawatha,
Utah bituminous coal. The reference peak at
67 ppm is from p-dioxane added as internal reference. Spectrum required approximately 1 hr
to obtain.

317

FUTURE APPLICATIONS

It is difficult to predict all future directions for ^{13}C nmr. It is not, however, difficult to anticipate some trends: (1) *in vivo* ^{13}C studies, including metabolic analyses, (2) ^{13}C nmr imaging using ^{13}C-labeled materials, (3) "routine" high-resolution ^{13}C nmr analyses of solid organic samples, (4) *quantitative* analysis of complex organic systems with very low impurity levels, and (5) significant new structural studies of native biopolymers, including structural proteins, nucleic acids, and complex lipids. Already, studies have appeared of intact cells[51] and in one favorable case, of an entire fruit.[52]

In keeping with tradition,[53] it is appropriate to finally include a spectrum of an hallucinogenic drug. Figure 10.12 shows the ^{13}C spectrum reported in 1976[54] of whole nutmeg.

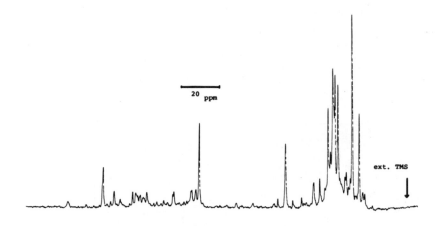

FIGURE 10.12.[54]   *Carbon-13 nmr spectrum of whole nutmeg (obtained from Spice Island Inc., South San Francisco, California) that was polished to fit in a nmr tube.*

REFERENCES

1. O. A. Reutov, T. N. Shatkina, E. T. Lippmaa, and T. I. Pehk, *Proc. Acad. Sci., USSR, Chem. Sect.*, **181**, 770 (1968); *Tetrahedron*, **25**, 5757 (1969); O. A. Reutov, T. N. Shatkine, E. V. Leont'eva, E. T. Lippmaa, and T. I. Pehk, *Proc. Acad. Sci. USSR, Chem. Sect.*, **183**, 1053; H. A. Staab and M. Haenel, *Chem. Ber.*, **103**, 1095 (1970).
2. E. Hedaya and M. E. Kent, *J. Am. Chem. Soc.*, **93**, 3283 (1971); J. B. Stothers, I. S. Y. Wang, D. Ouchi, and E. W. Warnhoff, *ibid.*, **93**, 6702 (1971); R. M. Roberts and T. L. Gibson, *ibid.*, **93**, 7340 (1971).
3. These studies are summarized in *Topics in Carbon-13 NMR Spectroscopy*, Vol. 1, G. C. Levy, Ed., Wiley-Interscience, New York, 1974, Chapter 6.
4. S. H. Pines and A. W. Douglas, *J. Am. Chem. Soc.*, **98**, 8119 (1976).
5. K. Ugurbil, T. R. Brown, J. A. den Hollander, P. Glynn, and R. G. Shulman, *Proc. Nat. Acad. Sci. (USA)*, **75**, 3742 (1978).
6. A. I. Scott and co-workers, unpublished results.
7. W. B. Moniz, C. F. Poranski, Jr., and S. A. Sojka, in *Topics in Carbon-13 NMR Spectroscopy*, Vol. 3, G. C. Levy, Ed., Wiley-Interscience, New York, 1979, Chapter 6 and papers cited therein.
8. E. M. Schulman, R. D. Bertrand, D. M. Grant, A. R. Lepley, and C. Walling, *J. Am. Chem. Soc.*, **94**, 5972 (1972).
9. (a) E. Lippmaa, T. Pehk, A. L. Buchachenko, and S. V. Rykov, *Chem. Phys. Lett.*, **5**, 521 (1970); (b) R. Kaptein, R. Freeman, H. D. W. Hill, and J. Bargon, *J. Chem. Soc., Chem. Commun.*, **1973**, 953.
10. S. Schaublin, A. Wokaun, and R. R. Ernst, *J. Magn. Resonance*, **27**, 273 (1977).
11. F. A. L. Anet and R. Anet, in *Dynamic Nuclear Magnetic Resonance Spectroscopy*, L. M. Jackman and F. A. Cotton, Eds., Academic, New York, 1975.
12. O. A. Gansow, J. Killough, and A. R. Burke, *J. Am. Chem. Soc.*, **93**, 4297 (1971).
13. S. Masamune, K. Hojo, G. Bigam, and D. L. Rabenstein, *J. Am. Chem. Soc.*, **93**, 4966 (1971).
14. (a) D. K. Dalling, D. M. Grant, and L. F. Johnson, *J. Am. Chem. Soc.*, **93**, 3678 (1971); (b) H. J. Schneider, R. Price, and T. Keller, *Angew. Chemie*, **83**, 759 (1971).
15. F. A. L. Anet and T. N. Rawdah, *J. Am. Chem. Soc.*, **101**, 1887 (1979); F. A. L. Anet and I Yavari, *ibid.*, **100**, 7814 (1978); *ibid.*, **99**, 7640 (1977); and earlier papers.
16. (a) Yu. K. Grishin, N. M. Sergeyev, D. A. Subbotin, and Yu. A. Ustynyuk, *Molec. Phys.*, **25**, 297 (1973); (b) C. Piccinni-Leopardi, O. Fabre, and J. Reisse, *Org. Magn. Resonance*, **8**, 233 (1976); (c) A. Blanchette, F. Sauriol-Lord, and M. St.-Jacques, *J. Am. Chem. Soc.*, **100**, 4055 (1978).

17. F. A. L. Anet, in *Topics in Carbon-13 NMR Spectroscopy*, Vol. 3, G. C. Levy, Ed., Wiley-Interscience, New York, 1979, Chapter 1.

18. G. C. Levy, J. T. Bailey, and D. A. Wright, *J. Magn. Resonance*, **37**, 353 (1980).

19. (a) F. A. Cotton, R. J. Haines, B. E. Hanson, and J. C. Sekutowski, *Inorg. Chem.*, **17**, 2010 (1978); (b) F. A. Cotton et al., *J. Am. Chem. Soc.*, **99**, 3673 (1977); (c) *ibid.*, **99**, 3293 (1977).

20. O. A. Gansow, A. R. Burke, and W. D. Vernon, *J. Am. Chem. Soc.*, **98**, 5817 (1976); G. L. Geoffroy and W. L. Gladfelter, *ibid.*, **99**, 6775 (1977).

21. (a) J. Dadok and R. F. Sprecher, *J. Magn. Resonance*, **13**, 243 (1974); (b) R. K. Gupta, J. A. Ferretti, and E. D. Becker, *J. Magn. Resonance*, **13**, 275 (1974).

22. R. R. Ernst, *J. Magn. Resonance*, **3**, 10 (1970); R. Kaiser, *J. Magn. Resonance*, **3**, 28 (1970).

23. D. Ziessow and B. Blümich, *Ber. der Bunsen-Gesellschaft Phys. Chem.*, **78**, 1168 (1974).

24. H. Hill, *Topics in Carbon-13 NMR Spectroscopy*, Vol. 3, G. C. Levy, Ed., Wiley-Interscience, New York, 1979, Chapter 1, 84-101 and references cited therein.

25. G. A. Morris and R. Freeman, *J. Am. Chem. Soc.*, **100**, 6763 (1978).

26. (a) R. D. Bertrand, W. B. Moniz, and A. N. Garroway, *J. Am. Chem. Soc.*, **100**, 5227 (1978); (b) G. A. Morris and R. Freeman, *ibid.*, **101**, 760 (1979); (c) G. A. Morris, *ibid.*, **102**, 428 (1980).

27. (a) H. J. Jakobsen, S. A. Linde, and S. Sørensen, *J. Magn. Resonance*, **15**, 385 (1974); H. J. Jakobsen and H. Beldsøe, *J. Magn. Resonance*, **26**, 183 (1977); (b) G. A. Morris and R. Freeman, *J. Am. Chem. Soc.*, **101**, 760 (1979).

28. (a) D. Terpstra, in *Topics in Carbon-13 NMR Spectroscopy*, Vol. 3, G. C. Levy, Ed., Wiley-Interscience, New York, 1979, Chapter 1, Section 6; (b) R. Freeman and G. Morris, *Bull. Magn. Res.*, **1**, 5 (1979).

29. G. Bodenhausen, D. Phil, thesis, Oxford University, 1977.

30. E. R. Andrew, *Phil. Transact.* in press; P. Brunner and R. R. Ernst, *J. Magn. Resonance*, **33**, 83 (1979).

31. P. C. Lauterbur, *Nature*, **242**, 190 (1973); P. C. Lauterbur, D. M. Kramer, W. V. House, and C. N. Chen, *J. Am. Chem. Soc.*, **97**, 6866 (1975).

32. A. Pines and J. J. Chang, *J. Am. Chem. Soc.*, **96**, 5590 (1974).

33. M. Schwartz, P. E. Fagerness, C. H. Wang, and D. M. Grant, *J. Chem. Phys.*, **63**, 2524 (1975).

34. C. A. Fyfe, J. R. Lyerla, and C. S. Yannoni, *J. Magn. Resonance*, **31**, 315 (1978).

35. I. D. Gay, *J. Phys. Chem.*, **38**, 38 (1974); D. Michel, W. Meiler, and H. Pfeifer, *J. Molec. Catal.*, **1**, 35 (1975); J. J. Chang, A. Pines, J. J. Fripiat, and H. A. Resing, *Surface*

Sci., **47**, 661 (1975); I. D. Gay and J. F. Kriz, *J. Phys. Chem.*, **79**, 2145 (1975); J. F. Kriz and I. D. Gay, *ibid.*, **80**, 2951 (1976).

36. (a) M. D. Sefcik, *J. Am. Chem. Soc.*, **101**, 2164 (1979); (b) M. D. Sefcik, J. Schaefer, and E. O. Stejskal, *ACS Symposium Series*, Vol. **40**, J. R. Katzer, Ed., 1977, Chapter 29; (c) J. B. Nagy, M. Gigot, A. Gourgue, and E. G. Derouane, *J. Molec. Catal.*, **2**, 265 (1977).

37. (a) Abstracts, 20th and 21st Experimental NMR Conferences, 1979, 1980; (b) J. Schaefer and E. O. Stejskal, *Topics in Carbon-13 NMR Spectroscopy*, Vol. 3, G. C. Levy, Ed., Wiley-Interscience, New York, 1979, Chapter 4.

38. S. R. Hartmann and E. L. Hahn, *Phys. Rev.*, **128**, 2042 (1962).

39. A. Pines, M. G. Gibby, and J. S. Waugh, *J. Chem. Phys.*, **59**, 569 (1973).

40. C. A. Fyfe, J. R. Lyerla, and C. S. Yannoni, *J. Am. Chem. Soc.*, **100**, 5635 (1978).

41. J. R. Lyerla, C. S. Yannoni, D. Bruck, and C. A. Fyfe, *J. Am. Chem. Soc.*, **101**, 4770 (1979).

42. CXP-300 spectrum, courtesy of Bruker Instruments, Inc.

43. M. Alla and E. Lippmaa, *Chem. Phys. Lett.*, **37**, 260 (1976).

44. (a) S. Pausak, A. Pines, and J. S. Waugh, *J. Chem. Phys.*, **59**, 591 (1973); (b) V. R. Cross and J. S. Waugh, *J. Magn. Resonance*, **25**, 225 (1977); (c) R. G. Griffin and D. J. Reuben, *J. Chem. Phys.*, **63**, 1272 (1975); (d) D. L. Vanderhart, *J. Chem. Phys.*, **64**, 830 (1976).

45. E. Lippmaa, M. Alla, and T. Tuherm, 19th Congress Ampere, Heidelberg, September 1976; E. O. Stejskal, J. Schaefer, and R. A. McKay, *J. Magn. Resonance*, **25**, 569 (1977).

46. (a) D. L. Vanderhart and H. L. Retcofsky, *Fuel*, **55**, 202 (1976); (b) G. E. Maciel, V. J. Bartuska, and F. P. Miknis, *Fuel*, **58**, 391 (1979).

47. P. Fischer, J. W. Stadelhofer, and M. Zander, *Fuel*, **57**, 345 (1978).

48. (a) G. E. Maciel, V. J. Bartuska, and F. P. Miknis, *Fuel*, **58**, 155 (1979); (b) D. Vitorović, D. Vučelić, M. J. Gašić, N. Juranić, and S. Macura, *Org. Geochem.*, **1**, 89 (1978); (c) F. P. Miknis, G. E. Maciel, and V. J. Bartuska, *Org. Geochem.*, **1**, 169 (1979).

49. J. N. Shoolery and W. L. Budd1e, *Anal. Chem.*, **48**, 1458 (1976).

50. (a) R. J. Pugmire, D. M. Grant, K. W. Zilm, L. L. Anderson, A. G. Oblad, and R. E. Wood, *Fuel*, **56**, 295 (1977); (b) D. M. Cantor, *Anal. Chem.*, **50**, 1185 (1978).

51. A. Lapidot and C. S. Irving, *Biochemistry*, **18**, 1788 (1979); see also R. S. Norton, *Bull. Magn. Resonance*, review in press.

52. M. Kainosho, *Tetrahedron Lett.*, 4279 (1976).

53. See first edition, p. 82.

54. M. Kainosho and H. Konishi, *Tetrahedron Lett.*, 4757 (1976).

Appendix

Answers to selected problems are provided below.

The spectral assignments given for the problem spectra are *suggested* assignments. Some pairs of assignments may be inter-changeable. Numbers appearing on structural formulas correspond to the spectral peaks indicated in the spectra. (Notes on spectral coding are given on p. 45.)

(2.1) $HO-\overset{\overset{O}{\|}}{C}-\underset{1}{\overset{4}{\underset{2}{\bigcirc}}}\overset{3}{}\overset{5}{-}\overset{6}{C}\equiv N$

(2.2) $^{4}CH_3\overset{\overset{O}{\|}}{\underset{1}{C}}-O-\underset{2}{CH}=\underset{3}{CH_2}$

(2.3) $\underset{H}{\overset{H}{}}\underset{3\ 2}{C=C}\underset{1}{\overset{7}{\underset{CH_3}{}}}$
$C=O$
$\underset{4}{O}\quad\overset{5}{}$
$\underset{6}{CH_2-C-CH_2}\quad\overset{H}{}$
$\underset{O}{\overset{\diagup}{}}$

(3.2) $\underset{4}{CH_3}-\underset{3}{CH_2}-CH_2-\underset{1}{CH_2}-CH_2-\underset{2}{CH_2}-CH_2-CH_3$

(3.5) $(\underset{3}{CH_3})_3\underset{2}{C}\underset{1}{CH_2}\underset{5}{CH}(\underset{4}{CH_3})_2$

(3.7) $\underset{2}{CH_3}\overset{\overset{H}{}}{\underset{}{-N-}}\underset{1}{CH}(\underset{3}{CH_3})_2$

323

(3.8)

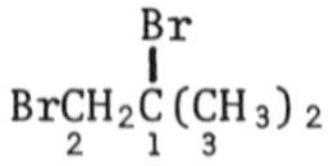

$$\underset{2 \;\; 1 \;\; 3}{BrCH_2\overset{\overset{\displaystyle Br}{|}}{C}(CH_3)_2}$$

(3.9)

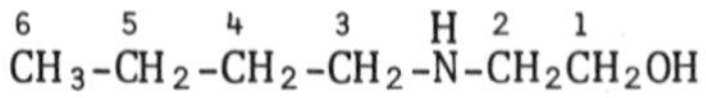

$$\underset{6 \quad\; 5 \quad\;\; 4 \quad\;\; 3 \quad\;\; H\; 2 \;\; 1}{CH_3-CH_2-CH_2-CH_2-N-CH_2CH_2OH}$$

(3.12)

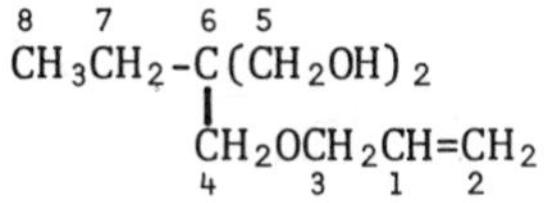

$$\underset{8 \;\;\; 7 \quad\; 6\; 5}{CH_3CH_2-\underset{\underset{4 \qquad 3 \quad 1 \;\; 2}{CH_2OCH_2CH=CH_2}}{\overset{|}{C}}(CH_2OH)_2}$$

(3.13)

$$\underset{6\;\;\; 4 \quad 3 \quad 5 \quad 1 \;\; 2}{CH_3CH_2CH_2CH_2C\equiv CH}$$

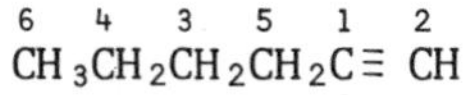

Note: internal acetylenic carbon much smaller than protonated carbons.

(4.1)

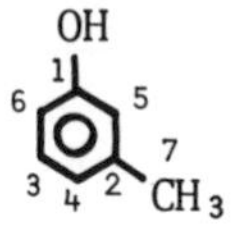

(4.2)

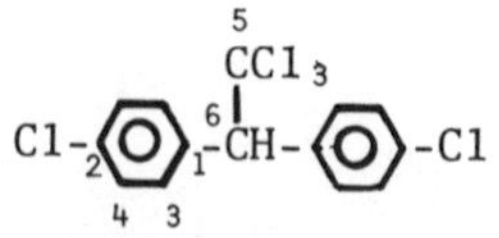

(DDT)

(4.3)   *ortho* (a)- and *meta* (b)-terphenyl

(4.4)

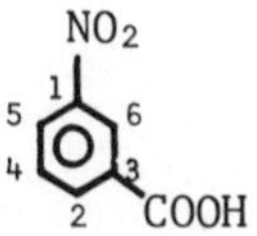

(4.6)

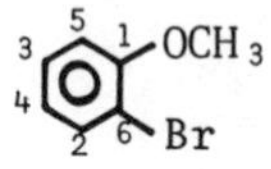

(5.1) $CH_3CH_2O\overset{O}{\overset{\|}{C}}C \equiv \underset{2\ 1}{C}\underset{3}{C}O\underset{4}{CH_2CH_3}$

(5.2)

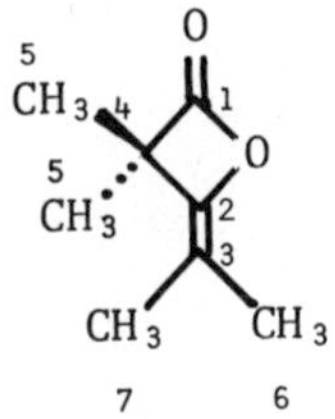

(5.3)   Nicotinamide

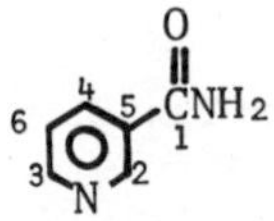

(5.4)   Ph-N(CH$_3$)CHO   predominantly:

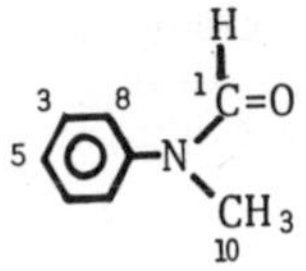

In the predominant isomer the CH$_3$ carbon is at higher field as a result of a steric compression shift caused by interaction with the *cisoid* carbonyl oxygen.

(8.1)   105 sec. The remaining nondipolar contribution is probably spin rotation or chemical shift anisotropy relaxation. Temperature or magnetic field studies will differentiate these two mechanisms.

(8.2)   Qualitative evaluation of the relaxation processes for C$_4$H$_7$XYZ should be considered first. Methyl C-4 is relaxed completely (within experimental error) by ^{13}C-^{1}H dipole-dipole interactions, as evidenced by the observed maximum NOE. Methyl C-1, however, has appreciable contributions from another mechanism.
    The relaxation data for C-1 at 60° indicate that the spin-rotation (SR) mechanism is operative (the other mechanisms (DD, CSA) result in longer T_1s at higher temperatures). At 30° the relaxation of C-1 is probably due solely to the SR and DD mechanisms. The SR contribution can be calculated from the NOE:

C-1

% SR relaxation $= \dfrac{0.9}{2.0} = 45\%$

$R_1^{SR} = (\dfrac{1}{9.0 \text{ sec}})(0.45) = 0.0499 \text{ sec}^{-1}$

$T_1^{SR} = 20 \text{ sec}$

$R_1^{DD} = (\dfrac{1}{9.0 \text{ sec}})(0.55) = 0.0611 \text{ sec}^{-1}$

$T_1^{DD} = 16.4 \text{ sec}$

Carbon-3 has a very short relaxation time (4.2 sec at 30°) and also a low NOE (0.6). This cannot be due to SR relaxation; C-3 is clearly not a free rotor. Nucleus Z must be a quadrupolar nucleus with the right characteristics for relaxing C-3 by the scalar mechanism. As with C-1, it is likely that just two mechanisms are causing relaxation of C-3, but here they are the dipole-dipole and scalar mechanisms. The individual contributions are:

C-3

$R_1^{DD} = 0.714 \text{ sec}^{-1}$

$T_1^{DD} = 14 \text{ sec}$

$R_1^{SC} = 0.167 \text{ sec}^{-1}$

$T_1^{SC} = 6 \text{ sec}$

30% DD and 70% SC relaxation

A long relaxation time is observed for the nonprotonated carbon C-2. The NOE indicates that C-2 is predominantly relaxed by ^{13}C-^{1}H dipole-dipole interactions with considerable contributions from other mechanisms. The T_1 for the degassed sample at 30° indicates that dipole-dipole interaction with dissolved oxygen contributes to the relaxation of C-2 in the undegassed sample:

$$R_1^{O_2(DD)} = \frac{1}{53 \text{ sec}} - \frac{1}{62 \text{ sec}}$$

$$= 0.0189 \text{ sec}^{-1} - 0.0161 \text{ sec}^{-1}$$

$$= 0.0028 \text{ sec}^{-1}$$

$$T_1^{O_2(DD)} = \frac{1}{0.0028 \text{ sec}^{-1}} \cong 357 \text{ sec}$$

The percentage contribution of $O_2(DD)$ relaxation at $30°$:

$$= (100) \frac{R_1^{O_2(DD)}}{R_1^{T}} = (100) \frac{0.0028 \text{ sec}^{-1}}{0.0189 \text{ sec}-1}$$

$$\cong 14.8\%$$

The contribution of $^{13}C-^{1}H$ dipole-dipole relaxation can be determined directly from the NOE:

$$R_1^{DD} = (R_1^{T}) (\frac{NOE}{2.0})$$

$$= (0.0189 \text{ sec}^{-1}) (\frac{1.4}{2.0})$$

$$R_1^{DD} = 0.0132 \text{ sec}^{-1}$$

$$(T_1^{DD} = 75.6 \text{ sec})$$

The contribution of chemical shift anisotropy to relaxation of C-2 can be calculated from the 75 MHz T_1 data:

$$\frac{75 \text{ MHz}}{25 \text{ MHz}} = 3$$

therefore T_1^{CSA} at 75 MHz $= (1/3)^2$ or $1/9 \ T_1^{CSA}$ at 25 MHz (similarly $R_1^{CSA,75 \text{ MHz}} = 9R_1^{CSA,25 \text{ MHz}}$).

R_1^{DD} at 75 MHz remains 0.0132 sec^{-1}

$R_1^{O_2(DD)}$ at 75 MHz remains 0.0028 sec^{-1}

$$R_1^{CSA,75 \text{ MHz}} = R_1^{Tot.,75 \text{ MHz}} - R_1^{DD} - R_1^{O_2(DD)} - R_1^{other}$$

Assuming that $R_1^{other} \approx 0$

$$R_1^{CSA, 75 \text{ MHz}} = 0.0385 - 0.0132 - 0.0028$$

then

$$R_1^{CSA, 25 \text{ MHz}} \cong \frac{0.0225}{9}$$

$$\cong 0.0025 \text{ sec}^{-1}$$

$$T_1^{CSA, 25 \text{ MHz}} \cong 400 \text{ sec}$$

The relaxation contributions for C-2 at 25 MHz in the unde-gassed sample:

$$R_1^{Total} = R_1^{O_2 (DD)} + R_1^{CSA} + R_1^{other*} \qquad *\text{Presumably } R_1^{SR}$$

$$0.0189 = 0.0028 \text{ sec}^{-1} + 0.0132 \text{ sec}^{-1} + 0.0025 \text{ sec}^{-1} + R_1^{other}$$

$$R_1^{other} \approx 0.0004 \text{ sec}^{-1}$$

$$R_1^{other} \approx 0.0004 \text{ sec}^{-1}$$

or $T_1^{other} \approx 2500 \text{ sec}$

(justifying our assumption that $R_1^{other} \sim 0$).

(9.1)

OH(OAc)

(9.2)   vanillin

(9.3)

Note that peak 2 is at very high field for an aromatic carbon with an attached methoxy group. This results from the two *ortho* methoxy groups. Peak 4 represents a carbon *ortho* to one methoxy and *para* to another.

Index